# Lecture Notes in Artificial Intelligence    12733

Subseries of Lecture Notes in Computer Science

More information about this subseries at http://www.springer.com/series/1244

Agnès Braud · Aleksey Buzmakov ·
Tom Hanika · Florence Le Ber (Eds.)

# Formal Concept Analysis

16th International Conference, ICFCA 2021
Strasbourg, France, June 29 – July 2, 2021
Proceedings

 Springer

*Editors*
Agnès Braud 🆔
ICube
University of Strasbourg
Illkirch, France

Tom Hanika 🆔
University of Kassel
Kassel, Germany

Aleksey Buzmakov 🆔
NRU Higher School of Economics
Perm, Russia

Florence Le Ber 🆔
ICube
University of Strasbourg
Illkirch, France

ISSN 0302-9743          ISSN 1611-3349    (electronic)
Lecture Notes in Artificial Intelligence
ISBN 978-3-030-77866-8       ISBN 978-3-030-77867-5    (eBook)
https://doi.org/10.1007/978-3-030-77867-5

LNCS Sublibrary: SL7 – Artificial Intelligence

This Springer imprint is published by the registered company Springer Nature Switzerland AG
The registered company address is: Gewerbestrasse 11, 6330 Cham, Switzerland

# Preface

This volume features the contributions accepted for the 16th International Conference on Formal Concept Analysis (ICFCA 2021), held during June 29 – July 2, 2021, at Université de Strasbourg, France. Formal concept analysis (FCA) is a mathematical field that emerged about 40 years ago and is rooted in lattice and order theory. Although being of a theoretical nature, FCA proved to be of interest to various applied fields such as knowledge discovery, machine learning, database theory, information visualization, and many others.

The goal of the International Conference on Formal Concept Analysis is to offer researchers from FCA and related backgrounds the possibility to present and discuss their research. Since its first edition in 2003 in Darmstadt, Germany, ICFCA has been held annually in several countries in Europe, Africa, America, and Australia. In 2015, ICFCA became biennial to alternate with the Conference on Concept Lattices and Their Applications (CLA).

The field of FCA originated in the 1980s in Darmstadt as a subfield of mathematical order theory, with prior developments in other research groups. Its original motivation was to consider complete lattices as lattices of concepts, drawing motivation from philosophy and mathematics alike. FCA has since then developed into a wide research area with applications far beyond its original motivation, for example, in logic, knowledge representation, unsupervised learning, data mining, human learning, and psychology.

There were 32 papers submitted to this year's ICFCA by authors in fifteen countries. Each submission was reviewed by at least two different members of the Program Committee and at least one member of the Editorial Board. The review process was single blind. Fourteen high-quality papers were chosen for publication in this volume, amounting to an acceptance rate of 43%. In addition, five papers were deemed mature enough to be discussed at the conference and were therefore included as short papers in this volume.

The research part of this volume is divided in five different sections. First, in "Theory" we compiled works that discuss advances on theoretical aspects of FCA. Second, the section "Rules" consists of contributions devoted to implications and association rules. The third section "Methods and Applications" is composed of results that are concerned with new algorithms and their applications. "Exploration and Visualisation" introduces different approaches to data exploration and in the final section we collected the five accepted short works.

In addition to the regular contributions, this volume also contains the abstracts of the four invited talks by outstanding researchers we were delighted to have at ICFCA 2021. In detail, we were pleased to host the following talks:

– "What do the Sources Say? Exploring Heterogeneous Journalistic Data as a Graph" by Ioana Manolescu (France)

- "Ontologies for On-Demand Design of Data-Centric Systems" by Magdalena Ortiz (Austria)
- "Towards human-guided rule learning" by Matthijs van Leeuwen (Netherlands)
- "Sustainable AI – What does it take for continued success in deployed applications?" by Stefan Wrobel (Germany)

We are deeply thankful to all authors who submitted their contributions to ICFCA 2021 as a platform for discussing their work. Our strong gratitude goes to the members of the Editorial Board and Program Committee, as well as to all the additional reviewers whose timely and thorough reviews made the fruitful discussions of the high-quality papers during the conference possible. Furthermore, we would like to express our sincere thanks to the local organizers who were always quick to solve any questions or problems that arose, and their hard work made for a great event.

We are very grateful to Springer for supporting the International Conference on Formal Concept Analysis as well as to the Université de Strasbourg and the Centre National de la Recherche Scientifique (CNRS) for hosting the event. Finally, we would like to highlight the great help of easychair.org for organizing the review process for ICFCA 2021 and of sciencesconf.org for hosting the website and their technical support, especially collecting the camera-ready papers.

June 2021

Agnès Braud<br>
Aleksey Buzmakov<br>
Tom Hanika<br>
Florence Le Ber

# Organization

## Executive Committee

### Conference Chair

| | |
|---|---|
| Florence Le Ber | Université de Strasbourg, France |
| Agnès Braud | Université de Strasbourg, France |

### Local Organization

| | |
|---|---|
| Xavier Dolques | Université de Strasbourg, France |
| Nicolas Lachiche | Université de Strasbourg, France |
| Aurélie Leborgne | Université de Strasbourg, France |
| Stella Marc-Zwecker | Université de Strasbourg, France |
| Peggy Rupp | CNRS, Strasbourg, France |

## Program and Conference Proceedings

### Program Chairs

| | |
|---|---|
| Aleksey Buzmakov | Higher School of Economics, Russia |
| Tom Hanika | Universität Kassel, Germany |

### Editorial Board

| | |
|---|---|
| Jaume Baixeries | Polytechnic University of Catalonia, Spain |
| Karell Bertet | Université de La Rochelle, France |
| Peggy Cellier | IRISA, Rennes, France |
| Sebastien Ferré | Université de Rennes 1, France |
| Bernhard Ganter | Technische Universität Dresden, Germany |
| Dmitry Ignatov | Higher School of Economics, Russia |
| Mehdi Kaytoue | INSA Lyon, France |
| Sergei Kuznetsov | Higher School of Economics, Russia |
| Leonard Kwuida | Bern University of Applied Sciences, Switzerland |
| Rokia Missaoui | Université du Québec en Outaouais, Canada |
| Amedeo Napoli | LORIA, Nancy, France |
| Sergei Obiedkov | Higher School of Economics, Russia |
| Manuel Ojeda-Aciego | University of Malaga, Spain |
| Uta Priss | Ostfalia University of Applied Sciences, Germany |
| Sebastian Rudolph | Technische Universität Dresden, Germany |
| Christian Sacarea | Babes-Bolyai University of Cluj-Napoca, Romania |
| Stefan E. Schmidt | Technische Universität Dresden, Germany |
| Barış Sertkaya | Frankfurt University of Applied Sciences, Germany |
| Gerd Stumme | Universität Kassel, Germany |

# Abstracts of Invited Talks

# What do the Sources Say? Exploring Heterogeneous Journalistic Data as a Graph

Ioana Manolescu

Inria Saclay, Île-de-France, France

**Abstract.** Professional journalism is of utmost importance nowadays. It is a main feature distinguishing dictatorships from democracies, and a mirror sorely needed by society to look upon itself and understand its functioning. In turn, understanding is necessary for making informed decisions, such as political choices.

With the world turning increasingly digital, journalists need to analyze very large amounts of data, while having no control over the structure, organization, and format of the data. Since 2013, my team has been working to understand data journalism and computational fact-checking use cases, to identify and develop tools adapted for this challenging setting. I will describe our SourcesSay project (2020–2024), in which extremely heterogeneous data sources are integrated as graphs, on top of which journalistic applications can be supported through flexible graph queries. I will explain the data source integration module, the role played by Information Extraction and Entity Disambiguation, as well as novel techniques to explore and simplify these graphs.

# Ontologies for On-Demand Design of Data-Centric Systems

Magdalena Ortiz

Faculty of Informatics, TU Wien
ortiz@kr.tuwien.ac.at

Over the last decade, ontologies have found impactful applications in data management, where they help improve access to data that is incomplete, heterogenous, or poorly structured [3, 4]. An ontology can act as a mediator between users and data sources to facilitate query formulation, and allow us to obtain more complete query answers by leveraging domain knowledge to infer implicit facts. Huge research efforts have been put into understanding the problems associated to query evaluation leveraging diverse ontology languages, many algorithms have been developed, and off-the-shelf engines for knowledge-enriched query answering exist, see e.g, [1] and its references.

We present our work advocating a novel use of ontologies [2], not only to access the data stored in systems, but also to facilitate the correct organization of data at design time. We propose a process called *focusing* to harness existing ontologies for the on-demand design of the schema of knowledge-enriched databases. Focusing solutions specify which terms of an ontology are relevant to a specific application, and explicate desired assumptions about their completeness and dynamicity. We present automated inferences services for obtaining and validating focusing solutions, and for answering queries the resulting knowledge-enriched databases. The definitions admit different ontology and query languages for specifying the scope of the system, and are accompanied by concrete decidability and complexity results for selected representative combinations.

**Acknowledgements.** Based on joint work with M, Šimkus, F. Murlak, Y. Ibáñez-Garca V. Gutiérrez-Basulto, and T. Gogacz. Supported by the Austrian Science Fund (FWF) projects P30360 and P30873.

# References

1. Bienvenu, M., Ortiz, M.: Ontology-mediated query answering with data-tractable description logics. In: ReasoningWeb. LNCS, vol. 9203, pp. 218–307. Springer (2015)
2. Gogacz, T., Gutiérrez-Basulto, V., Ibáñez-García, Y., Murlak, F., Ortiz, M., Simkus, M.: Ontology focusing: knowledge-enriched databases on demand. In: ECAI. Frontiers in AI and Applications, vol. 325, pp. 745–752. IOS Press (2020)
3. Schneider, T., Simkus, M.: Ontologies and data management: A Brief Survey Künstliche Intell. **34**(3), 329–353 (2020)
4. Xiao, G., et al.: Ontology-based data access: A survey. In: IJCAI, pp. 5511–5519 (2018). ijcai.org

# Towards Human-Guided Rule Learning

Matthijs van Leeuwen

Leiden University, the Netherlands

**Abstract.** Interpretable machine learning approaches such as predictive rule learning have recently witnessed a strong increase in attention, both within and outside the scientific community. Within the field of data mining, the discovery of descriptive rules has long been studied under the name of subgroup discovery. Although predictive and descriptive rule learning have subtle yet important differences, they both suffer from two drawbacks that make them unsuitable for use in many real-world scenarios. First, hyperparameter optimisation is typically cumbersome and/or requires large amounts of data, and second, results obtained by purely data-driven approaches are often unsatisfactory to domain experts.

In this talk I will argue that domain experts often have relevant knowledge not present in the data, which suggests a need for human-guided rule learning that integrates knowledge-driven and data-driven modelling. A first step in this direction is to eliminate the need for extensive hyperparameter tuning. To this end we propose a model selection framework for rule learning that 1) allows for virtually parameter-free learning, naturally trading off model complexity with goodness of fit; and 2) unifies predictive and descriptive rule learning, more specifically (multi-class) classification, regression, and subgroup discovery. The framework we propose is based on the minimum description length (MDL) principle, we consider both (non-overlapping) rule lists and (overlapping) rule sets as models, and we introduce heuristic algorithms for finding good models.

In the last part of the talk, I will give a glimpse of the next steps towards human-guided rule learning, which concern exploiting expert knowledge to further improve rule learning. Specifically, I will describe initial results obtained within the iMODEL project, in which we develop theory and algorithms for interactive model selection, involving the human in the loop to obtain results that are more relevant to domain experts.

# Sustainable AI – What Does It Take for Continued Success in Deployed Applications?

Stefan Wrobel[1,2]

[1] University of Bonn, Germany
[2] Fraunhofer IAIS, Germany

**Abstract.** Advances in machine learning research have been so impressive that one would be tempted to believe that today most practical problems could easily be solved purely with data and machine learning. However, in the real world, the requirements demanded from a deployed application go far beyond achieving an acceptably low error for a trained model. Deployed applications must guarantee sustained success with respect to their functionality, their business viability, and their ethical acceptability. In this talk, we will analyze the challenges faced in practice with respect to these three dimensions of sustainability, pointing out risks and common misunderstandings and highlighting the role of hybrid modeling. We will then discuss our lessons learned from a number of real world projects for companies with respect to approaches for engineering and operating ML systems. The talk will conclude with a perspective on the demands placed on AI systems by customers and society, presenting our methodology for testing and ultimately certifying such systems.

# Contents

**Exploration and Visualisation**

**Short Papers**

# Theory

# Representing Partition Lattices
# Through FCA

Mike Behrisch[1] , Alain Chavarri Villarello[2], and Edith Vargas-García[2](✉)

[1] Institute of Discrete Mathematics and Geometry,
Technische Universität Wien, Vienna, Austria
behrisch@logic.at
[2] Department of Mathematics, ITAM Río Hondo, Mexico City, Mexico
{achavar3,edith.vargas}@itam.mx

**Abstract.** We investigate the standard context, denoted by $\mathbb{K}(\mathcal{L}_n)$, of the lattice $\mathcal{L}_n$ of partitions of a positive integer $n$ under the dominance order. Motivated by the discrete dynamical model to study integer partitions by Latapy and Duong Phan and by the characterization of the supremum and (infimum) irreducible partitions of $n$ by Brylawski, we show how to construct the join-irreducible elements of $\mathcal{L}_{n+1}$ from $\mathcal{L}_n$. We employ this construction to count the number of join-irreducible elements of $\mathcal{L}_n$, and confirm that the number of objects (and attributes) of $\mathbb{K}(\mathcal{L}_n)$ has order $\Theta(n^2)$. We also discuss the embeddability of $\mathbb{K}(\mathcal{L}_n)$ into $\mathbb{K}(\mathcal{L}_{n+1})$ with special emphasis on $n = 9$.

**Keywords:** Join-irreducibility · Standard context · Integer partition

## 1 Introduction

The study of *partitions* of an integer started to gain attention in 1674 when Leibniz investigated [8, p. 37] the number of ways one can write a positive integer $n$ as a sum of positive integers in decreasing order, which he called 'divulsiones', today known as (unrestricted) partitions, see [11]. He observed that there are 5 partitions of the number 4, namely the partitions $4, 3 + 1, 2 + 2$, $2 + 1 + 1$ and $1 + 1 + 1 + 1$; for the number 5 there are 7 partitions, for 6 there are 11 partitions etc., and so he asked about the number $p(n)$ of partitions of a positive integer $n$. In 1918 G. H. Hardy and S. Ramanujan in [6] published an asymptotic formula to count $p(n)$. To our knowledge, until now, there is no 'closed-form expression' known to express $p(n)$ for any integer $n$.

From [1], it is known that the set of all partitions of a positive integer $n$ with *the dominance ordering* (defined in the next section) is a *lattice*, denoted by $\mathcal{L}_n$. Moreover, Brylawski proposed a dynamical approach to study this lattice, which is explained in a more intuitive form by Latapy and Duong Phan in [7]. Motivated by their method to construct $\mathcal{L}_{n+1}$ from $\mathcal{L}_n$, we restrict their approach to

The third author gratefully acknowledges financial support by the Asociación Mexicana de Cultura A.C.

A. Braud et al. (Eds.): ICFCA 2021, LNAI 12733, pp. 3–19, 2021.
https://doi.org/10.1007/978-3-030-77867-5_1

the join-irreducible elements of $\mathcal{L}_n$ and show how to construct the join-irreducible elements of $\mathcal{L}_{n+1}$ from those of $\mathcal{L}_n$. Our second main result is to give a recursive formula for the number of join-irreducibles of $\mathcal{L}_n$, and, since partition conjugation is a lattice antiautomorphism making $\mathcal{L}_n$ autodual (see [1]), we also derive the number of meet-irreducibles. Then we count the number of objects (and attributes) of the standard context $\mathbb{K}(\mathcal{L}_n)$ and prove that this number has order $\Theta(n^2)$. Retrospectively, we learned that an alternative formula for it had been found independently by Bernhard Ganter [3, Proposition 3] (not using our recursion based on [7]). Further, we show how to incrementally obtain $\mathbb{K}(\mathcal{L}_{n+1})$ from $\mathbb{K}(\mathcal{L}_n)$, giving a polynomial-time algorithm for $\mathbb{K}(\mathcal{L}_n)$ of time complexity $\mathcal{O}(n^5)$. Finally, picking up a question from [3], we argue that $\mathbb{K}(\mathcal{L}_9)$ cannot be embedded into $\mathbb{K}(\mathcal{L}_{10})$. This is known for symmetric embeddings from [3, Proposition 5], but it is even true without the symmetry requirement.

The sections of this paper should be read in consecutive order. The following one introduces some notation and prepares basic definitions and facts concerning lattice theory and formal concept analysis. In Sect. 3 we explore the relation between the join-irreducible elements of the lattice $\mathcal{L}_{n+1}$ and those of the lattice $\mathcal{L}_n$, and the final section concludes the task by counting the supremum irreducible elements of $\mathcal{L}_n$ and discussing the embeddability question for $\mathbb{K}(\mathcal{L}_n)$.

## 2   Preliminaries

### 2.1   Lattices and Partitions

Throughout the text $\mathbb{N} := \{0, 1, 2, \ldots\}$ denotes the set of *natural numbers* and $\mathbb{N}_+ := \{1, 2, \ldots\}$ denotes the set of *positive integers*. Moreover, if $P$ is a set and $\leq$ is a binary relation on $P$, which is *reflexive, antisymmetric* and *transitive*, then $\mathbb{P} := (P, \leq)$ is a *partially ordered set*. We often identify $\mathbb{P}$ and $P$; for example, we write $x \in \mathbb{P}$ instead of $x \in P$ when this is convenient. For such a partially ordered set $\mathbb{P}$ and elements $a, b \in \mathbb{P}$, we say that $a$ is *covered* by $b$ (or $b$ *covers* $a$), and write $a \prec b$, if $a < b$ and $a \leq z < b$ implies $z = a$. Partially ordered sets $\mathbb{P}$ and $\mathbb{Q}$ are *(order-)isomorphic* if there is a bijective map $\varphi \colon \mathbb{P} \to \mathbb{Q}$ such that $a \leq b$ in $\mathbb{P}$ if and only if $\varphi(a) \leq \varphi(b)$ in $\mathbb{Q}$. Every such $\varphi$ is called an *order-isomorphism*.

A *lattice* is a partially ordered set $\mathbb{P}$ such that any two elements $a$ and $b$ have a least upper bound, called *supremum* of $a$ and $b$, denoted by $a \vee b$, and a greatest lower bound, called *infimum* of $a$ and $b$, denoted by $a \wedge b$. Moreover, if supremum $\bigvee S$ and infimum $\bigwedge S$ exist for all $S \subseteq \mathbb{P}$, then $\mathbb{P}$ is called a *complete lattice*. Let $\mathbb{L}$ be a lattice and $M \subseteq L$. Then $M$ is a (carrier set of a) *sublattice* $\mathbb{M}$ of $\mathbb{L}$ if whenever $a, b \in M$, then $a \vee b \in M$ and $a \wedge b \in M$.

For a lattice $\mathbb{L}$, we say that $a \in \mathbb{L}$ is *join-irreducible*, denoted by $\vee$-irreducible, if it is not *the minimum element* $0_{\mathbb{L}}$ (provided such an element exists) and for every $b, c \in \mathbb{L}$ such that $a = b \vee c$ we can conclude that $a = b$ or $a = c$. In particular for a finite lattice, the join-irreducible elements are those which cover precisely one element. *Meet-irreducible* ($\wedge$-irreducible) elements are defined dually, and in finite lattices they can be described by being covered by precisely one element. For a lattice $\mathbb{L}$, we denote by $\mathcal{J}(\mathbb{L})$ and by $\mathcal{M}(\mathbb{L})$ the set of *all join-irreducible elements*, and of *all meet-irreducible elements* of $\mathbb{L}$, respectively. As an example, the meet-irreducible elements of the lattices $\mathbb{N}_5$ and $\mathbb{M}_3$ appear shaded in Fig. 1.

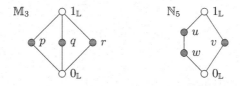

**Fig. 1.** Lattices $\mathbb{M}_3$ and $\mathbb{N}_5$ with meet-irreducible elements appearing shaded

Our aim is to study the join-irreducible elements in the lattice of positive integer partitions. We thus define a partition formally.

**Definition 1.** *An (ordered) partition of a positive integer $n \in \mathbb{N}_+$ is an n-tuple $\boldsymbol{\alpha} := (a_1, \ldots, a_n)$ of natural numbers such that*

$$a_1 \geq a_2 \geq \ldots \geq a_n \geq 0 \qquad and \qquad n = a_1 + a_2 + \ldots + a_n.$$

*If there is $k \in \{1, \ldots, n\}$ such that $a_k > 0$ and $a_i = 0$ for all $i > k$, the partition $\boldsymbol{\alpha}$ can be written in the form $(a_1, \ldots, a_k)$, whereby we delete the zeros at the end.*

For example, $(3, 2, 2, 1, 0, 0, 0, 0)$ is a partition of 8 because $3 \geq 2 \geq 2 \geq 1 \geq 0$ and $3+2+2+1 = 8$. By deleting the zeros, we write more succinctly $(3, 2, 2, 1)$ for this partition. Graphically, we can illustrate a partition using a diagram that has a ladder shape, which is known as a Ferrers diagram (cf. Fig. 2). We have rotated all our Ferrers diagrams by $90°$ counterclockwise as compared to the convention prevailing in the literature. The reason for this is given after Definition 4. From a partition $\boldsymbol{\alpha}$ of $n$, it is possible to obtain *the conjugated partition* $\boldsymbol{\alpha}^*$ in the sense of [1] by reading its Ferrers diagram by rows, from bottom to top. This operation can also be seen as reflecting the Ferrers diagram along a diagonal axis. For instance, the partition $(3, 2, 2, 1)$ of $8 = 3 + 2 + 2 + 1$ has the Ferrers diagram presented in Fig. 2, and its conjugate consists of 4 grains from the first row, 3 grains from the second, and 1 grain from the third. So we get the partition $(4, 3, 1)$; its Ferrers diagram is also shown in Fig. 2 below.

Ferrers diagram of $(3, 2, 2, 1)$.          The conjugated partition of $(3, 2, 2, 1)$ is $(4, 3, 1)$.

**Fig. 2.** Ferrers diagrams

The *set of all partitions* of $n \in \mathbb{N}_+$, denoted by $\mathrm{Part}(n)$, can be ordered in different ways. One of them is the dominance ordering, defined as follows.

**Definition 2** ([1]). *Let $\alpha = (a_1, \ldots, a_k)$ and $\beta = (b_1, \ldots, b_m)$ in Part(n) be partitions of $n \in \mathbb{N}_+$. We define the* dominance ordering *between $\alpha$ and $\beta$ by*

$$\alpha \geq \beta \quad \text{if and only if} \quad \sum_{i=1}^{j} a_i \geq \sum_{i=1}^{j} b_i \text{ for all } j \geq 1.$$

In [1, Proposition 2.2] it is shown that the set Part($n$) with the dominance ordering is a lattice. We denote by $\mathcal{L}_n = (\text{Part}(n), \leq)$ *the lattice of all partitions* of $n \in \mathbb{N}_+$ with the dominance ordering. Characterizing the covering relation is central for the construction of a finite lattice, and the following theorem provides this characterization in the case of $\mathcal{L}_n$.

**Theorem 3 (Brylawski** [1]). *In the lattice $\mathcal{L}_n$, the partition $\alpha = (a_1, \ldots, a_n)$ covers $\beta = (b_1, \ldots, b_n)$, denoted $\beta \prec \alpha$, if and only if either of the following two cases (not necessarily disjoint) is satisfied:*

1. *There exists $j \in \{1, \ldots, n\}$ such that $a_j = b_j + 1$, $a_{j+1} = b_{j+1} - 1$ and $a_i = b_i$ for all $i \in \{1, \ldots, n\} \setminus \{j, j+1\}$.*
2. *There exist $j, h \in \{1, \ldots, n\}$ such that $a_j = b_j + 1$, $a_h = b_h - 1$, $b_j = b_h$ and $a_i = b_i$ for all $i \in \{1, \ldots, n\} \setminus \{j, h\}$.*

In [7] Matthieu Latapy and Ha Duong Phan give a dynamic approach to study the lattice $\mathcal{L}_n$ so that the dominance ordering and the covering relation can be visualized more intuitively. To this end they use the following definitions.

**Definition 4** ([7]). *For a partition $\alpha = (a_1, \ldots, a_n)$, the* height difference *of $\alpha$ at $j \in \{1, \ldots, n\}$, denoted by $d_j(\alpha)$, is the integer $a_j - a_{j+1}$, where $a_{n+1} := 0$. For $j \in \{1, \ldots, n\}$ we say (cf. Fig. 3) that the partition $\alpha = (a_1, \ldots, a_n)$ has:*

1. *a* cliff *at $j$ if $d_j(\alpha) \geq 2$.*
2. *a* slippery plateau *at $j$ if there exists some $k > j$ such that $d_k(\alpha) = 1$ and $d_i(\alpha) = 0$ for all $i \in \{j, j+1, \ldots, k-1\}$. The integer $k-j$ is called the* length *of the slippery plateau at $j$.*
3. *a* non-slippery plateau *at $j$ if there is some $k > j$ such that $\alpha$ has a cliff at $k$ and $d_i(\alpha) = 0$ for all $i \in \{j, j+1, \ldots, k-1\}$. The integer $k-j$ is the* length *of the non-slippery plateau at $j$.*
4. *a* slippery step *at $j$ if $\alpha' = (a_1, \ldots, a_{j-1}, a_j - 1, a_{j+1}, \ldots, a_n)$ is a partition with a slippery plateau at $j$.*
5. *a* non-slippery step *at $j$ if $\alpha' = (a_1, \ldots, a_{j-1}, a_j - 1, a_{j+1}, \ldots, a_n)$ is a partition with a non-slippery plateau at $j$.*

The covering relation of Theorem 3 is also described in [7] using a correspondence between the two cases of Theorem 3 and *two transition rules*:

1. If $\alpha$ has a cliff at $j$, one grain can fall from column $j$ to column $j+1$.

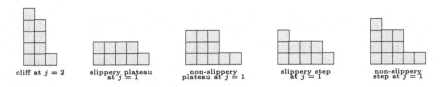

cliff at $j = 2$     slippery plateau at $j = 1$     non-slippery plateau at $j = 1$     slippery step at $j = 1$     non-slippery step at $j = 1$

**Fig. 3.** The slippery plateau and the slippery step have length 3, the non-slippery plateau and the non-slippery step have length 2.

2. If $\alpha$ has a slippery step of length $l$ at $j$, one grain can slip from column $j$ to column $h = j + l + 1$.

Given a partition $\alpha \in \mathcal{L}_n$, if one applies to $\alpha$ the first or second transition rule, every time that $\alpha$ has a cliff or slippery step, then we obtain all partitions $\beta$ such that $\beta \prec \alpha$. If $\beta$ is obtained from $\alpha$ by applying one of the two transition rules, then $\beta$ is called *directly reachable* from $\alpha$. This is denoted by $\alpha \xrightarrow{j} \beta$, where $j$ is the column from which the grain falls or slips. The set $\mathcal{D}(\alpha) = \{\beta \in \mathcal{L}_n \mid \exists j : \alpha \xrightarrow{j} \beta\} = \{\beta \in \mathcal{L}_n \mid \beta \prec \alpha\}$ denotes the set of *all partitions directly reachable* from $\alpha$.

*Example 5.* Consider the partition $\alpha = (5, 3, 2, 1, 1)$ in $\mathcal{L}_{12}$. It has a cliff at $j = 1$, and also has a slippery step at $j = 2$ and at $j = 3$. We obtain $\beta_1$, $\beta_2$ and $\beta_3$ after applying the corresponding transition rules $\alpha \xrightarrow{1} \beta_1$, $\alpha \xrightarrow{2} \beta_2$, $\alpha \xrightarrow{3} \beta_3$, the Ferrers diagrams of which are:

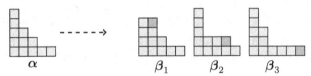

Therefore, $\mathcal{D}(\alpha) = \{(4, 4, 2, 1, 1), (5, 2, 2, 2, 1), (5, 3, 1, 1, 1, 1)\}$.

To study the standard context of the lattice $\mathcal{L}_n$, we quickly recall the notion of context and standard context as known from formal concept analysis [5].

## 2.2 Notions of Formal Concept Analysis

Formal concept analysis (FCA) is a method for data analysis based on the notion of a *formal concept*. FCA was born around 1980, when a research group in Darmstadt, Germany, headed by Rudolf Wille, began to develop a framework for lattice theory applications, [4]: For sets $G$ and $M$ and any binary relation $I \subseteq G \times M$ between $G$ and $M$ the triple $\mathbb{K} = (G, M, I)$ is called a *formal context*, and $I$ its *incidence relation*. The elements of $G$ are called *objects* and those of $M$ are *attributes*. The incidence relation induces a Galois connection between $G$ and $M$ in the natural way. Corresponding pairs $(A, B)$ of Galois closed sets are called *formal concepts*, and they can be ordered by inclusion in the first component.

The ordered set of all formal concepts of $\mathbb{K}$ is denoted by $\mathfrak{B}(\mathbb{K})$ and forms a complete lattice, called the *concept lattice* of $\mathbb{K}$. One of the fundamental theorems of formal concept analysis is the statement, shown in the monograph [5] by Bernhard Ganter and Rudolf Wille, that the concept lattices are, up to isomorphism, exactly the complete lattices. Every concept lattice is complete, and every complete lattice is isomorphic to some concept lattice. This applies in particular to finite lattices $\mathbb{L}$, such as $\mathcal{L}_n$ for $n \in \mathbb{N}_+$, which are always complete and are usually described by their *standard context* $(\mathcal{J}(\mathbb{L}), \mathcal{M}(\mathbb{L}), \leq)$: One easily verifies that every concept lattice of a finite context and every finite lattice is *doubly founded*, see [5] for the definition. Moreover, in [5] it is proved that if $\mathbb{L}$ is a doubly founded complete lattice, then $\mathbb{L}$ is isomorphic to $\mathfrak{B}(\mathcal{J}(\mathbb{L}), \mathcal{M}(\mathbb{L}), \leq)$.

Studying the lattice $\mathcal{L}_n$ of positive integer partitions through its standard context immediately leads to the problem of describing the join-irreducible elements of $\mathcal{L}_n$. Again due to finiteness, this poses the question which $\alpha \in \mathcal{L}_n$ have the property that the conditions in Theorem 3 are fulfilled by exactly one other partition $\beta$, in other words, where $\mathcal{D}(\alpha)$ is a singleton. We find it non-obvious how to describe and count the set of such partitions $\alpha$ directly. Instead we shall try to understand how to derive $\mathcal{J}(\mathcal{L}_{n+1})$ from $\mathcal{J}(\mathcal{L}_n)$ for $n \in \mathbb{N}_+$.

## 3     Relation Between the Lattices $\mathcal{L}_n$ and $\mathcal{L}_{n+1}$

From a partition $\alpha = (a_1, \ldots, a_n)$ of $n \in \mathbb{N}_+$ we obtain a new partition of $n+1$, if we add one grain to its first column, i.e., $\alpha^{\downarrow 1} := (a_1 + 1, a_2, \ldots a_n, 0)$. For $i \in \{1, \ldots, n\}$ we denote by $\alpha^{\downarrow i}$ the tuple $(a_1, a_2, \ldots, a_i + 1, \ldots, a_n, 0)$, which is not always a partition of $n + 1$, and by $\alpha^{\downarrow n+1}$ the tuple $(a_1, a_2, \ldots, a_n, 1)$. If $A \subseteq \mathcal{L}_n$, then $A^{\downarrow i} = \{\alpha^{\downarrow i} \mid \alpha \in A\}$. In Fig. 4 the lattice diagrams of $\mathcal{L}_6$ and $\mathcal{L}_7$ are shown, the elements of $\mathcal{L}_6^{\downarrow 1}$ appear shaded in $\mathcal{L}_7$.

Subsequently we prove that there is an isomorphic copy of $\mathcal{L}_n$ in $\mathcal{L}_{n+1}$.

**Lemma 6.** $\mathcal{L}_n$ *is isomorphic to* $\mathcal{L}_n^{\downarrow 1}$ *for all* $n \geq 1$.

*Proof.* Let $\varphi \colon \mathcal{L}_n \longrightarrow \mathcal{L}_n^{\downarrow 1}$ be given by $\varphi(\alpha) = \alpha^{\downarrow 1}$. Clearly $\varphi$ is injective. Since $\varphi(\mathcal{L}_n) = \mathcal{L}_n^{\downarrow 1}$, we get that $\varphi$ is bijective. To show that $\varphi$ is an order-isomorphism, let $\alpha = (a_1, \ldots, a_n)$ and $\beta = (b_1, \ldots, b_n)$ be in $\mathcal{L}_n$. Then $\alpha^{\downarrow 1} = (a'_1, \ldots, a'_{n+1})$, $\beta^{\downarrow 1} = (b'_1, \ldots, b'_{n+1})$, where $a'_1 = a_1 + 1$, $b'_1 = b_1 + 1$, $a'_{n+1} = b'_{n+1} = 0$ and $a'_i = a_i$, $b'_i = b_i$ for all $i \in \{2 \ldots, n\}$, and it follows that

$$\alpha \leq \beta \iff \sum_{i=1}^{j} a_i \leq \sum_{i=1}^{j} b_i \text{ for all } j \in \{1, \ldots, n\}$$

$$\iff 1 + \sum_{i=1}^{j} a_i \leq 1 + \sum_{i=1}^{j} b_i \text{ for all } j \in \{1, \ldots, n\}$$

$$\Longleftrightarrow \sum_{i=1}^{j} a_i' \le \sum_{i=1}^{j} b_i' \text{ for all } j \in \{1, \dots, n+1\} \Longleftrightarrow \alpha^{\downarrow_1} \le \beta^{\downarrow_1}.$$

$\square$

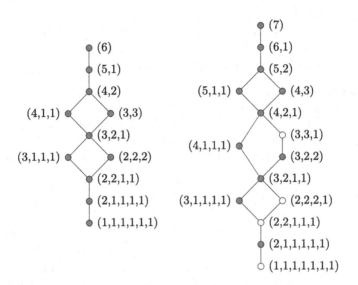

**Fig. 4.** Lattices $\mathcal{L}_6$ and $\mathcal{L}_7$

**Proposition 7** ([7]). *The set $\mathcal{L}_n^{\downarrow_1}$ forms a sublattice of $\mathcal{L}_{n+1}$ for all $n \ge 1$.*

We write $\mathbb{K} \rightarrowtail \mathbb{L}$ to indicate that the lattice $\mathbb{L}$ has a sublattice isomorphic to the lattice $\mathbb{K}$. This embedding relation is transitive, that is, if $\mathbb{K} \rightarrowtail \mathbb{L} \rightarrowtail \mathbb{M}$, then $\mathbb{K} \rightarrowtail \mathbb{M}$. Moreover, from Lemma 6 and Proposition 7 we obtain $\mathcal{L}_n \rightarrowtail \mathcal{L}_{n+1}$ for all $n \in \mathbb{N}_+$.

**Lemma 8.** *If $n \ge 7$, then $\mathcal{L}_n$ is non-modular and non-distributive.*

*Proof.* In Fig. 4 the set $\{(4,2,1), (4,1,1,1), (3,3,1), (3,2,2), (3,2,1,1)\} \subseteq \mathcal{L}_7$ is a sublattice isomorphic to $N_5$. Consequently, for $n \ge 7$ we have by Lemma 6 and Proposition 7 that $N_5 \rightarrowtail \mathcal{L}_7 \rightarrowtail \mathcal{L}_8 \rightarrowtail \dots \rightarrowtail \mathcal{L}_{n-1} \rightarrowtail \mathcal{L}_n$. By transitivity it follows that $N_5 \rightarrowtail \mathcal{L}_n$. Hence, applying the $M_3$-$N_5$-Theorem in [2, p. 89] to $\mathcal{L}_n$, we obtain that it is non-modular and non-distributive. $\square$

If $1 \le n \le 5$, the lattices $\mathcal{L}_n$ form a chain under the dominance ordering. From this and Fig. 4 we see that for $n \in \{1, \dots, 6\}$, the lattice $\mathcal{L}_n$ does not have sublattices isomorphic to $N_5$ or $M_3$, hence it is distributive and thus modular.

In what follows, the process to obtain the lattice $\mathcal{L}_{n+1}$ from $\mathcal{L}_n$ described in [7] is shown. For their construction the authors of [7] analyzed the directly reachable partitions from a partition $\alpha \in \mathcal{L}_n$, considering the following sets.

Let $n \in \mathbb{N}_+$. The *set of all partitions* of $n$ with a *cliff* at 1 is denoted by $\mathrm{C}(n)$, with a *slippery step* at 1 by $\mathrm{S}(n)$, with a *non-slippery step* at 1 by $\mathrm{NS}(n)$, with a *slippery plateau of length* $l \geq 1$ *at* 1 by $\mathrm{P}_l(n)$, and with a *non-slippery plateau at* 1 by $\mathrm{NP}(n)$.

In order to generate the elements of $\mathcal{L}_{n+1}$ from $\mathcal{L}_n$, we follow the two steps given in [7]. For each $\alpha \in \mathcal{L}_n$ we perform:

**Step 1** Add one grain to the first column, that is, construct $\alpha^{\downarrow 1}$.
**Step 2** If $\alpha \in \mathrm{S}(n)$ or $\alpha \in \mathrm{NS}(n)$, then construct $\alpha^{\downarrow 2}$.
  If $\alpha \in \mathrm{P}_l(n)$ for some $l \geq 1$, then construct $\alpha^{\downarrow l+2}$.

Via these two steps every partition $\alpha \in \mathcal{L}_n$ generates at most two elements of $\mathcal{L}_{n+1}$, called the *sons* of $\alpha$. The element $\alpha^{\downarrow 1}$, described in the first step, is called the *left son* and the one in the second step is the *right son* (if it exists). If $\beta$ is a son of $\alpha$, we call $\alpha$ the *father* of $\beta$.

By the following theorem from [7], every element in $\mathcal{L}_{n+1}$ has a father in $\mathcal{L}_n$.

**Theorem 9 (Latapy and Phan [7]).** *For all $n \geq 1$, we have:*

$$\mathcal{L}_{n+1} = \mathcal{L}_n^{\downarrow 1} \sqcup \mathrm{S}(n)^{\downarrow 2} \sqcup \mathrm{NS}(n)^{\downarrow 2} \sqcup \bigsqcup_{1 \leq l < n} \mathrm{P}_l(n)^{\downarrow l+2}.$$

Because these unions are disjoint and because the map $\alpha \mapsto \alpha^{\downarrow i}$ is injective, we have that every $\beta \in \mathcal{L}_{n+1}$ possesses a unique father in $\mathcal{L}_n$. Using this close connection between $\mathcal{L}_n$ and $\mathcal{L}_{n+1}$, in the next section we shall generate the elements of $\mathcal{J}(\mathcal{L}_{n+1})$ from elements of $\mathcal{J}(\mathcal{L}_n)$ via the two steps described above. This will allow us to determine the number of objects and attributes of the standard context of $\mathcal{L}_n$.

## 4   Standard Context of $\mathcal{L}_n$

In this section we study the standard context of $\mathcal{L}_n$, which is a restricted version of the context $(\mathrm{Part}(n), \mathrm{Part}(n), \leq)$. This restriction provides us with a smaller number of objects and attributes, and, most importantly, from these proper subsets of $\mathcal{L}_n$ one can recover the full information about the structure of $\mathcal{L}_n$. This is so since for a finite lattice $\mathbb{L}$ one can prove that $\mathbb{L}$ is isomorphic to the concept lattice $\underline{\mathfrak{B}}(\mathcal{J}(\mathbb{L}), \mathcal{M}(\mathbb{L}), \leq \cap (\mathcal{J}(\mathbb{L}) \times \mathcal{M}(\mathbb{L})))$, and hence our motivation to investigate the standard context of $\mathcal{L}_n$ in more detail.

*Example 10.* We calculate the standard context of $\mathcal{L}_6$, the diagram of which is shown in Fig. 4. We have that

$$\mathcal{J}(\mathcal{L}_6) = \{(2,1,1,1,1), (2,2,1,1), (2,2,2), (3,1,1,1), (3,3), (4,1,1), (5,1), (6)\}$$
$$\mathcal{M}(\mathcal{L}_6) = \{(1,1,1,1,1,1), (2,1,1,1,1), (2,2,2), (3,1,1,1), (3,3), (4,1,1),$$
$$(4,2), (5,1)\}.$$

Thus, the standard context, denoted by $\mathbb{K}(\mathcal{L}_6)$, is:

| $\mathbb{K}$ | 111111 | 21111 | 222 | 3111 | 33 | 411 | 42 | 51 |
|---|---|---|---|---|---|---|---|---|
| 21111 | | × | × | × | × | × | × | × |
| 2211 | | | × | × | × | × | × | × |
| 222 | | | × | | × | × | × | × |
| 3111 | | | | × | × | × | × | × |
| 33 | | | | | × | | × | × |
| 411 | | | | | | × | × | × |
| 51 | | | | | | | | × |
| 6 | | | | | | | | |

Observe that while the context $(\mathcal{L}_6, \mathcal{L}_6, \leq)$ has 11 objects and 11 attributes the standard context has only 8 of each. This advantage in size becomes much more pronounced for larger values of $n$ than just $n = 6$, as we shall see below.

Determining the number of elements in $\mathcal{L}_n$ is a problem that has captivated mathematicians over the years. Whether our approach via formal concept analysis and the standard context $\mathbb{K}(\mathcal{L}_n)$ is to offer any advantage over working with $\mathcal{L}_n$ directly, remains, however, unclear until we show that $\mathbb{K}(\mathcal{L}_n)$ stays much more manageable in size than $\mathcal{L}_n$ when $n$ increases. Therefore, it is natural to ask about the number of elements which are $\vee$-irreducible or $\wedge$-irreducible. To study the growth of the standard context, it is, in fact, sufficient to know how the number $|\mathcal{J}(\mathcal{L}_n)|$ is growing, since partition conjugation * is a lattice antiautomorphism [1], and hence we have $\mathcal{J}(\mathcal{L}_n)^* = \mathcal{M}(\mathcal{L}_n)$ and thus $|\mathcal{J}(\mathcal{L}_n)| = |\mathcal{M}(\mathcal{L}_n)|$. In the following table we present $|\mathcal{J}(\mathcal{L}_n)|$ for $1 \leq n \leq 8$. The apparent pattern in the last column suggests a relation between $|\mathcal{J}(\mathcal{L}_n)|$ and $|\mathcal{J}(\mathcal{L}_{n+1})|$.

| $n$ | $|\mathcal{L}_n|$ | $|\mathcal{M}(\mathcal{L}_n)|$ | $|\mathcal{J}(\mathcal{L}_n)|$ | $|\mathcal{J}(\mathcal{L}_{n+1})| - |\mathcal{J}(\mathcal{L}_n)|$ |
|---|---|---|---|---|
| 1 | 1 | 0 | 0 | 1 |
| 2 | 2 | 1 | 1 | 1 |
| 3 | 3 | 2 | 2 | 2 |
| 4 | 5 | 4 | 4 | 2 |
| 5 | 7 | 6 | 6 | 2 |
| 6 | 11 | 8 | 8 | 3 |
| 7 | 15 | 11 | 11 | 3 |
| 8 | 22 | 14 | 14 | 3 |

For a finite lattice the join-irreducible elements can be characterized as those that cover precisely one element. In $\mathcal{L}_n$ these are those that have exactly one cliff (and no slippery step) or exactly one slippery step (and no cliff). In [1] Brylawski characterized the $\vee$-irreducible elements as follows.

**Lemma 11 ([1, Corollary 2.5]).** *For $n \geq 1$ the join-irreducible partitions from $\mathcal{J}(\mathcal{L}_n)$ can be categorized into four types where always $m, l, s \geq 1$:*

**Type A:** $(\overset{m}{k, k, \ldots, k})$ *for* $k \geq 2$.

**Type B:** $(\overset{m}{k, k, \ldots, k}, \overset{m+l}{k-1, k-1, \ldots, k-1})$ *for* $k \geq 2$.

**Type C:** $(\overset{m}{k, k, \ldots, k}, \overset{m+l}{1, 1, \ldots, 1})$ *for* $k \geq 3$.

**Type D:** $(\overset{m}{k, \ldots, k}, \overset{m+l}{k-1, \ldots, k-1}, \overset{m+l+s}{1, 1, \ldots, 1})$ *for* $k \geq 4$.

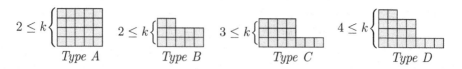

$2 \leq k \Big\{$     *Type A*     $2 \leq k \Big\{$     *Type B*     $3 \leq k \Big\{$     *Type C*     $4 \leq k \Big\{$     *Type D*

When we apply to the elements of $\mathcal{J}(\mathcal{L}_n)$ the two steps described in Sect. 3 to generate elements of $\mathcal{L}_{n+1}$ from $\mathcal{L}_n$, we obtain the following proposition.

**Proposition 12.** *Let $n \geq 3$. For every partition $\boldsymbol{\alpha} \in \mathcal{J}(\mathcal{L}_n) \setminus \{(2, 1, \ldots, 1)\}$ among its (at most two) sons there is exactly one that belongs to $\mathcal{J}(\mathcal{L}_{n+1})$; moreover, the type B partition $(2, 1, \ldots, 1) \in \mathcal{J}(\mathcal{L}_n)$ has two sons in $\mathcal{J}(\mathcal{L}_{n+1})$.*

*Proof.* Every $\boldsymbol{\alpha} \in \mathcal{J}(\mathcal{L}_n)$ belongs to one of the four types from Lemma 11. So we apply to $\boldsymbol{\alpha}$ the two steps from Sect. 3 and analyze the resulting partitions.

**Type A:** If $\boldsymbol{\alpha}$ is a partition of type A, then we consider two subcases:
- If $m = 1$, then $\boldsymbol{\alpha} = (k)$ has a cliff at 1. Hence, it has the left son $\boldsymbol{\alpha}^{\downarrow_1} = (k + 1) = (n + 1)$, which is a partition of type A, but it has no right son. So we have $\boldsymbol{\alpha}^{\downarrow_1} \in \mathcal{J}(\mathcal{L}_{n+1})$.
- If $m \geq 2$, then $\boldsymbol{\alpha} = (k, k, \ldots, \overset{m}{k})$ has a non-slippery plateau at 1, hence it has only the left son $\boldsymbol{\alpha}^{\downarrow_1} = (k + 1, k, \ldots, \overset{m}{k})$, which is a partition of type B. Thus, $\boldsymbol{\alpha}^{\downarrow_1} \in \mathcal{J}(\mathcal{L}_{n+1})$.

**Type B:** If $\boldsymbol{\alpha}$ is a partition of type B, then we consider three subcases:
- If $m = 1, l \geq 1, k \geq 3$, then $\boldsymbol{\alpha} = (k, k - 1, \ldots, \overset{1+l}{k - 1})$ has a non-slippery step at 1. Hence, this partition has two distinct sons. The left one is $\boldsymbol{\alpha}^{\downarrow_1} = (k + 1, k - 1, \ldots, \overset{1+l}{k - 1})$, which is not $\vee$-irreducible, and the right one is $\boldsymbol{\alpha}^{\downarrow_2} = (k, k, k - 1, \ldots, \overset{1+l}{k - 1})$, which belongs to $\mathcal{J}(\mathcal{L}_{n+1})$ because it is of type B (if $l \geq 2$) or of type A (if $l = 1$).
- If $m = 1, l \geq 1, k = 2$, the special $\boldsymbol{\alpha} = (2, 1, \ldots, \overset{1+l}{1})$ has a slippery step at 1. Hence, the left son $\boldsymbol{\alpha}^{\downarrow_1} = (3, 1, \ldots, \overset{1+l}{1})$ belongs to $\mathcal{J}(\mathcal{L}_{n+1})$ since it is of type C, and the right son $\boldsymbol{\alpha}^{\downarrow_2} = (2, 2, 1, \ldots, \overset{1+l}{1})$ *also* belongs to $\mathcal{J}(\mathcal{L}_{n+1})$ for it is of type B (if $l \geq 2$) or of type A (if $l = 1$).
- If $m \geq 2, l \geq 1, k \geq 2$, then $\boldsymbol{\alpha} = (k, k, \ldots, \overset{m}{k}, k - 1, k - 1, \ldots, \overset{m+l}{k - 1})$ has a slippery plateau at 1 of length $m - 1$, thus it has two sons. The left son $\boldsymbol{\alpha}^{\downarrow_1} = (k + 1, k, \ldots, \overset{m}{k}, k - 1, \ldots, \overset{m+1}{k - 1})$ is not $\vee$-irreducible because for $k \geq 3$ it does not belong to any type described by Brylawski, and if $k = 2$, then $\boldsymbol{\alpha}^{\downarrow_1} = (3, 2, \ldots, \overset{m}{2}, 1, \ldots, \overset{m+1}{1})$ is not $\vee$-irreducible either. The right son $\boldsymbol{\alpha}^{\downarrow_{m+1}} = (k, \ldots, \overset{m}{k}, \overset{m+1}{k}, k - 1, \ldots, \overset{m+l}{k - 1})$ belongs to $\mathcal{J}(\mathcal{L}_{n+1})$ because it is of type B (if $l \geq 1$) or of type A (if $l = 1$).

**Type C:** If $\boldsymbol{\alpha}$ is a partition of type C, then we consider two subcases:
- If $m = 1, l \geq 1, k \geq 3$, then $\boldsymbol{\alpha}$ has a cliff at 1, and it has only the left son $\boldsymbol{\alpha}^{\downarrow_1} = (k + 1, 1, \ldots, \overset{1+l}{1})$, belonging to $\mathcal{J}(\mathcal{L}_{n+1})$ because it is of type C.

- If $m \geq 2, l \geq 1, k \geq 3$, then $\alpha$ has a non-slippery plateau at 1. Hence, it only has the left son $\alpha^{\downarrow_1} = (k+1, \overset{m}{k, \ldots, k}, 1, \ldots, \overset{m+l}{1})$, which is of type D. Thus, $\alpha^{\downarrow_1} \in \mathcal{J}(\mathcal{L}_{n+1})$.

**Type D:** If $\alpha$ is a partition of type D, then we consider two subcases:

- If $m \geq 2, l, s \geq 1, k \geq 4$, then $\alpha$ has a slippery plateau at 1 of length $m - 1$, hence it has two sons.

  The left son $\alpha^{\downarrow_1} = (k+1, \overset{m}{k, \ldots, k}, k-1, \ldots, \overset{m+l}{k-1}, 1, \ldots, \overset{m+l+s}{1})$ is not $\vee$-irreducible. The right son

$$\alpha^{\downarrow_{m+1}} = (\overset{m}{k, \ldots, k}, \overset{m+1}{k}, k-1 \ldots, \overset{m+l}{k-1}, 1, \ldots, \overset{m+l+s}{1})$$

  belongs to $\mathcal{J}(\mathcal{L}_{n+1})$, because it is of type D (if $l \geq 2$) or of type C (if $l = 1$).

- If $m = 1, l, s \geq 1, k \geq 4$, then $\alpha$ has a non-slippery step at 1. Hence, the left son $\alpha^{\downarrow_1} = (k+1, k-1, \ldots, \overset{m+l}{k-1}, 1, \ldots, \overset{m+l+s}{1})$ is not $\vee$-irreducible, and the right son $\alpha^{\downarrow_2} = (k, k, k-1, \ldots, \overset{m+l}{k-1}, 1, \ldots, \overset{m+l+s}{1})$ belongs to $\mathcal{J}(\mathcal{L}_{n+1})$ because it is of type D (if $l \geq 2$) or of type C (if $l = 1$).

This shows that every element $\alpha \in \mathcal{J}(\mathcal{L}_n) \setminus \{(2, 1, \ldots, 1)\}$ has precisely one son which belongs to $\mathcal{J}(\mathcal{L}_{n+1})$ and the only partition that has two sons belonging to $\mathcal{J}(\mathcal{L}_{n+1})$ is $(2, 1, \ldots, 1)$. □

Exploiting Proposition 12, we can now define a map $\eta \colon \mathcal{J}(\mathcal{L}_n) \to \mathcal{J}(\mathcal{L}_{n+1})$ as follows: if $\alpha \in \mathcal{J}(\mathcal{L}_n) \setminus \{(2, 1, \ldots, 1)\}$, denote by $\eta(\alpha)$ the unique son of $\alpha$ that belongs to $\mathcal{J}(\mathcal{L}_{n+1})$; moreover, let $\eta(2, 1, \ldots, 1) := (2, 2, 1, \ldots, 1)$. Although the partition $(2, 1, \ldots, 1)$ has two sons that belong to $\mathcal{J}(\mathcal{L}_{n+1})$, it is convenient to ignore the left son in the definition of $\eta$. From the proof of Proposition 12, we can work out an explicit expression for $\eta$. For any $\alpha \in \mathcal{J}(\mathcal{L}_n)$ we have

$$\eta(\alpha) = \begin{cases} \alpha^{\downarrow_1} & \text{if } \alpha \in \mathrm{C}(n) \cup \mathrm{NP}(n), \\ \alpha^{\downarrow_2} & \text{if } \alpha \in \mathrm{S}(n) \cup \mathrm{NS}(n), \\ \alpha^{\downarrow_{l+2}} & \text{if } \alpha \in \mathrm{P}_l(n) \text{ for some } 1 \leq l < n. \end{cases} \tag{1}$$

Since $\eta$ is injective we have $|\eta(\mathcal{J}(\mathcal{L}_n))| = |\mathcal{J}(\mathcal{L}_n)|$. Moreover it is clear that $|\mathcal{J}(\mathcal{L}_{n+1})| = |\eta(\mathcal{J}(\mathcal{L}_n))| + |\mathcal{J}(\mathcal{L}_{n+1}) \setminus \eta(\mathcal{J}(\mathcal{L}_n))|$. Thus, to calculate $|\mathcal{J}(\mathcal{L}_{n+1})|$, we have to identify those elements of $\mathcal{J}(\mathcal{L}_{n+1})$ that are not in the image of $\eta$.

**Lemma 13.** *For $n \geq 3$ we have $\mathcal{J}(\mathcal{L}_{n+1}) \setminus \eta(\mathcal{J}(\mathcal{L}_n)) = E_1 \cup E_2$, where $E_1$ and $E_2$ are the following exceptional sets of partitions of $n + 1$:*

$$E_1 := \{(2, 1, \ldots, \overset{n}{1})\}, \qquad E_2 := \{(\overset{m}{3, \ldots, 3}, 1, \ldots, \overset{m+l}{1}) \mid m \geq 1, l \geq 1\}.$$

*Proof.* Let $\beta \in \mathcal{J}(\mathcal{L}_{n+1})$. Then, similarly to Proposition 12, we consider several cases, according to the four types of $\vee$-irreducible partitions described by Brylawski, cf. Lemma 11.

**Type A:** If $\beta$ is of the form $(k, k, \ldots, \overset{m}{k})$, with $k \geq 2$ and $m \geq 1$, then we consider two subcases:
- If $m \geq 2$, then $\beta$ is a son of $(k \ldots, k, k \overset{m}{-} 1) \in \mathcal{J}(\mathcal{L}_n)$.
- If $m = 1$, then $\beta = (n+1)$ is a son of $(n) \in \mathcal{J}(\mathcal{L}_n)$.

**Type B:** If $\beta$ is of the form $(k, \ldots, \overset{m}{k}, k-1, \ldots, \overset{m+l}{k-1})$, with $k \geq 2$ and $m, l \geq 1$, then we consider three subcases:
- If $m = 1$, $l \geq 1$, and $k \geq 3$, then $\beta$ is a son of $(k-1, \ldots, k-1) \in \mathcal{J}(\mathcal{L}_n)$.
- If $m = 1$, $l \geq 1$, and $k = 2$, then $\beta = (2, 1, \ldots, 1)$ is a son of $(1, 1, \ldots, 1)$, which does not belong to $\mathcal{J}(\mathcal{L}_n)$, but $\beta$ does not arise from any of the four cases in the proof of Proposition 12.
- If $m \geq 2$, $l \geq 1$, and $k \geq 2$, then $\beta$ is a son of $(k, \ldots, \overset{m-1}{k}, k-1, \ldots, k-1)$ from $\mathcal{J}(\mathcal{L}_n)$.

**Type C:** If $\beta$ is of the form $(k, \ldots, \overset{m}{k}, 1, \ldots, \overset{m+l}{1})$, with $k \geq 2$ and $m, l \geq 1$, then we consider three subcases:
- If $m = 1$, $l \geq 1$, and $k \geq 3$, then $\beta$ is the left son of $(k-1, 1, \ldots, \overset{1+l}{1})$. For $k \geq 4$ it belongs to $\mathcal{J}(\mathcal{L}_n)$, so $\beta \in \eta(\mathcal{J}(\mathcal{L}_n))$. However, $\beta \notin \eta(\mathcal{J}(\mathcal{L}_n))$ in the case where $k = 3$ as $\eta$ chooses the right son of $(2, 1 \ldots, 1)$.
- If $m \geq 2$, $l \geq 1$, and $k \geq 4$, then $\beta$ is the right son of the type D partition $(k, \ldots, \overset{m-1}{k}, \overset{m}{k-1}, 1, \ldots, \overset{m+l}{1}) \in \mathcal{J}(\mathcal{L}_n)$.
- If $m \geq 2$, $l \geq 1$, and $k = 3$, then $\beta = (3, \ldots, \overset{m}{3}, 1, \ldots, \overset{m+l}{1})$ is the right son of $(3, \ldots, \overset{m-1}{3}, \overset{m}{2}, 1, \ldots, \overset{m+l}{1}) \notin \mathcal{J}(\mathcal{L}_n)$, but $\beta$ is not obtained from any of the four cases (in the proof) of Proposition 12.

**Type D:** If $\beta$ is of the form $(k, \ldots, \overset{m}{k}, k-1, \ldots, \overset{m+l}{k-1}, 1, \ldots, \overset{m+l+s}{1})$ with $k \geq 4$ and $m, l, s \geq 1$, then we consider two subcases:
- If $m = 1$ and $l, s \geq 1$, then $\beta$ is the left son of $(k-1, \ldots, k-1, \overset{1+l}{1}, 1, \ldots, \overset{1+l+s}{1})$, which belongs to $\mathcal{J}(\mathcal{L}_n)$.
- If $m \geq 2$ and $l, s \geq 1$, then $\beta$ is the right son of the type D partition $(k, \ldots, \overset{m-1}{k}, \overset{m}{k-1}, \ldots, \overset{m+l}{k-1}, 1, \ldots, \overset{m+l+s}{1}) \in \mathcal{J}(\mathcal{L}_n)$.

Thus, almost all elements of $J(\mathcal{L}_{n+1})$ are sons of some element in $J(\mathcal{L}_n)$ except for those of the form $(3, 3, \ldots, \overset{m}{3}, 1, \ldots, \overset{m+l}{1})$, with $m \geq 2$ and $l \geq 1$, or the partition $(2, 1, \ldots, 1)$. Additionally, the partition $(3, 1, \ldots, 1)$, which we could get from a father in $\mathcal{J}(\mathcal{L}_n)$, was excluded from the image of $\eta$ by definition. Therefore, we conclude that the elements of $\mathcal{J}(\mathcal{L}_{n+1})$ that are not in the image of $\eta$ are all of the form $(3, \ldots, \overset{m}{3}, 1, \ldots, \overset{m+l}{1})$ for $m, l \geq 1$ or they are the partition $(2, 1, \ldots, 1)$. $\qquad \square$

We can now describe how to construct the set $\mathcal{J}(\mathcal{L}_{n+1})$ from $\mathcal{J}(\mathcal{L}_n)$.

**Theorem 14.** *Let $n \geq 3$. Then*

$$\mathcal{J}(\mathcal{L}_{n+1}) = (\mathcal{J}(\mathcal{L}_n) \cap C(n))^{\downarrow_1} \sqcup (\mathcal{J}(\mathcal{L}_n) \cap \mathrm{NP}(n))^{\downarrow_1} \sqcup (\mathcal{J}(\mathcal{L}_n) \cap S(n))^{\downarrow_2}$$
$$\sqcup (\mathcal{J}(\mathcal{L}_n) \cap \mathrm{NS}(n))^{\downarrow_2} \sqcup \bigsqcup_{1 \leq l < n} (\mathcal{J}(\mathcal{L}_n) \cap P_l(n))^{\downarrow_{l+2}} \sqcup E_1(n+1) \sqcup E_2(n+1),$$

*where $E_1(k) = \{(2, 1, \ldots, 1)\}, E_2(k) = \{(3, \ldots, \overset{m}{3}, 1, \ldots, \overset{m+l}{1}) \mid m \geq 1, l \geq 1\} \subseteq \mathcal{L}_k$.*

*Proof.* Obviously, $\mathcal{J}(\mathcal{L}_{n+1}) = \eta(\mathcal{J}(\mathcal{L}_n)) \sqcup (\mathcal{J}(\mathcal{L}_{n+1}) \setminus \eta(\mathcal{J}(\mathcal{L}_n)))$, and, from the definition of $\eta$ given in (1), we have

$$\eta(\mathcal{J}(\mathcal{L}_n)) = (\mathcal{J}(\mathcal{L}_n) \cap C(n))^{\downarrow_1} \sqcup (\mathcal{J}(\mathcal{L}_n) \cap \mathrm{NP}(n))^{\downarrow_1} \sqcup (\mathcal{J}(\mathcal{L}_n) \cap S(n))^{\downarrow_2}$$
$$\sqcup (\mathcal{J}(\mathcal{L}_n) \cap \mathrm{NS}(n))^{\downarrow_2} \sqcup \bigsqcup_{1 \leq l < n} (\mathcal{J}(\mathcal{L}_n) \cap P_l(n))^{\downarrow_{l+2}}.$$

By Lemma 13, it follows that $\mathcal{J}(\mathcal{L}_{n+1}) \setminus \eta(\mathcal{J}(\mathcal{L}_n)) = E_1(n+1) \sqcup E_2(n+1)$. $\square$

Counting the elements of $\mathcal{J}(\mathcal{L}_{n+1})$ was the main motivation to study the relationship between $\mathcal{J}(\mathcal{L}_n)$ and $\mathcal{J}(\mathcal{L}_{n+1})$. For this we need one more result.

**Lemma 15.** *For $n \geq 3$ we have $|E_2(n+1)| = \lfloor n/3 \rfloor$, i.e., the number of partitions of $n+1$ of the form $(3, \ldots, \overset{x}{3}, 1, \ldots, \overset{x+y}{1})$ with $x \geq 1$ and $y \geq 1$ is $\lfloor n/3 \rfloor$.*

*Proof.* We have a one-to-one correspondence between each partition of the form $(3, \ldots, \overset{x}{3}, 1, \ldots, \overset{x+y}{1})$ and each integer solution of $3x + y = n + 1$ subject to the inequalities $x \geq 1$, $y \geq 1$. If we subtract 1 from both sides in the last equation and change the variable, then we obtain

$$3x + y = n \text{ subject to } x \geq 1, y \geq 0. \tag{2}$$

Thus, the number of solutions of (2) will be the same as the number of partitions in $E_2(n+1)$. Moreover, we have $3 \cdot \lfloor n/3 \rfloor + r = n$, where $0 \leq r < 3$. Now if $x \in \{1, \ldots, \lfloor n/3 \rfloor\}$, then $n - 3x \geq n - 3\lfloor n/3 \rfloor = r \geq 0$. Letting $y = n - 3x$, we have that $(x, y)$ is a solution of (2). But if $x > \lfloor n/3 \rfloor$, then $x \geq \lfloor n/3 \rfloor + 1$, implying $y = n - 3x \leq n - 3(\lfloor n/3 \rfloor + 1) = r - 3 < 0$, whence $(x, y)$ is no solution. So all solutions of (2) are $\{(x, y) \mid 1 \leq x \leq \lfloor n/3 \rfloor, y = n - 3x\}$; thus, there are $\lfloor n/3 \rfloor$ of them, and therefore $\lfloor n/3 \rfloor$ partitions in $E_2(n+1)$. $\square$

The following theorem reveals the pattern that appeared in the last column of the table shown after Example 10.

**Theorem 16.** *Starting from $|\mathcal{J}(\mathcal{L}_1)| = 0$, for every $n \geq 1$ we have the recursion*

$$|\mathcal{J}(\mathcal{L}_{n+1})| = |\mathcal{J}(\mathcal{L}_n)| + \left\lfloor \frac{n}{3} \right\rfloor + 1.$$

*Proof.* For $n = 1$ we have $|\mathcal{J}(\mathcal{L}_2)| = |\mathcal{J}(\mathcal{L}_1)| + \lfloor 1/3 \rfloor + 1$ since $|\mathcal{J}(\mathcal{L}_1)| = 0$; for $n = 2$ we get $|\mathcal{J}(\mathcal{L}_3)| = |\mathcal{J}(\mathcal{L}_2)| + \lfloor 2/3 \rfloor + 1$. For $n \geq 3$, since $\eta$ is injective, we have $|\eta(\mathcal{J}(\mathcal{L}_n))| = |\mathcal{J}(\mathcal{L}_n)|$, and Lemma 13 states that

$$|\mathcal{J}(\mathcal{L}_{n+1}) \setminus \eta(\mathcal{J}(\mathcal{L}_n))| = |E_1(n+1) \sqcup E_2(n+1)| = |E_1(n+1)| + |E_2(n+1)|.$$

Moreover, $|E_1(n+1)| = 1$, and Lemma 15 yields $|E_2(n+1)| = \lfloor n/3 \rfloor$. Thus,

$$|\mathcal{J}(\mathcal{L}_{n+1})| = |\eta(\mathcal{J}(\mathcal{L}_n))| + |\mathcal{J}(\mathcal{L}_{n+1}) \setminus \eta(\mathcal{J}(\mathcal{L}_n))| = |\mathcal{J}(\mathcal{L}_n)| + \left\lfloor \tfrac{n}{3} \right\rfloor + 1.$$

$\square$

From the last theorem, we can get a closed formula for $|\mathcal{J}(\mathcal{L}_{n+1})|$, which gives us a clearer picture of the cardinality of $|\mathcal{J}(\mathcal{L}_{n+1})|$ in terms of $n$.

**Corollary 17 (cf [3, Proposition 3]).** *For all $n \in \mathbb{N} \setminus \{0\}$ we have*

$$|\mathcal{J}(\mathcal{L}_{n+1})| = n \left( \left\lfloor \tfrac{n}{3} \right\rfloor + 1 \right) - \tfrac{3}{2} \left\lfloor \tfrac{n}{3} \right\rfloor^2 - \tfrac{1}{2} \left\lfloor \tfrac{n}{3} \right\rfloor. \tag{3}$$

*Proof.* By Theorem 16 we have $|\mathcal{J}(\mathcal{L}_{i+1})| - |\mathcal{J}(\mathcal{L}_i)| = \lfloor i/3 \rfloor + 1$ for $i \geq 1$. Thus,

$$|\mathcal{J}(\mathcal{L}_{n+1})| = |\mathcal{J}(\mathcal{L}_{n+1})| - |\mathcal{J}(\mathcal{L}_1)| = \sum_{i=1}^{n} (|\mathcal{J}(\mathcal{L}_{i+1})| - |\mathcal{J}(\mathcal{L}_i)|) = n + \sum_{i=1}^{n} \left\lfloor \tfrac{i}{3} \right\rfloor$$

since $|\mathcal{J}(\mathcal{L}_1)| = 0$. To calculate $\sum_{i=1}^{n} \lfloor i/3 \rfloor$, put $q = \lfloor n/3 \rfloor$ and $r = n - 3q$. Hence, $n = 3q + r$ with $0 \leq r < 3$, and we have $\sum_{i=1}^{n} \lfloor i/3 \rfloor = \sum_{i=0}^{n} \lfloor i/3 \rfloor = u + v$, where

$$u = \sum_{i=0}^{3q} \left\lfloor \tfrac{i}{3} \right\rfloor = 3 \cdot 0 + 3 \cdot 1 + \ldots + 3 \cdot (q-1) + q = 3\frac{q(q-1)}{2} + q = \tfrac{3}{2}q^2 - \tfrac{1}{2}q,$$

$$v = \sum_{i=1}^{r} \left\lfloor \tfrac{3q+i}{3} \right\rfloor = \underbrace{q + \ldots + q}_{r \text{ times}} = rq = (n - 3q)q = nq - 3q^2.$$

Thus, $|\mathcal{J}(\mathcal{L}_{n+1})| = n + \tfrac{3}{2}q^2 - \tfrac{1}{2}q + nq - 3q^2 = n(q+1) - \tfrac{3}{2}q^2 - \tfrac{1}{2}q.$ $\square$

To obtain the standard context $\mathbb{K}(\mathcal{L}_{n+1})$ from $\mathbb{K}(\mathcal{L}_n)$ we do the following. First, we construct the objects, i.e., we construct $\mathcal{J}(\mathcal{L}_{n+1})$ from $\mathcal{J}(\mathcal{L}_n)$ as Theorem 14 shows. Second, we calculate $\mathcal{J}(\mathcal{L}_{n+1})^*$ in order to obtain the attributes of $\mathbb{K}(\mathcal{L}_{n+1})$. Finally, we fill the cross table using the dominance ordering as the incidence relation between objects and attributes. This can be done efficiently:

**Corollary 18.** *The set $\mathcal{J}(\mathcal{L}_n)$ can be obtained in time $\Theta(n^3)$, and the standard context $\mathbb{K}(\mathcal{L}_n)$ can be produced in time $\mathcal{O}(n^5)$.*

*Proof.* With constant effort we create $\mathcal{J}(\mathcal{L}_{k_0})$ for an initial value $k_0 \in \mathbb{N}_+$ and sort its partitions according to the classes occurring in Theorem 14. We then iterate over the $\Theta(k^2)$ partitions in $\mathcal{J}(\mathcal{L}_k)$ (see Corollary 17) plus the $\Theta(k)$ exceptional partitions (see Lemma 15) to obtain $\mathcal{J}(\mathcal{L}_{k+1})$ from $\mathcal{J}(\mathcal{L}_k)$ using the recursion in Theorem 14. As part of this process we divide the resulting partitions of $k + 1$ into the classes of Theorem 14 to prepare for the next step. We do this for $k_0 \leq k < n$, taking $\sum_{k=k_0}^{n-1} \Theta(k^2)$, i.e., $\Theta(n^3)$ time units.

We further iterate once over the $\Theta(n^2)$ objects in $\mathcal{J}(\mathcal{L}_n)$ to get the attributes $\mathcal{J}(\mathcal{L}_n)^*$. Then for each object we iterate over these $\Theta(n^2)$ attributes, decide whether there is a cross in $\mathcal{O}(n)$, and so create $\mathbb{K}(\mathcal{L}_n)$ in $\mathcal{O}(n^5)$ time units.  □

Note that any algorithm for $\mathbb{K}(\mathcal{L}_n)$ has a lower bound of $o(n^4)$ time units as already writing a non-trivial context of size $\Theta(n^2) \times \Theta(n^2)$ needs $o(n^4)$ individual steps. A non-recursive algorithm can be obtained from [3, Proposition 2].

The results leading to Theorems 14, 16 and Corollary 17 are based on the second author's bachelor's thesis, which was completed in February 2020 and defended in June. We were surprised to learn that meanwhile the sizes of $\mathcal{J}(\mathcal{L}_n)$ had been discovered independently on the other side of the globe [3]. Seeing [3], the question of embeddability of $\mathbb{K}(\mathcal{L}_n)$ into $\mathbb{K}(\mathcal{L}_{n+1})$ caught our attention, as it was answered negatively for symmetric context embeddings ($n = 9$ being the smallest impossible case) but left open in general. We were intrigued to construct a non-symmetric embedding based on $\eta \colon \mathcal{J}(\mathcal{L}_n) \to \mathcal{J}(\mathcal{L}_{n+1})$, cp. (1); however our attempts were bound to fail:

**Proposition 19.** *There is no context embedding of $\mathbb{K}(\mathcal{L}_9)$ into $\mathbb{K}(\mathcal{L}_{10})$.*

*Proof.* A context embedding is a pair of injective maps $\alpha \colon \mathcal{J}(\mathcal{L}_n) \to \mathcal{J}(\mathcal{L}_{n+1})$, $\beta \colon \mathcal{M}(\mathcal{L}_n) \to \mathcal{M}(\mathcal{L}_{n+1})$ that send crosses to crosses and empty cells of $\mathbb{K}(\mathcal{L}_n)$ to empty cells. As $\mathbb{K}(\mathcal{L}_n)$ is finite, this fact can be written as a long conjunction over $G \times M := \mathcal{J}(\mathcal{L}_n) \times \mathcal{M}(\mathcal{L}_n)$; also the injectivity requirement can be added by saying that values of distinct objects (attributes resp.) have to be distinct:

$$\bigwedge_{\substack{(g,m)\in G\times M \\ g\leq m}} \alpha(g) \leq \beta(m) \wedge \bigwedge_{\substack{(g,m)\in G\times M \\ g\not\leq m}} \neg(\alpha(g) \leq \beta(m)) \wedge \bigwedge_{\substack{g,h\in G \\ g\sqsubset_G h}} \alpha(g) \neq \alpha(h) \wedge \bigwedge_{\substack{l,m\in M \\ l\sqsubset_M m}} \beta(l) \neq \beta(m),$$

where $\sqsubset_G$ and $\sqsubset_M$ are arbitrary strict linear orders on $G$ and $M$, resp. Now the search for a context embedding obviously is an instance of a constraint satisfaction problem where the variables are the elements of $G$ and $M$, and the solutions (value assignments in $\mathcal{J}(\mathcal{L}_{n+1})$ and $\mathcal{M}(\mathcal{L}_{n+1})$, resp.) are the maps of the embedding. The instance is given by the following conjunctive formula

$$\varphi = \bigwedge_{\substack{(g,m)\in G\times M \\ g\leq m}} g \leq m \wedge \bigwedge_{\substack{(g,m)\in G\times M \\ g\not\leq m}} \neg(g \leq m) \wedge \bigwedge_{\substack{g,h\in G \\ g\sqsubset_G h}} g \neq h \wedge \bigwedge_{\substack{l,m\in M \\ l\sqsubset_M m}} l \neq m,$$

and for $n = 9$ we coded it in the SMT-LIB2.0 language and fed it to the Z3 sat solver [9,10]. Instead of an embedding the answer was that $\varphi$ was unsatisfiable over the structure $(\mathcal{J}(\mathcal{L}_{10}) \times \mathcal{M}(\mathcal{L}_{10}), \leq)$. We also had the solver generate a formal proof of unsatisfiability; alas, it is 21 694 lines long and remains inaccessible to humans in its present form.  □

Studying the standard context $\mathbb{K}(\mathcal{L}_n)$ is of interest because its concept lattice $\mathfrak{B}(\mathbb{K}(\mathcal{L}_n))$ is isomorphic to $\mathcal{L}_n$. From equation (3) we have that the number of objects (and attributes) of $\mathbb{K}(\mathcal{L}_n)$ has order $\Theta(n^2)$. This means that, as $n$ increases, the number of objects (and attributes) in the context $\mathbb{K}(\mathcal{L}_n)$ grows much more slowly than the number $p(n)$ of unrestricted partitions of $n$, which satisfies the asymptotic formula $p(n) \sim \frac{1}{4n\sqrt{3}} e^{\pi\sqrt{2n/3}}$ [6].

It is quite remarkable (though from a formal concept analysis perspective not too surprising) that from a fraction of the partitions of $n$, we can construct a context $\mathbb{K}(\mathcal{L}_n)$ the size of which is 'only' of order $\mathcal{O}(n^4)$ but whose number of formal concepts is precisely the total number of partitions of $n$.

Asymptotic expansions [12,13] (and even exact numbers for many integers $n$, see [11]) for the sizes $p(n)$ of the lattices $\mathfrak{B}(\mathbb{K}(\mathcal{L}_n)) \cong \mathcal{L}_n$ are known. Moreover, by Corollary 17, the standard contexts $\mathbb{K}(\mathcal{L}_n)$ satisfy precise size estimates $\mathcal{O}(n^4)$ and can be computed efficiently, that is, in polynomial time $\mathcal{O}(n^5)$, see Corollary 18. Therefore, in addition to being a novel, perhaps slightly esoteric, means for the computation of $p(n)$, the sequence $(\mathbb{K}(\mathcal{L}_n))_{n\in\mathbb{N}_+}$ of contexts might also prove itself to be a promising playground for testing conjectures or the efficiency of new formal concept analytic algorithms (regarding e.g. the computation of concept lattices or stem bases etc.).

**Acknowledgements.** The authors are grateful to Prof. Bernhard Ganter for pointing out the topic, for helpful advice and encouraging remarks. They also would like to thank Dr. Christian Meschke for his constant support. Moreover, they appreciate the constructive comments and suggestions given by the anonymous referees, which improved the presentation of the material.

# References

1. Brylawski, T.: The lattice of integer partitions. Discrete Math. **6**(3), 201–219 (1973). https://doi.org/10.1016/0012-365X(73)90094-0
2. Davey, B.A., Priestley, H.A.: Introduction to Lattices and Order, 2nd edn. Cambridge University Press, New York (2002). https://doi.org/10.1017/CBO9780511809088
3. Ganter, B.: Notes on integer partitions. In: 15th International Conference on Concept Lattices and Their Applications, Tallinn, pp. 19–31. CEUR-WS.org (2020)
4. Ganter, B., Wille, R.: Applied lattice theory: formal concept analysis. In: Grätzer, G. (ed.) General Lattice Theory, 2nd edn. Birkhäuser, Basel (1998)
5. Ganter, B., Wille, R.: Formal Concept Analysis. Mathematical Foundations. Springer, Berlin (1999). https://doi.org/10.1007/978-3-642-59830-2
6. Hardy, G.H., Ramanujan, S.: Asymptotic formulæ in combinatory analysis. Proc. London Math. Soc. (2) **17**(1), 75–115 (1918). https://doi.org/10.1112/plms/s2-17.1.75
7. Latapy, M., Phan, T.H.D.: The lattice of integer partitions and its infinite extension. Discrete Math. **309**(6), 1357–1367 (2009). https://doi.org/10.1016/j.disc.2008.02.002
8. Mahnke, D.: Leibniz auf der Suche nach einer allgemeinen Primzahlgleichung. Bibl. Math. (3) **XIII**, 29–61 (1912–1913). https://www.ophen.org/pub-102519

9. Microsoft Research: Z3 Theorem Prover (2021). https://github.com/z3prover/z3 or https://rise4fun.com/Z3/

10. de Moura, L., Bjørner, N.: Z3: an efficient SMT solver. In: Ramakrishnan, C.R., Rehof, J. (eds.) TACAS 2008. LNCS, vol. 4963, pp. 337–340. Springer, Heidelberg (2008). https://doi.org/10.1007/978-3-540-78800-3_24

11. OEIS: Sequence A000041. In: Sloane, N.J.A. (ed.) The On-line Encyclopedia of Integer Sequences. OEIS Foundation (2020). https://oeis.org/A000041. Accessed 13 Dec 2020

12. Rademacher, H.: On the partition function $p(n)$. Proc. London Math. Soc. (2) **43**(4), 241–254 (1937). https://doi.org/10.1112/plms/s2-43.4.241

13. Rademacher, H.: On the expansion of the partition function in a series. Ann. Math. (2) **44**(3), 416–422 (1943). https://doi.org/10.2307/1968973

# Fixed-Point Semantics for Barebone Relational Concept Analysis

Jérôme Euzenat[(⊠)]

Univ. Grenoble Alpes, Inria, CNRS, Grenoble INP, LIG, 38000 Grenoble, France
Jerome.Euzenat@inria.fr

**Abstract.** Relational concept analysis (RCA) extends formal concept analysis (FCA) by taking into account binary relations between formal contexts. It has been designed for inducing description logic TBoxes from ABoxes, but can be used more generally. It is especially useful when there exist circular dependencies between objects. In this case, it extracts a unique stable concept lattice family grounded on the initial formal contexts. However, other stable families may exist whose structure depends on the same relational context. These may be useful in applications that need to extract a richer structure than the minimal grounded one. This issue is first illustrated in a reduced version of RCA, which only retains the relational structure. We then redefine the semantics of RCA on this reduced version in terms of concept lattice families closed by a fixed-point operation induced by this relational structure. We show that these families admit a least and greatest fixed point and that the well-grounded RCA semantics is characterised by the least fixed point. We then study the structure of other fixed points and characterise the interesting lattices as the self-supported fixed points.

## 1 Motivation

Formal concept analysis (FCA [7]) is a useful tool for inducing a classification structure from data. Relational concept analysis (RCA [13]) is one of its extensions allowing to take advantage of relationships between objects to extract dependent concept lattices. One of its strong point is its ability to deal with circular dependencies between objects.

Although the result returned by RCA is solid and useful, it may not be the only possible result. The relational structure, when containing circuits, has the capability to induce richer lattice structures. Indeed, in the absence of information or of reason to separate objects, RCA classifies them within the same concept. On the contrary, in the absence of information or of reason to aggregate objects, it is possible to keep them in different concepts. A good compromise may sometimes reside in between these two extremes. As a data mining procedure, RCA can be useful in returning all possible structures and not necessarily the safest ones.

This is not really a problem in the target RCA application: extracting the core classes of a description logic ontology. However, this may be a problem

A. Braud et al. (Eds.): ICFCA 2021, LNAI 12733, pp. 20–37, 2021.
https://doi.org/10.1007/978-3-030-77867-5_2

for other applications. This work was initially motivated by one such application of RCA: We developed a link key candidate extraction algorithm on top of relational concept analysis [2]. Link keys are rules for identifying the same individuals from different data sources. In this context, the concepts of extracted lattices are link key candidates which will be selected on the basis of two independent measures [1]. As a data mining task, RCA is more useful if it generates all the possible link key candidates.

Hereafter, we illustrate the considered problem on $RCA^0$, a minimal version of RCA. Although $RCA^0$ is simply a convenient way to illustrate the problem it requires solutions that will apply to RCA as a whole.

Understanding the nature of the problem and its relation with RCA lead to consider its semantics. The current semantics of RCA [14] focusses on the grounding of the process. We redefine this semantics on properties directly characterising the solutions.

We first consider the core function involved in the classical RCA algorithm and identify acceptable results as the fixed points of this function. We show, in the case of $RCA^0$, that the classical RCA semantics corresponds to extracting its least fixed point.

We also provide a direct way to generate the greatest fixed point. However, although RCA extracts the minimal fixed point in its simplest form, this is not the case of the greatest fixed point: it would make reference to non-existent concepts. Hence we discuss the notion of self-supported concept lattice, so that the acceptable RCA results would be self-supported fixed points.

FCA is a domain of fixed points, hence it is easy to get lost among the various fixed points involved: ($a$) In description logics, on which RCA relies, the semantics of concepts is given by fixed points when circularities occur [11]; ($b$) FCA's goal is to compute fixed points: concepts are the result of a closure operator which is also a fixed point [4]; ($c$) finally, when confronted to cycles, the RCA concept lattice is the fixed point of the function that grows a lattice family from the previous one. The present work is concerned with the latter kind of fixed points.

In the remainder, we present the context as well as related work (Sect. 2). We illustrate the considered problem on a minimal example (Sect. 3). We then provide a fixed-point semantics for RCA based on a context-expansion function (Sect. 4) which allows to characterise the classical RCA semantics. However, this semantics being not fully satisfactory, we introduce the complementary notion of self-supported concept lattices (Sect. 5). We finally discuss concrete processing issues (Sect. 6).

## 2    Preliminaries and Related Work

We mix preliminaries with related works for reasons of space, but also because the paper directly builds on this related work.

## 2.1   Formal Concept Analysis

Formal Concept Analysis (FCA) [7] starts with a binary context $\langle G, M, I \rangle$ where $G$ denotes a set of objects, $M$ a set of attributes, and $I \subseteq G \times M$ a binary relation between $G$ and $M$, called the incidence relation. The statement $gIm$ is interpreted as "object $g$ has attribute $m$". Two operators $\cdot^\uparrow$ and $\cdot^\downarrow$ define a Galois connection between the powersets $\langle 2^G, \subseteq \rangle$ and $\langle 2^M, \subseteq \rangle$, with $A \subseteq G$ and $B \subseteq M$:

$$A^\uparrow = \{m \in M \mid gIm \text{ for all } g \in A\}$$
$$B^\downarrow = \{g \in G \mid gIm \text{ for all } m \in B\}$$

The operators $\cdot^\uparrow$ and $\cdot^\downarrow$ are decreasing, i.e. if $A_1 \subseteq A_2$ then $A_2^\uparrow \subseteq A_1^\uparrow$ and if $B_1 \subseteq B_2$ then $B_2^\downarrow \subseteq B_1^\downarrow$. Intuitively, the less objects there are, the more attributes they share, and dually, the less attributes there are, the more objects have these attributes. It can be checked that $A \subseteq A^{\uparrow\downarrow}$ and that $B \subseteq B^{\downarrow\uparrow}$, that $A^\uparrow = A^{\uparrow\downarrow\uparrow}$ and that $B^\downarrow = B^{\downarrow\uparrow\downarrow}$.

For $A \subseteq G$, $B \subseteq M$, a pair $\langle A, B \rangle$, such that $A^\uparrow = B$ and $B^\downarrow = A$, is called a formal concept, where $A$ is the extent and $B$ the intent of $\langle A, B \rangle$. Moreover, for a formal concept $\langle A, B \rangle$, $A$ and $B$ are closed sets for the closure operators $\cdot^{\uparrow\downarrow}$ and $\cdot^{\downarrow\uparrow}$, respectively, i.e. $A^{\uparrow\downarrow} = A$ and $B^{\downarrow\uparrow} = B$.

Concepts are partially ordered by $\langle A_1, B_1 \rangle \leq \langle A_2, B_2 \rangle \Leftrightarrow A_1 \subseteq A_2$ or equivalently $B_2 \subseteq B_1$. With respect to this partial order, the set of all formal concepts forms a complete lattice called the concept lattice of $\langle G, M, I \rangle$.

Formal concept analysis can be considered as a function that associates to a formal context $\langle G, M, I \rangle$ its concept lattice $\langle C, \leq \rangle = FCA(\langle G, M, I \rangle)$ (or $\mathfrak{B}(G, M, I)$ [7]). By abuse of language, when a variable $L$ denotes a concept lattice $\langle C, \leq \rangle$, $L$ will also be used to denote $C$.

## 2.2   Extending FCA

Formal concept analysis is defined on relatively simple structures hence many extensions have been designed. They allow FCA to ($a$) deal with more complex input structure, and/or ($b$) generate more expressive and interpretable knowledge structures.

**Scaling.** Scaling is one type of extension of type ($a$). A scaling operation $\varsigma : \mathscr{X} \mapsto 2^D$ generates boolean attributes named after a language $D$ from a structure $\Sigma \in \mathscr{X}$. In FCA, $D = M$ and $I$ is provided by its matrix. In scaled contexts, this language can be interpreted so that the incidence relation $I$ is immediately derived from the attribute $m$ following:

$$gIm \text{ iff } \Sigma \models m(g)$$

Hence, adding attributes to a context under such a structure may be performed as:

$$K_{M'}^\Sigma(\langle G, M, I \rangle) = \langle G, M \cup M', I \cup \{\langle g, m \rangle \in G \times M' \mid \Sigma \models m(g)\}\rangle$$

Applying a scaling operation $\varsigma$ to a formal context $K$ following a structure $\Sigma$ can be thus decomposed into (i) determining the set of attributes $\varsigma(\Sigma)$ to add, and (ii) extending the context with such attributes:

$$\sigma_\varsigma(K, \Sigma) = K^{\Sigma}_{\varsigma(\Sigma)}(K)$$

Many conceptual scaling operations have been discussed in [7] for dealing with non boolean variables in formal contexts. In general, $\Sigma$ is void, $D$ is expressed as predicates, e.g. $\cdot = v$ for nominal scaling or $\cdot \leq n$ for ordinal scaling, and $\models$ is the evaluation of the predicate for the value.

Logical scaling [12] has been introduced for more versatile languages such as description logics and SQL. It introduces query results within formal contexts. In this case, $\Sigma$ is a logical theory or database tables, $D$ the set of formulas of the logic or instantiated queries and $\models$ is entailment or query evaluation.

Relational scaling operations considered in [13] are based on a struture $\Sigma = \langle R, C \rangle$ made of a family of relations $R = \{r_y\}_{y \in Y}$, i.e. relations $r_y \subseteq G_x \times G_z$ between two sets of objects, and a family $C = \{C_x\}_{x \in X}$ of sets of concepts whose extent is a subset of $G_x$. Its language $D_{\varsigma,R,C}$ is the set of attribute descriptions involving $\varsigma$, $R$ and $C$. For example, qualified existential scaling ($\exists$) adds attributes $\exists r.c$ for $r \in R$, $r \subseteq G_x \times G_z$, $c \in C_z$ and $\models$ checks that

$$\langle R, L \rangle \models gI\exists r.c \text{ iff } \exists g'; \langle g, g' \rangle \in r \wedge g' \in extent(c)$$

Various relational scaling operations are used in RCA such as existential, strict and wide universal, min and max cardinality, which all follow the classical role restriction semantics of description logics [3].

**Other Extensions.** Pattern structures [6,9] provide a more structured attribute language without scaling. However, its use is not directly related to the problem of context dependencies considered here as the attributes do not refer to concepts.

On the contrary, other approaches [5,10] aim at extracting conceptual structures from $n$-ary relations without resorting to scaling. Their concepts have intents that can be thought of as conjunctive queries and extents as tuples of objects, i.e. answers to these queries. Hence, instead of being classes, i.e. monadic predicates, concepts correspond to general polyadic predicates. For that purpose, they rely on more expressive input, e.g. in Graph-FCA [5] the incidence relation is a hypergraph between objects, and produce a more expressive representation. A comparison of RCA and Graph-FCA is provided in [8]. Graph-FCA adopts a different approach than RCA but should, in principle, suffer from the same problem as the one illustrated here. However, intents would need to refer to concepts so created, i.e. named subqueries. This remains to be studied.

## 2.3   Relational Concept Analysis

Relational Concept Analysis (RCA) [13] extends FCA to the processing of relational datasets and allows inter-object relations to be materialised and incorporated into formal concept intents. RCA is a way to induce a description logic TBox from a simple ABox [3], using specific scaling operations. It may also be though of as a general way to deal with circular references using different scaling operations.

RCA applies to a relational context $\langle K^0, R \rangle$, composed of a set of formal contexts $K^0 = \{\langle G_x, M_x^0, I_x^0 \rangle\}_{x \in X}$ and a set of binary relations $R = \{r_y\}_{y \in Y}$. A relation $r_y \subseteq G_x \times G_z$ connects two object sets, a domain $G_x$ $(dom(r_y) = G_x, x \in X)$ and a range $G_z$ $(ran(r_y) = G_z, z \in X)$.

RCA applies relational scaling operations from a set $\Omega$ to each $K_x^i \in K^i$ and all relations $r_y \subseteq G_x \times G_z$ from the set of concepts in corresponding $L_z = FCA(K_z^i)$.

For performing its operations, RCA thus relies on $FCA$ and $\sigma_\varsigma$. More precisely it uses $FCA^*$ and $\sigma_\Omega^*$ defined as:

$$FCA^*(\{\langle G_x, M_x, I_x \rangle\}_{x \in X}) = \{FCA(\langle G_x, M_x, I_x \rangle)\}_{x \in X}$$

$$\sigma_\Omega^*(\{\langle G_x, M_x, I_x \rangle\}_{x \in X}, R, \{L_x\}_{x \in X}) = \left\{ \bigoplus_{r_y \in R \ | \ r \subseteq G_x \times G_z}^{\varsigma \in \Omega} \sigma_\varsigma(\langle G_x, M_x, I_x \rangle, r_y, L_z) \right\}_{x \in X}$$

such that $\bigoplus_{r_y \in R \ | \ r \subseteq G_x \times G_z}^{\varsigma \in \Omega}$ scales, with all operations in $\Omega$, the given context with all the relations starting from $x$ (to any $z$).

RCA starts from the initial formal context family $K^0$ and thus iterates the application of the two operations:

$$K^{i+1} = \sigma_\Omega^*(K^i, R, FCA^*(K^i))$$

until reaching closure, i.e. reaching $n$ such that $K^{n+1} = K^n$. Then, $RCA_\Omega(K^0, R) = FCA^*(K^n)$.

By abuse of notation, we note $\langle G, M, I \rangle \subseteq \langle G, M', I' \rangle$ whenever $M \subseteq M'$ and $I = I' \cap (G \times M)$. In this case, because $I$ is the incidence relation between the same $G$ and $M \subseteq M'$, the relation only depends on $M$ and $M'$. This is generalised to formal context families $\{\langle G_x, M_x, I_x \rangle\}_{x \in X} \subseteq \{\langle G_x, M_x', I_x' \rangle\}_{x \in X}$ whenever $\forall x \in X, M_x \subseteq M_x'$.

The RCA process always reaches a closed formal context family for reason of finiteness [13] and the sequence $(K^i)_{i=0}^n$ is non-contracting, i.e. $\forall i \geq 0, K^i \subseteq K^{i+1}$ [14]. The well-grounded semantics of RCA [14] further establishes that RCA indeed finds $the$ $K^n$ satisfying these constraints through correctness (the concepts of $FCA^*(K^n)$ are grounded in $K^0$ through $R$) and completess (all such concepts are in $K^n$).

## 2.4   RCA⁰

To keep the paper short and simple, we restrict it to $RCA^0$, a special case of RCA. It is restricted in two ways:

- It contains only one formal context ($|X| = 1$),
- which has no attributes ($M_x^0 = \varnothing$).

Additionally, we will consider below only qualified existential scaling ($\Omega = \{\exists\}$).

Because $RCA^0$ is a restriction of RCA, we will use the same notation as defined above, thought it operates on simpler structures.

Although $RCA^0$ seems very simple, FCA can be encoded into $RCA^0$. Introducing $RCA^0$ is sufficient to hint at the problem that we want to illustrate[1].

## 3   RCA May Accept Different Concept Lattice Families: Illustration

As an $RCA^0$ example, consider the following ABox:

$A = \{\top(a), \top(b), \top(c), \top(d), p(a, b), p(b, a), p(c, d), p(d, c), p(a, a), p(b, b)\}$

This can be encoded as an empty formal context and the relation of Fig. 1 (left). The empty context will generate the single lattice of Fig. 1 (right) (names are assigned to concepts according to their extent).

| $p$ | $a$ | $b$ | $c$ | $d$ |
|---|---|---|---|---|
| $a$ | × | × |   |   |
| $b$ | × | × |   |   |
| $c$ |   |   |   | × |
| $d$ |   |   | × |   |

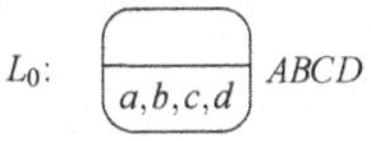

$L_0$:  [ a,b,c,d ] ABCD

**Fig. 1.** Relation (left) and initial concept lattice (right).

Scaling with $\exists$ and $p$ provides the attribute $\exists p.ABCD$ which generates the new context of Fig. 2 (left), leading to the lattice of Fig. 2 (right) which is the one returned by RCA.

|   | $\exists p.ABCD$ |
|---|---|
| $a$ | × |
| $c$ | × |
| $b$ | × |
| $d$ | × |

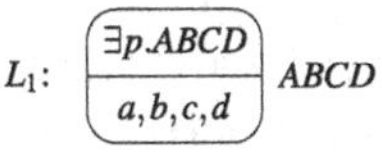

$L_1$:  [ $\exists p.ABCD$ / a,b,c,d ] ABCD

**Fig. 2.** Scaled context (left) and final concept lattice $L_1$ (right).

However, the concept lattices of Fig. 3 are other valid lattices worth considering.

---

[1] An anonymous reviewer complements the remarks of Sect. 2.2 noting that $RCA^0$ is also very related to Graph-FCA as they both have only one context and using existential scaling.

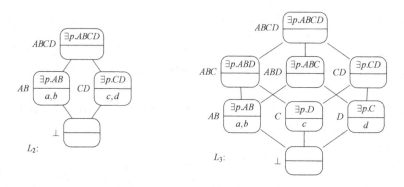

**Fig. 3.** Alternative concept lattices ($L_2$ and $L_3$).

They correspond to different knowledge bases:

$T_1 = \{ABCD \sqsubseteq \exists p.ABCD\}$

$A_1 = \{ABCD(a), ABCD(b), ABCD(c), ABCD(d), p(a,b), p(b,a), p(c,d), p(d,c), p(a,a), p(b,b)\}$

and

$$T_2 = \{AB \sqsubseteq \top \sqcap \exists p.AB, CD \sqsubseteq \top \sqcap \exists p.CD, ABCD \sqsubseteq \exists p.ABCD\}$$
$$A_2 = \{AB(a), AB(b), CD(c), CD(d), p(a,b), p(b,a), p(c,d), p(d,c), p(a,a), p(b,b)\}$$

and

$$T_3 = \{AB \sqsubseteq ABC \sqcap ABD \sqcap \exists p.AB, C \sqsubseteq ABC \sqcap CD \sqcap \exists p.D, D \sqsubseteq ABD \sqcap CD \sqcap \exists p.C,$$
$$ABC \sqsubseteq ABCD \sqcap \exists p.ABD, ABD \sqsubseteq ABCD \sqcap \exists p.ABC, CD \sqsubseteq ABCD \sqcap \exists p.CD,$$
$$ABCD \sqsubseteq \exists p.ABCD\}$$
$$A_3 = \{AB(a), AB(b), C(c), D(d), p(a,b), p(b,a), p(c,d), p(d,c), p(a,a), p(b,b)\}$$

In addition to extracting the TBox, these extend the ABox. However, in RCA and FCA, objects are also assigned to the created concepts. In this case, this assignment has consequences on the scaled attributes taken into account and hence the resulting lattice.

As in classical RCA, each concept of these lattices is closed with respect to the specific formal context scaled by $\exists$ and $p$ from the concepts of the lattice. Moreover, the lattices are self-supported in the sense that their attributes refer only to their concepts.

The problem applies to RCA as a whole as $RCA^0$ is included in RCA. Hence the question: Why does RCA returns only one lattice, and which one? Answering it requires to reconsider the RCA semantics.

## 4    Semantics and Properties: A Context Approach

The alternative lattices presented in Sect. 3 are legitimate because, independently of the attributes, they rely exclusively on the structure of the relations between formal contexts. This structure is already used in the well-grounded RCA semantics, but they have not been fully exploited.

The answer will require to further define 'legitimate', in terms of fixed points of a specific function, and characterise the semantics of RCA as indeed grounded, in terms of these fixed points.

### 4.1    The Lattice $\mathscr{K}$ of RCA$^0$ Contexts

We first define the space of formal context families considered by RCA. They are determined by three elements given once and for all: $K^0 = \{\langle G, M^0, I^0 \rangle\}_{x \in X}$, $R = \{r_y\}_{y \in Y}$, and $\Omega$. This is even more specific for RCA$^0$ with $K^0 = \langle G, \varnothing, \varnothing \rangle$ and $\Omega = \{\varsigma_\exists\}$, but for most of this section we will ignore it.

The contexts considered by RCA are formal context families obtained by the scaled initial context using the scaling operations. Given a finite set of objects $G$, the set of concepts that can be created from such contexts is finite and moreover each concept can be identified by its extent. Hence, we will consider that this induces a set of concept names $N(G) = 2^G$ valid for any such concept lattice; the extent of a so named concept will be the set of objects in its name. Given a finite set of relations $R$ and scaling operations $\Omega$, this determines the finite set $D_{\Omega,R,N(G)} = \bigcup_{r \in R}^{\varsigma \in \Omega} D_{\varsigma,r,N(G)}$ of possible scaled attributes in RCA$^0$.

Hence, the formal contexts considered by RCA are those obtained by adding subsets of $D_{\Omega,R,N(G)}$:

$$\mathscr{K}_{\langle G,M^0,I^0 \rangle, R, \Omega} = \{K_M^{\langle R,N(G) \rangle}(\langle G, M^0, I^0 \rangle) \mid M \subseteq D_{\Omega,R,N(G)}\}$$

with $K_M^{\langle R,N(G) \rangle}(.)$ the operation defined in Sect. 2.2.

Given $K, K' \in \mathscr{K}_{\langle G,M^0,I^0 \rangle, R, \Omega}$ such that $K = \langle G, M^0 \cup M, I^0 \cup I \rangle$ and $K' = \langle G, M^0 \cup M', I^0 \cup I' \rangle$, $K \vee K'$ and $K \wedge K'$ are defined as:

$$K \vee K' = \langle G, M^0 \cup (M \cup M'), I^0 \cup (I \cup I') \rangle \qquad \text{(join)}$$
$$K \wedge K' = \langle G, M^0 \cup (M \cap M'), I^0 \cup (I \cap I') \rangle \qquad \text{(meet)}$$

It is clear that $\mathscr{K}_{K^0,R,\Omega}$ is closed by meet and join.

*Property 1.* $\langle \mathscr{K}_{K^0,R,\Omega}, \vee, \wedge \rangle$ is a complete lattice

*Proof.* $\vee$ and $\wedge$ satisfy commutativity, associativity and the absorption laws directly from the union and intersection on sets, so this is a lattice. It is complete because finite. □

*Property 2.* $\forall K, K' \in \mathscr{K}_{K^0,R,\Omega}, K \subseteq K'$ iff $K = K \wedge K'$

*Proof.* This property also comes directly from its set theoretic counterpart application to $M$ and $M'$: $K \subseteq K' \Leftrightarrow M \subseteq M' \Leftrightarrow M = M \cap M' \Leftrightarrow K = K \wedge K'$ □

## 4.2  The Context Expansion Function $F$

We reformulate RCA as based on a main single function, $F_{K^0,R,\Omega}$, the context expansion function attached to a relational context $\langle K^0, R \rangle$ and a set $\Omega$ of scaling operations.

**Definition 1 (Context expansion function).** *Given a relational context $\langle K^0, R \rangle$ and a set of relational scaling operations $\Omega$, the function $F_{K^0,R,\Omega}$ : $\mathscr{K}_{K^0,R,\Omega} \mapsto \mathscr{K}_{K^0,R,\Omega}$ is defined by:*

$$F_{K^0,R,\Omega}(K) = \sigma_{\Omega}^*(K, R, FCA^*(K)))$$

The function expression is independent from $K^0$, $K^0$ is used to restrict the domain of the function so that its elements cover $K^0$. From now on, we will abbreviate $\mathscr{K}_{K^0,R,\Omega}$ as $\mathscr{K}$ and $F_{K^0,R,\Omega}$ as $F$. This is legitimate because, for a given relational context, $K^0$, $R$ and $\Omega$ do not change. $F$ is an extensive and monotone internal operation for $\mathscr{K}$:

*Property 3.* $\forall K \in \mathscr{K}$, $F(K) \in \mathscr{K}$

*Proof.* Scaling only adds attributes from $D_{\Omega,R,N(G)}$.    □

*Property 4 (F is extensive and monotone).* The function $F$ attached to a relational context and a set of scaling operator satisfies:

$$K \subseteq F(K) \qquad \text{(extensivity)}$$
$$K \subseteq K' \Rightarrow F(K) \subseteq F(K') \qquad \text{(monotony)}$$

*Proof.* extensivity holds because $F$ eventually adds to each formal context in $K$ new attributes scaled from $FCA(K)$. The set of attributes can thus not be smaller. monotony holds because $K \subseteq K'$ means that $M \subseteq M'$. This entails that the set of concepts of $FCA(K)$ is included in that of $FCA(K')$, hence the set of attributes $A$ scaled from $K$ is included in the set $A'$ scaled from $K'$. Since, they are added to $M$ and $M'$, then $M \cup A \subseteq M' \cup A'$, hence $F(K) \subseteq F(K')$.    □

Extensivity corresponds to the non-contracting property of the well-grounded semantics [14] and monotony is also called order-preservation.

## 4.3  Fixed Points of $F$

Given $F$, it is possible to define its sets of fixed points, i.e. the sets of formal contexts closed for $F$, as:

**Definition 2 (fixed point).** *A formal context $K \in \mathscr{K}$ is a fixed point for a context expansion function $F$, if $F(K) = K$. We call fp(F) the set of fixed points for $F$.*

Since $\mathscr{K}$ is a complete lattice and $F$ is order-preserving (or monotone) on $\mathscr{K}$, then the Knaster-Tarski theorem applies:

**Theorem 1 (Knaster-Tarski theorem [15]).** *Let $\mathcal{K}$ be a complete lattice and let $F : \mathcal{K} \mapsto \mathcal{K}$ be an order-preserving function. Then the set of fixed points of $F$ in $L$ is also a complete lattice.*

In particular, this warrants that there exists least and greatest fixed points of $F$ in $\mathcal{K}$ (called $lfp(F)$ and $gfp(F)$).

In FCA, and subsequently in RCA without circular dependencies, the images by $FCA^*$ of all fixed points of $F$ are isomorphic. Even with RCA and circular dependencies (between the objects or between the contexts), this is often the case. But the example of Sect. 3 shows that, even in $RCA^0$, there may be several fixed points for $F$ whose lattice is non isomorphic. Hence the question: which fixed point is returned by RCA's well-grounded semantics [14]?

### 4.4 The Well-Grounded Semantics of RCA is the Least Fixed-Point Semantics

RCA may be redefined as

$$RCA_\Omega(K^0, R) = FCA^*(F^\infty(K^0))$$

RCA iterates $F$ from $K^0$ until closure, and ultimately applies $FCA^*$. Since $K^0$ belongs to $\mathcal{K}$, then it computes a fixed point of $F$. This is the least fixed point.

**Proposition 1 (The RCA algorithm computes the least fixed point).** *Given $F$ the context expansion function associated to $K^0$, $R$ and $\Omega$,*

$$RCA_\Omega(K^0, R) = FCA^*(lfp(F_{K^0,R,\Omega}))$$

*Proof.* $RCA_\Omega(K^0, R) = FCA^*(F^n(K^0))$ for some $n$ at which $F(F^n(K^0)) = F^n(K^0)$ [13]. Let $K^\infty = F^n(K^0)$, $K^\infty \in fp(F)$ (Definition 2). $\forall K \in fp(F)$, $K \in \mathcal{K}$, thus $K^0 \subseteq K$ because all the contexts in $\mathcal{K}$ contain $M^0$. By monotony (Property 4), $K^\infty = F^n(K^0) \subseteq F^n(K) = K$, because $K$ is a fixed point. Thus, $K^\infty$ is a fixed point more specific than all fixed points: it is the least fixed point. □

### 4.5 Computing the Greatest Fixed Point

A natural question is how to obtain the greatest fixed point. In fact, under this approach this is (theoretically) surprisingly easy.

**Proposition 2.** $gfp(F_{\langle G,M^0,I^0\rangle,R,\Omega}) = K^{\langle R,N(G)\rangle}_{D_{\Omega,R,N(G)}}(\langle G, M^0, I^0\rangle)$

*Proof.* This context is the greatest element of $\mathcal{K}$ as it contains all attributes of $D_{\Omega,R,N(G)}$. It is also a fixed point because $F$ extensive and internal. □

The lattice corresponding to the greatest fixed point will be $L = FCA^*(gfp(F_{K^0,R,\Omega}))$.

This result is easy but very uncomfortable. The obtained lattice may contain many useless attributes. Indeed, $\exists r.c$ is well defined by the incidence relation, but it is of no use to RCA if $c$ does not belong to $L$.

In the example of Sect. 3, the attribute $\exists p.A$ belongs to $D_{\Omega,R,N(G)}$ though $A$ does not belong to the maximal lattice $L_3$, because it is not closed. The fact that both $a$ and $b$ satisfy this attribute makes that it will find its place in the intent of $AB$. If one considers the lattice in isolation, this is perfectly valid because the scaled context is well-defined: $\exists p.A$ is just an attribute among others satisfied by $a$ and $b$. However, if the lattice is transformed in a description logic TBox, this is not correct to refer to an undefined class. This is not the result that we expected: we need the results to be self-supported.

This problem is even more embarrassing if one wants to enumerate all fixed points, which are as many solutions to the RCA problem: many of these will feature such non-supported attributes.

## 5    Self-supported Fixed Points

We first quickly approach this problem from the concept lattice standpoint, it is better understood with both contexts and lattices together[2]. We then define self-supported concept lattices and consider their interaction with fixed points.

### 5.1    The Lattice $\mathscr{L}$ of RCA$^0$ Lattices and the Lattice Expansion Function $E$

From $\mathscr{K}_{K^0,R,\Omega}$, one can define $\mathscr{L}_{K^0,R,\Omega}$ as the finite set of images of $\mathscr{K}_{K^0,R,\Omega}$ by FCA. These are concept lattices obtained by applying FCA on $K^0$ extended with a subset of $D_{\Omega,R,N(G)}$:

$$\mathscr{L}_{\langle G,M^0,I^0\rangle,R,\Omega} = \{FCA(\langle G, M^0 \cup M, I^0 \cup I\rangle) \mid M \subseteq D_{\Omega,R,N(G)}\}$$

We define a specific type of homomorphisms between two concept lattices when concepts are simply mapped into concepts with the same extent and possibly increased intent.

**Definition 3 (Lattice homomorphism).** *A concept lattice homomorphism $h : \langle C, \leq \rangle \mapsto \langle C', \leq' \rangle$ is a function which maps each concept $c \in C$ into a corresponding concept $h(c) \in C'$ such that:*

- $\forall c \in C,\ intent(c) \subseteq intent(h(c))$, and
- $\forall c \in C,\ extent(c) = extent(h(c))$, and
- $\forall c, d \in C,\ c \leq d \Rightarrow h(c) \leq' h(d)$.

---

[2] Instead of developing both $\mathscr{K}$ and $\mathscr{L}$ independently and maintaining an equivalence between them, it would have been possible to use a more FCA-like structure associating the corresponding contexts and lattices.

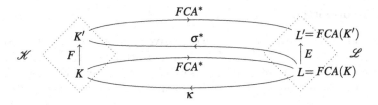

**Fig. 4.** Relations between $F$ and $E$ through the alternation of $FCA^*$ and $\sigma_\Omega^*$.

We note $L \preceq L'$ if there exists a homomorphism from $L$ to $L'$. In principle, $L \simeq L'$ if $L \preceq L'$ and $L' \preceq L$, but here, $\simeq$ is $=$. The order between concept lattices is straigthforwardly extended to families of concept lattices such that: $\{L_x\}_{x \in X} \preceq \{L'_x\}_{x \in X}$ iff $\forall x \in X, L_x \preceq L'_x$.

There exists an implicit function $\kappa : \mathscr{L}_{K^0,R,\Omega} \mapsto \mathscr{K}_{K^0,R,\Omega}$ such that $\forall L \in \mathscr{L}_{K^0,R,\Omega}, L = FCA(\kappa(L))$. Since $\simeq$ is the same as $=$ which identifies lattices containing concept having exactly the same intent and extent. $\kappa(L)$ can be induced by collecting the attributes present in $L$ intents to build the unique $M$, from which the corresponding $I$ is obtained [7].

We define $E_{K^0,R,\Omega}$, the lattice expansion function attached to a relational context $\langle K^0, R \rangle$ and a set $\Omega$ of scaling operators.

**Definition 4 (Lattice expansion function).** *Given a relational context $\langle K^0, R \rangle$ and a set of relational scaling operations $\Omega$ the function $E_{K^0,R,\Omega} : \mathscr{L}_{K^0,R,\Omega} \mapsto \mathscr{L}_{K^0,R,\Omega}$ is defined by:*

$$E_{K^0,R,\Omega}(L) = FCA^*(\sigma_\Omega^*(\kappa(L), R, L))$$

Here again, $K^0$ is only used to constrain the domain of the function, not its expression. From now on, we will abbreviate $\mathscr{L}_{K^0,R,\Omega}$ as $\mathscr{L}$ and $E_{K^0,R,\Omega}$ as $E$.

Instead of considering that $RCA(K^0) = FCA^*(F^\infty(K^0))$, it is possible to consider that $RCA(K^0) = E^\infty(FCA^*(K^0))$. Hence, RCA may be redefined as

$$RCA_\Omega(K^0, R) = E^\infty(FCA^*(K^0))$$

RCA iterates $E$ from $FCA^*(K^0)$ until closure. The definition of $E$ amounts to first scaling and then applying FCA, though $F$ does the opposite (see Fig. 4).

In consequence, $E$ is the function corresponding to $F$ in the sense that $E = FCA \circ F \circ \kappa$ and $FCA^* \circ E = F \circ FCA^*$ (see Fig. 4). Actually, the results obtained for $\mathscr{K}$ and $F$, hold exactly for $\mathscr{L}$ and $E$:

- $\langle \mathscr{L}, \preceq \rangle$ is a complete lattice;
- $E$ is an internal, monotone and extensive operation of $\mathscr{L}$;
- $RCA_\Omega(K^0, R) = \mathrm{lfp}(E_{K^0,R,\Omega})$.

$E$ inherits exactly all properties of $F$: the desirable ones and the problematic ones. So, apparently no progress has been made.

## 5.2   Self-supported Lattices

The problem is that both $F$ and $E$ are extensive functions. Hence, it is possible, starting from anywhere in $\mathscr{K}$ or $\mathscr{L}$, to consider attributes that do not refer to concepts and these attributes will be preserved. As a consequence, there are fixed points with these unwanted attributes and they are also found in the greatest fixed point.

One may consider identifying such attributes from the greatest fixed point and forbidding them. However, these meaningless attributes are contextual: one supported attribute in the greatest fixed point, may be non supported in a smaller lattice. This is a base difficulty for enumerating these fixed points.

Instead, we consider only self-supported lattices, i.e. lattices whose intents only refer to their own concepts.

**Definition 5 (Self-supported lattices).** *Let $\mathscr{L}$ a set of concept lattices, its set of self-supported lattices is*

$$S(\mathscr{L}) = \{L \in \mathscr{L}_{K^0,R,\Omega} \mid \forall c \in L, intent(c) \subseteq D_{\Omega,R,L}\}$$

The set of interesting lattices that may be returned by RCA$^0$ can be circumscribed as $fp(E) \cap S(\mathscr{L})$ as these are stable and self-supported. Moreover, by construction of $\mathscr{K}$ and $\mathscr{L}$, they cover $K^0$.

$E$ has the advantage of preserving self-supportivity.

**Proposition 3 ($E$ is internal to $S(\mathscr{L})$).** $\forall L \in S(\mathscr{L})$, $E(L) \in S(\mathscr{L})$.

*Proof.* If $L \in S(\mathscr{L})$, all attributes in intents of $L$ are supported by concepts in $L$. $E = FCA^* \circ \sigma_\Omega^*$. $\sigma_\Omega^*$ first adds to $\kappa(L)$ attributes which are supported by $L$. $L \preceq E(L)$, so these concepts are still in $E(L)$. Hence, the attributes in $\kappa(L)$ and those scaled by $\sigma_\Omega^*$ are still supported by $E(L)$.                    □

But the definition of $S$ does not provide a direct way to transform a non self-supported lattice into a self-supported one: the suppression of non self-supported attributes from intents could result in non-concepts (with non closed-extent). One possible way to solve this problem consists of extracting only the attributes currently in the lattice and to apply $FCA^*$ to the resulting context.

For that purpose, we introduce a filtering function $\pi : \mathscr{L} \mapsto \mathscr{K}$ which suppresses *from the induced context ($\kappa(L)$)* those attributes non supported *by the lattice*:

$$\pi(L) = \langle G, M \setminus D_{\Omega,R,N(G)\setminus L}, I \setminus \{\langle g,m \rangle \mid m \in D_{\Omega,R,N(G)\setminus L}\}\rangle$$

such that $\kappa(L) = \langle G, M, I \rangle$.

One can define $Q : \mathscr{L} \mapsto \mathscr{L}$, such that

$$Q(L) = FCA^*(\pi(L))$$

or $P : \mathscr{K} \mapsto \mathscr{K}$, such that $P(K) = \pi(FCA^*(K))$, see Fig. 5.

Contrary to $E$, $Q$ is anti-extensive and monotone:

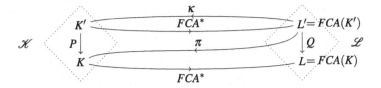

**Fig. 5.** Relations between $P$ and $Q$ through the alternation of $FCA^*$ and $\pi$.

**Proposition 4 ($Q$ is anti-extensive and monotone).** *The function $Q$ satisfies:*

$$Q(L) \preceq L \qquad \text{(anti-extensivity)}$$
$$L \preceq L' \Rightarrow Q(L) \preceq Q(L') \qquad \text{(monotony)}$$

*Proof.* **anti-extensivity** $\pi(L) \subseteq \kappa(L)$ because $\pi$ simply suppresses attributes from $\kappa(L)$. Hence, $FCA^*(\pi(L)) \preceq FCA^*(\kappa(L))$ because the latter contain all concepts of the former (identified by extent) eventually featuring the removed attributes. Moreover, $FCA^*(\kappa(L)) = L$ by definition, thus $Q(L) = FCA^*(\pi(L)) \preceq FCA^*(\kappa(L)) = L$.

**monotony** If $L \preceq L'$, then $\kappa(L) \subseteq \kappa(L')$, otherwise $FCA^*$ would not generate a smaller lattice. In addition, $L \preceq L'$ entails $N(G) \setminus L \supseteq N(G) \setminus L'$ which entails $D_{\Omega,R,N(G)\setminus L} \supseteq D_{\Omega,R,N(G)\setminus L'}$, which finally together leads to $M \setminus D_{\Omega,R,N(G)\setminus L} \subseteq M' \setminus D_{\Omega,R,N(G)\setminus L'}$. Then, $\pi(L) \subseteq \pi(L')$ because a smaller context supported by a smaller lattice cannot result in a larger context. Hence, $Q(L) = FCA^*(\pi(L)) \preceq FCA^*(\pi(L')) = Q(L')$. □

It would be possible to redefine $S(\mathscr{L})$ as $fp(Q)$. Like with $E$, it is possible to apply the Knaster-Tarski theorem to show that $\langle fp(Q), \preceq \rangle$ is a complete lattice.

But like $E$, $Q$ is not a closure operator as it is not idempotent. However, with the same arguments as [13], it can be argued that the repeated application of $Q$ converges to a self-supported concept lattice.

**Proposition 5.** $\forall L \in \mathscr{L}, \exists n; Q^n(L) = Q^{n+1}(L)$ *and* $Q^n(L) \in S(\mathscr{L})$.

*Proof.* First, $L$ is a finite concept lattice. Moreover, $Q(L) \preceq L$, hence it not possible to build an infinite chain of non converging application of $Q$ since at each iteration, either $\pi$ suppresses no attribute (and then closure has been reached), or it suppresses at least one attribute and then a strictly smaller context is reached. Ultimately, the least fixed point $lfp(Q) = FCA^*(K^0)$ is reached. It is a fixed point because $\kappa(FCA^*(K^0)) = K^0$ contains no scaled attribute and thus is self-supported. When closure is reached, this is because $\pi$ does not find any non-supported attribute in the lattice intents. This means that all of them are supported by the lattice. □

By convention, we note $Q^\infty$ the closure function associated with $Q$.

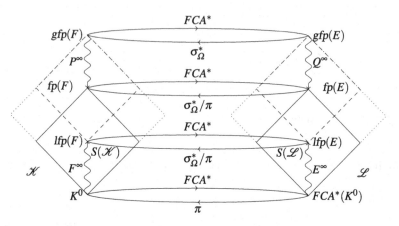

**Fig. 6.** The $\mathscr{L}$ (resp. $\mathscr{K}$) lattice and effects of $E$ and $Q$ (resp. $F$ and $P$) for characterising $fp(E)$ and $S(\mathscr{L})$ (resp. $fp(F)$ and $S(\mathscr{K})$).

We end up with two operations, $E$ and $Q$, the former extensive and the latter anti-extensive, that may be transformed into closure operators. These functions are instrumental to provide the infimum and supremum of our desired lattices (see also Fig. 6):

**Proposition 6.** $lfp(E)$ and $Q^\infty(gfp(E))$ are respectively the infimum and supremum of $fp(E) \cap S(\mathscr{L})$ for $\preceq$.

*Proof.* $lfp(E)$ is the lower bound for $fp(E) \cap S(\mathscr{L})$ because it is the lower bound for $fp(E)$. It is the infimum of $fp(E) \cap S(\mathscr{L})$ for $\preceq$ because $FCA^*(K^0) \in S(\mathscr{L})$ and by Proposition 3 this property is preserved by $E$ and since $lfp(E) = E^\infty(FCA^*(K^0))$, it belongs to $S(\mathscr{L})$.

$Q^\infty(gfp(E))$ is the upper bound for $fp(E) \cap S(\mathscr{L})$ because $gfp(E)$ contains all possible closed concepts that can be built from $D_{\Omega,R,N(G)}$. Hence, those attributes not belonging to $\pi(gfp(E))$ cannot belong to any self-supported lattice. By Proposition 5, $Q^\infty(gfp(E)) \in S(\mathscr{L})$. If $Q^\infty(gfp(E)) \notin fp(E)$, this entails $Q^\infty(gfp(E)) \prec E(Q^\infty(gfp(E)))$ and moreover that $\exists n; E^n(Q^\infty(gfp(E))) \in fp(E)$ (because $E$ is extensive and the space is finite). But, by Proposition 3, $E$ preserves self-supportiveness. Thus, $E^n(Q^\infty(gfp(E))) \in fp(E) \cap S(\mathscr{L})$ and $Q^\infty(gfp(E)) \prec E^n(Q^\infty(gfp(E)))$, which is contradictory with the fact that $Q^\infty(gfp(E))$ is an upper-bound for all fixed points. Thus, $Q^\infty(gfp(E))$ is the supremum of $fp(E) \cap S(\mathscr{L})$ for $\preceq$.                               □

## 6   Discussion

Our initial goal was to define which concept lattices could be considered as the result of RCA on a relational context. RCA provides a practical algorithm (based on $F$ or $E$ and $FCA^*$) to find out the smallest of these: $lfp(E)$. We have characterised the greatest one: $gfp(Q)$ or $Q^\infty(gfp(E))$.

We end up with two functions, complementary in their structure, one expanding the context, the other contracting it. In the perspective of enumerating all self-supported fixed points, it is tempting to either start from $lfp(E)$ and use $E$ or start from $gfp(Q)$ and use $Q$. Unfortunately, these starting points being fixed points for these very functions, this leads nowhere. It is necessary to escape the fixed points. For instance, starting from $lfp(E)$, one could add non-supported attributes until they become supported. Performing this attribute by attribute is not very smart. The example of Sect. 3 shows possible fixed points: $L_1$, $L_2$ and $L_3$. They require to add 2 or 6 attributes to $L_1$. A smarter strategy would consist of analysing the sets of attributes that support each others, through the induction of concepts, and adding these one by one to $lfp(E)$ or suppressing them from $gfp(Q)$. There is a known bound to this set since none of the attributes not in the intents of $gfp(Q)$ can be added, and none of those in $lfp(E)$ can be suppressed. Finally, these sets may entertain dependencies (adding one set of attributes would immediately support another). This may be dealt with by computing such dependencies or by applying the required closure operator ($E^\infty$ or $Q^\infty$) after each addition.

Such a procedure seems to be achievable with $RCA^0$, it will be more difficult to set up with RCA due to dependencies across lattices.

## 7  Conclusions

Motivated by the requirement to extract more concepts with relational concept analysis, we gave a new, fixed-point based, semantics for $RCA^0$. The main contribution of this work is the formulation of the RCA semantics in terms of fixed points of the function ($F$ or $E$) at the core of $RCA^0$. Then it is shown that the well-grounded semantics of RCA corresponds to the least fixed-point semantics.

We also identified as self-supported fixed points those other fixed points of interest. The least fixed point being the smaller of these. This led to develop another function ($P$ or $Q$) which, together with $FCA^*$, allows extracting the greatest of them as an alternative to RCA.

This result does not mean that RCA is wrong. In FCA, conceptual scaling has been considered as a human-driven analysis tool: a knowledgeable person could provide attribute in this language for describing better the data to be analysed. In RCA, scaling is used as an extraction tool, with the drawback to potentially generate many attributes. By only extracting the least fixed point, RCA avoids generating too many of them.

In the context of extracting a TBox for a particular ABox, extracting the least fixed point is adequate since it may be relatively complex and it is a good starting point. But for other applications, such as link key candidate extraction, it is very important to have all possible fixed points because external measures are used for selecting the best one (which has no reason to be either the least or the greatest one).

The definitions and results of Sects. 4 and 5 have been restricted to $RCA^0$ for the sake of clarity. Although this remains to be proved, they should hold for RCA as a whole. Indeed, all definitions can be applied to families of contexts and lattices, the order between them being the product order induced by the piece-wise conjunction. All operations remain monotone and extensive (or anti-extensive) as soon as the selected scaling operations are. This is enough to preserve the results.

**Acknowledgements.** This work has been partially funded by the ANR Elker project (ANR-17-CE23-0007-01). The author thanks Philippe Besnard for pointing to the Knaster-Tarski theorem.

# References

1. Atencia, M., David, J., Euzenat, J.: Data interlinking through robust linkkey extraction. In: Proceedings of 21st European Conference on Artificial Intelligence (ECAI), Praha (CZ), pp. 15–20 (2014)
2. Atencia, M., David, J., Euzenat, J., Napoli, A., Vizzini, J.: Link key candidate extraction with relational concept analysis. Discret. Appl. Math. **273**, 2–20 (2020)
3. Baader, F., Calvanese, D., McGuinness, D., Nardi, D., Patel-Schneider, P. (eds.): The Description Logic Handbook: Theory, Implementations and Applications. Cambridge University Press, Cambridge (2003)
4. Belohlávek, R.: Introduction to formal concept analysis. Technical report, Univerzita Palackého, Olomouc (CZ) (2008)
5. Ferré, S., Cellier, P.: Graph-FCA: an extension of formal concept analysis to knowledge graphs. Discret. Appl. Math. **273**, 81–102 (2020)
6. Ganter, B., Kuznetsov, S.O.: Pattern structures and their projections. In: Delugach, H.S., Stumme, G. (eds.) ICCS-ConceptStruct 2001. LNCS (LNAI), vol. 2120, pp. 129–142. Springer, Heidelberg (2001). https://doi.org/10.1007/3-540-44583-8_10
7. Ganter, B., Wille, R.: Formal Concept Analysis: Mathematical Foundations. Springer, Heidelberg (1999). https://doi.org/10.1007/978-3-642-59830-2
8. Keip, P., Ferré, S., Gutierrez, A., Huchard, M., Silvie, P., Martin, P.: Practical comparison of FCA extensions to model indeterminate value of ternary data. In: Proceedings of 15th International Conference on Concept Lattices and Their Applications (CLA), Tallinn (EE). CEUR Workshop Proceedings, vol. 2668, pp. 197–208 (2020)
9. Kuznetsov, S.O.: Pattern structures for analyzing complex data. In: Sakai, H., Chakraborty, M.K., Hassanien, A.E., Ślęzak, D., Zhu, W. (eds.) RSFDGrC 2009. LNCS (LNAI), vol. 5908, pp. 33–44. Springer, Heidelberg (2009). https://doi.org/10.1007/978-3-642-10646-0_4
10. Kötters, J.: Concept lattices of a relational structure. In: Pfeiffer, H.D., Ignatov, D.I., Poelmans, J., Gadiraju, N. (eds.) ICCS-ConceptStruct 2013. LNCS (LNAI), vol. 7735, pp. 301–310. Springer, Heidelberg (2013). https://doi.org/10.1007/978-3-642-35786-2_23
11. Nebel, B.: Reasoning and Revision in Hybrid Representation Systems. Lecture Notes in Artificial Intelligence, vol. 422. Springer, Berlin (1990). https://doi.org/10.1007/BFb0016445
12. Prediger, S.: Logical scaling in formal concept analysis. In: Lukose, D., Delugach, H., Keeler, M., Searle, L., Sowa, J. (eds.) ICCS-ConceptStruct 1997. LNCS, vol. 1257, pp. 332–341. Springer, Heidelberg (1997). https://doi.org/10.1007/BFb0027881

13. Rouane Hacene, M., Huchard, M., Napoli, A., Valtchev, P.: Relational concept analysis: mining concept lattices from multi-relational data. Ann. Math. Artif. Intell. **67**(1), 81–108 (2013)
14. Rouane-Hacene, M., Huchard, M., Napoli, A., Valtchev, P.: Soundness and completeness of relational concept analysis. In: Cellier, P., Distel, F., Ganter, B. (eds.) ICFCA 2013. LNCS (LNAI), vol. 7880, pp. 228–243. Springer, Heidelberg (2013). https://doi.org/10.1007/978-3-642-38317-5_15
15. Tarski, A.: A lattice-theoretical fixpoint theorem and its applications. Pac. J. Math. **5**(2), 285–309 (1955)

# Boolean Substructures in Formal Concept Analysis

Maren Koyda[1,2(✉)] [iD] and Gerd Stumme[1,2] [iD]

[1] Knowledge and Data Engineering Group, University of Kassel, Kassel, Germany
{koyda,stumme}@cs.uni-kassel.de
[2] Interdisciplinary Research Center for Information System Design,
University of Kassel, Kassel, Germany

**Abstract.** It is known that a (concept) lattice contains an n-dimensional Boolean suborder if and only if the context contains an n-dimensional contra-nominal scale as subcontext. In this work, we investigate more closely the interplay between the Boolean subcontexts of a given finite context and the Boolean suborders of its concept lattice. To this end, we define mappings from the set of subcontexts of a context to the set of suborders of its concept lattice and vice versa and study their structural properties. In addition, we introduce closed-subcontexts as an extension of closed relations to investigate the set of all sublattices of a given lattice.

**Keywords:** Formal Concept Analysis · Contranominal scales · Boolean contexts · Boolean lattices · Sublattices · Subcontexts · Closed relations

## 1 Introduction

In the field of Formal Concept Analysis (FCA) the basic data structure is a so-called formal context. It consists of a set of objects, a set of attributes, and an incidence relation on those sets representing which object *has* which attribute. Each such context gives rise to concepts which consist of a maximal set of objects that all share the same maximal set of attributes. The concepts, ordered by subset relation, form a complete lattice.

One frequently occurring type of substructure (more precisely: suborder or sub(semi)lattice) of a concept lattice are Boolean algebras. In the formal context, they correspond to subcontexts that are isomorphic to a contranominal scale, i.e., a context of type $(\{1, \ldots, k\}, \{1, \ldots, k\}, \neq)$. This means in particular the existence of $k$ objects that just differ slightly on $k$ attributes. However, despite of the only slight difference, these Boolean subcontexts are responsible for an exponential growth of the concept lattice [3]. Such Boolean subcontexts occur in real-world data as well as in randomly generated formal contexts [5].

---

Authors are given in alphabetical order. No priority in authorship is implied.

© Springer Nature Switzerland AG 2021
A. Braud et al. (Eds.): ICFCA 2021, LNAI 12733, pp. 38–53, 2021.
https://doi.org/10.1007/978-3-030-77867-5_3

In this paper we investigate the connection between the Boolean substructures in the formal context and in its corresponding concept lattice. Based on closed subrelations of a formal context [14], that provide a method to characterize the complete sublattices of the corresponding concept lattice, we introduce *closed-subcontexts* and present a one-to-one correspondence to all sublattices. Through this, we merge the obvious two-step-approach of limiting the lattice to an interval and determining its complete sublattices in one structure. Since this construction is an – almost arbitrary and difficult to handle – mixture of subcontext and subrelation and in addition is not directly specific to the field of Boolean substructures, we investigate the connection between Boolean subcontexts and Boolean sublattices and suborders, respectively, in Sect. 6 in a direct way without having to manipulate the incidence relation. To this end, we lift two well-known order embeddings [7] to the level of subcontexts and suborders to find the Boolean suborders corresponding to a Boolean subcontext. In addition, we introduce a construction to generate the Boolean subcontext associated to a given Boolean suborder. We combine these methods to investigate to which degree the join and meet operators of the lattice are respected by those maps.

As our work is triggered by complexity issues in data analysis where only finite sets are considered, **all statements in this paper are about finite sets and structures only,** unless explicitly stated otherwise.

As for the structure of this paper, in Sect. 2 we recall some basic notions and give a brief introduction to the approaches our investigations are based on. Afterwards, in Sect. 3 we give a short overview of previous works applied to the investigation of substructures of formal contexts and concept lattices. In Sect. 4 we introduce some notions required for our investigation on Boolean substructures. We introduce closed-subcontexts in Sect. 5 to determine the set of all Boolean sublattices. Our second approach is presented in Sect. 6 where we use embeddings of Boolean structures in concept lattices and construct the subcontexts associated to Boolean suborders. In Sect. 7 we compare both approaches, and discuss the differences and their overlap. We conclude our work and give an outlook in Sect. 8.

To advanced readers, we recommend proceeding directly to Sect. 4 and Fig. 1 as it illustrates the connections investigated in this work.

## 2    Recap on FCA and Notations

### 2.1    Foundations

Following, we recall some basic notions from FCA. For a detailed introduction we refer to [7]. A formal context is triple $\mathbb{K} := (G, M, I)$, where $G$ is the finite *object set*, $M$ the finite *attribute set*, and $I \subseteq G \times M$ a binary *incidence relation*. Instead of writing $(g, m) \in I$ for an object $g \in G$ and an attribute $m \in M$, we also write $gIm$ and say *object $g$ has attribute $m$*. One kind of formal context is the family of *contranominal scales*, denoted by $\mathbb{N}^c(k) := (\{1, 2, ..., k\}, \{1, 2, ..., k\}, \neq)$.

On the power set of the objects and the power set of the attributes there are two operations given: $\cdot' \colon \mathcal{P}(G) \to \mathcal{P}(M),\ A \mapsto A' := \{m \in M \mid \forall g \in A \colon (g, m) \in$

$I$} and $\cdot'\colon \mathcal{P}(M) \to \mathcal{P}(G)$, $B \mapsto B' := \{g \in G \mid \forall m \in B\colon (g,m) \in I\}$ Instead of $A'$ we also write $A^I$ to specify which incidence relation is used for the operation. A *formal concept* $C = (A, B)$ of the context $(G, M, I)$ is a pair consisting of an object subset $A \subseteq G$, called *extent*, and an attribute subset $B \subseteq M$, called *intent*, that satisfies $A' = B$ and $B' = A$. An object set $O \subseteq G$ is called *minimal object generator* of a concept $(A, B)$ if $O'' = A$ and $P'' \neq A$ for every proper subsets $P \subsetneq O$. Analogous, the *minimal attribute generator* of a concept $(A, B)$ is defined. The set of all minimal object generators (or rather all minimal attribute generators) of $(A, B)$ is denoted by $minG_{obj}(A, B)$ $(minG_{att}(A, B))$. The set of all formal concepts $(\mathfrak{B}(\mathbb{K}))$ together with the order defined by $(A_1, B_1) \leq (A_2, B_2)$ iff $A_1 \subseteq A_2$ for two concepts $(A_1, B_1)$ and $(A_2, B_2)$ determines the *concept lattice* $\underline{\mathfrak{B}}(\mathbb{K}) := (\mathfrak{B}(\mathbb{K}), \leq)$. The concept lattice of $\mathbb{N}^c(k)$ is called *Boolean lattice of dimension* $k$ and is denoted by $\mathfrak{B}(k) := \underline{\mathfrak{B}}(\mathbb{N}^c(k))$.

There are two tools for basic structural investigations of a formal context $\mathbb{K} = (G, M, I)$ in FCA. An object $g \in G$ is called *clarifiable* if another object $g \neq h \in G$ with $g' = h'$ exists. Furthermore, an object $g \in G$ is called *reducible* if a set of objects $X \subseteq G$ with $g \nsubseteq X$ and $g' = X'$ exists. Otherwise $g$ is called *irreducible*. The same applies to the set of attributes. The concept lattice of a context $\mathbb{K}$ that has no clarifiable/ reducible objects and attributes is isomorphic to the lattice of any context that can be constructed by adding reducible or clarifiable objects or attributes to $\mathbb{K}$. The stepwise elimination of all clarifiable/ reducible attributes and objects of a formal context results in a *clarified/reduced* context, the *standard context* of $\underline{\mathfrak{B}}(\mathbb{K})$.

To study particular parts of a formal context the selection of a *subcontext* is useful. A *subcontext* $\mathbb{S} := (H, N, J)$ of a formal context $\mathbb{K} = (G, M, I)$ is a formal context with $H \subseteq G$, $N \subseteq M$ and $J = I \cap (H \times N)$. We write $\mathbb{S} \leq \mathbb{K}$ to describe $\mathbb{S}$ as a subcontext of $\mathbb{K}$ and use the notion $[H, N]$ instead of $(H, N, I \cap (H \times N))$. The set of all subcontexts of a formal context $\mathbb{K}$ is denoted by $\mathcal{S}(\mathbb{K})$.

$1_L$ and $0_L$ denote the top and the bottom element of a lattice $\underline{L}$. The elements covering $0_L$ are called *atoms* and the elements covered by $1_L$ *coatoms*. We denote by $At(\underline{L})$ and $CoAt(\underline{L})$, respectively, the set of all atoms and coatoms of $\underline{L}$. $\underline{S} = (S, \leq)$, a subset $S \subseteq L$ together with the same order relation as $\underline{L}$, is called *suborder* of $\underline{L}$. The set of all suborders of $\underline{L}$ is denoted by $\mathcal{SO}(\underline{L})$. If $(a, b \in S \Rightarrow (a \vee b) \in S)$ holds we call $\underline{S}$ *sub-$\vee$-semilattice* of $\underline{L}$. If $(a, b \in S \Rightarrow (a \wedge b) \in S)$ holds we call $\underline{S}$ *sub-$\wedge$-semilattice* of $\underline{L}$. A $\underline{S}$ that is both, a sub-$\vee$-semilattice and a sub-$\wedge$-semilattice, is called *sublattice* of $\underline{L}$. The set of all sublattices of $\underline{L}$ is denoted by $\mathcal{SL}(\underline{L})$. If $(T \subseteq S \Rightarrow (\bigvee T), (\bigwedge T) \in S)$ holds $\forall\, T \subseteq S$ we call $\underline{S}$ *complete sublattice* of $\underline{L}$. The requirement for completeness can be translated into $1_L$ and $0_L$ being included in $\underline{S}$ if $\underline{L}$ is a finite lattice.

## 2.2   Relating Substructures in FCA

Wille [14] presents closed relations to characterize complete sublattices of a concept lattice. A relation $J \subseteq I$ is called *closed relation* of a formal context $\mathbb{K} = (G, M, I)$ if every concept of the context $(G, M, J)$ is a concept of $\mathbb{K}$ as well. Closed relations are linked to the complete sublattices of $\underline{\mathfrak{B}}(\mathbb{K})$ [7, chap.

3.3]: The set of all closed subrelations of $\mathbb{K}$ and all complete sublattices of $\mathfrak{B}(\mathbb{K})$ have a one-to-one correspondence. The bijection $C(\underline{S}) := \bigcup \{A \times B | (A, B) \in \underline{S}\}$ maps the set of all complete sublattices to the set of all closed relations. By limiting the lattice to an interval, the described one-to-one correspondence can be found between the complete lattices of the interval and the closed relations of the formal context associated to the interval.

A connection of the concept lattices of a formal context $\mathbb{K} = (G, M, I)$ and its subcontext $\mathbb{S} = [H, N]$ is given by Ganter and Wille [7, Proposition 32] by the two maps $\varphi_1 : \mathfrak{B}[H, N] \to \mathfrak{B}(G, M, I)$, $(A, B) \mapsto (A'', A')$ and $\varphi_2 : \mathfrak{B}[H, N] \to \mathfrak{B}(G, M, I)$, $(A, B) \mapsto (B', B'')$. Both maps are *order embeddings*. This means for all $(A_1, B_1), (A_2, B_2) \in \mathfrak{B}[H, N]$ that $(A_1, B_1) \leq (A_2, B_2)$ in $\mathfrak{B}[H, N]$ if and only if $\varphi_i(A_1, B_1) \leq \varphi_i(A_2, B_2)$ in $\mathfrak{B}(G, M, I)$ for both $i \in \{1, 2\}$. Hence, every structure contained in $\mathfrak{B}(\mathbb{S})$ also appears in $\mathfrak{B}(\mathbb{K})$.

# 3 Related Work

In the field of Formal Concept Analysis, there are several approaches to analyze smaller parts of a formal context or a concept lattice, as well as to investigate the connection between the two data structures. In [2] local changes to a formal context and their effects on the corresponding concept lattice, namely the number of concepts, are explored. Albano [1] studies the impact of contranominal scales in a formal context to the size of the corresponding concept lattice by giving an upper bound for $\mathfrak{B}(k)$-free lattices. The approach of Wille [14] on the one-to-one correspondence between closed subrelations of a formal context complete sublattices of the associated concept lattice is the basis for the work of Kauer and Krupke [9]. They investigate the problem of constructing the closed subrelation referring to a complete sublattice generated by a given subset of elements while not computing the whole concept lattice. Based on granulation as introduced in [15] the authors of [12] analyze substructures of formal contexts and concept lattices by considering them as granules that provide different levels of accuracy.

Also, many common methods deal with the detection of substructures in the first place. They are based on the selection of structurally meaningful attributes and objects of a formal context. For this purpose, Hanika et al. [8] search for a *relevant* attribute set that reflects the original lattice structure and the distribution of the objects as good as possible. Considering many-valued contexts, Ganter and Kuznetsov [6] select features based on their scaling. Another approach is to generate a meaningful subset by selecting entire concepts directly of the formal context by measuring their individual value for the context and the associated concept lattice. A natural idea is the consideration of extent and intent size of the concepts. Based on this, Kuznetsov [10] proposed a stability measure for formal concepts, measuring the ratio of extent subsets generating the same intent. Another measure, the support, was used by Stumme et al. [13] to generate so-called *iceberg lattices*, which also have a use in the field of mining of frequent association rules.

Besides meaningful reduction, altering the dataset is a standard method in FCA, which is motivated by an attempt to reduce the complexity of the dataset

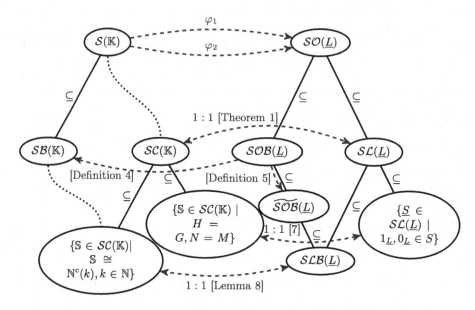

**Fig. 1.** Connections between the subcontexts of a formal context $\mathbb{K}$ and the suborders of the corresponding concept lattice $\underline{L} := \underline{\mathfrak{B}}(\mathbb{K})$. The set of all subsemilattices of $\underline{L}$ is denoted by $\widetilde{S\mathcal{OB}}(\underline{L})$.

or deal with noise. In this realm, Dias and Vierira investigate the replacement of *similar* objects by a single representative [4]. Approximate frequent itemsets have been investigated to handle noisy data [11], where the authors state an additional threshold for both rows and columns of the dataset.

Since we aim to investigate existing substructures of formal contexts and concept lattices, we turn away from those notions in general.

## 4   Boolean Subcontexts and Sublattices

In this work, we investigate Boolean substructures in formal contexts as well as in the corresponding concept lattices. Therefore, as illustrated in Fig. 1, we link the different substructures of a formal context with the substructures of the corresponding concept lattice. In this section we introduce the concrete definitions that serve as a foundation to analyze those connections.

**Definition 1.** *Let $\mathbb{K}$ be a formal context, $\mathbb{S} \leq \mathbb{K}$. $\mathbb{S}$ is called* Boolean subcontext *of dimension k of $\mathbb{K}$, if $\underline{\mathfrak{B}}(\mathbb{S}) \cong \underline{\mathfrak{B}}(k)$. $\mathbb{S}$ is called* reduced *if $\mathbb{S}$ is a reduced context. The set of all Boolean subcontexts of dimension $k$ of $\mathbb{K}$ and the set of all reduced Boolean subcontexts of dimension $k$ of $\mathbb{K}$ are denoted by $S\mathcal{B}_k(\mathbb{K})$ and $S\mathcal{RB}_k(\mathbb{K})$.*

Note that a reduced Boolean subcontext of dimension $k$ is isomorphic to the contranominal scale $\mathbb{N}^c(k)$.

| | a | b | c | d | e |
|---|---|---|---|---|---|
| 1 | × | × | | | |
| 2 | × | | × | | |
| 3 | | × | × | × | × |
| 4 | × | × | × | | |
| 5 | | × | | × | × |
| 6 | | | × | × | × |
| 7 | | | | × | |
| 8 | | | | | × |

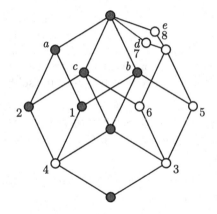

**Fig. 2.** Example of a formal context $\mathbb{K} = (G, M, I)$ with $G = \{1, 2, ..., 8\}$ and $M = \{a, b, ...e\}$ containing three reduced Boolean subcontexts and its corresponding concept lattice $\underline{\mathfrak{B}}(\mathbb{K})$. (Color figure online)

**Definition 2.** *Let $\underline{L}$ be a lattice and $\underline{S}$ a suborder of $\underline{L}$. $\underline{S}$ is called* Boolean suborder *of dimension k if $\underline{S} \cong \mathfrak{B}(k)$. If $\underline{S}$ is a sublattice of $\underline{L}$, $\underline{S}$ is called* Boolean sublattice *of dimension k. The set of all Boolean suborders of dimension $k$ of a lattice $\underline{L}$ is denoted by $\mathcal{SOB}_k(\underline{L})$. The set of all Boolean sublattices of dimension $k$ of a lattice $\underline{L}$ is denoted by $\mathcal{SLB}_k(\underline{L})$.*

If all dimensions are considered, the number $k$ is left out in the following.

Note that $\mathcal{SLB}_k(\underline{L})$ is a subset of $\mathcal{SOB}_k(\underline{L})$ and the standard context of a Boolean lattice $\underline{L}$ of dimension $k$ consists of a formal context $\mathbb{K} \cong \mathbb{N}^c(k)$ [7, Proposition 12]. Conversely, a formal context $\mathbb{K}$ consisting of a reduced Boolean subcontext of dimension $k$ and an arbitrary number of additional reducible attributes and objects has a corresponding concept lattice $\underline{\mathfrak{B}}(\mathbb{K}) \cong \mathfrak{B}(k)$.

For a better understanding of these structures, we introduce the example given in Fig. 2. We will refer back to this illustration throughout the paper.

*Example 1.* $\mathbb{S} = (\{4, 5, 6\}, \{b, c, d, e\}, J)$ with $J = I \cap (\{4, 5, 6\} \times \{b, c, d, e\})$ is a Boolean subcontext of dimension 3 of the formal context $\mathbb{K}$ given in Fig. 2. $\mathbb{S}$ is not reduced, since $d^J = e^J$ holds. However, $\mathbb{S}$ includes the reduced Boolean subcontexts $\mathbb{S}_1 = [\{4, 5, 6\}, \{b, c, d\}]$ and $\mathbb{S}_2 = [\{4, 5, 6\}, \{b, c, e\}]$. The third reduced Boolean subcontext in $\mathbb{K}$ is $\mathbb{S}_3 = [\{1, 2, 3\}, \{a, b, c\}]$. The concept lattice of $\mathbb{K}$ in Fig. 2 contains 15 Boolean suborders of dimension 3, two of which are also Boolean sublattices.

## 5   Closed-Subcontexts

At first, we leave the field of (Boolean) suborders and narrow our focus on (Boolean) sublattices. On the context side, we introduce so-called *closed-subcontexts* and show their one-to-one relationship to the sublattices of the concept lattice.

In [14], Wille introduced closed relations of a context to characterize the complete sublattices of its concept lattice. In finite lattices, complete sublattices differ from (non-complete) sublattices in that they always include the top element and the bottom element of the lattice. We adopt Wille's construction to match with (non necessarily complete) sublattices.

**Definition 3.** *Let* $\mathbb{K} = (G, M, I)$ *and* $\mathbb{S} = (H, N, J)$ *be two formal contexts. We call* $\mathbb{S}$ *closed-subcontext of* $\mathbb{K}$ *iff* $H \subseteq G$, $N \subseteq M$, $J \subseteq I \cap (H \times N)$ *and every concept of* $\mathbb{S}$ *is a concept of* $\mathbb{K}$ *as well. The set of all closed-subcontexts of* $\mathbb{K}$ *is denoted by* $SC(\mathbb{K})$.

The sublattices of $\mathfrak{B}(\mathbb{K})$ have a one-to-one correspondence to closed-subcontexts of $\mathbb{K}$ as follows.

**Theorem 1.** *Let* $\mathbb{K}$ *be a formal context and* $\underline{S}$ *be a sublattice of* $\mathfrak{B}(\mathbb{K})$. *Then*

$$\mathbb{K}_{\underline{S}} := ( \bigcup_{(A,B)\in\underline{S}} A, \bigcup_{(A,B)\in\underline{S}} B, \bigcup_{(A,B)\in\underline{S}} A \times B)$$

*is a closed-subcontext of* $\mathbb{K}$. *Conversely, for every closed-subcontext* $\mathbb{S}$ *of* $\mathbb{K}$, $\underline{\mathfrak{B}}(\mathbb{S})$ *is a sublattice of* $\mathfrak{B}(\mathbb{K})$.
*Furthermore, the map* $f(\underline{S}) := \mathbb{K}_{\underline{S}}$ *maps the set of sublattices of* $\mathfrak{B}(\mathbb{K})$ *bijectively onto the set of closed-subcontexts of* $\mathbb{K}$.

*Proof.* For each formal concept $(A, B) \in \underline{S}$ the formal concept $(A, B) \in \mathfrak{B}(\mathbb{K}_{\underline{S}})$ is due to construction a concept in $\mathbb{K}$. On the other side let $\mathbb{S} = (H, N, J)$ be a closed-subcontext of $\mathbb{K}$. The concept set of $\mathbb{S}$ is a subset of the concept set of $\mathbb{K}$ and therefore $\underline{\mathfrak{B}}(\mathbb{S})$ is a suborder of $\underline{\mathfrak{B}}(\mathbb{K})$. Let $(A_1, B_1), (A_2, B_2) \in \underline{\mathfrak{B}}(\mathbb{S})$. Let $(A_S, B_S)$ be the infimum of both in $\mathbb{S}$ and $(A_K, B_K)$ the infimum of both in $\mathbb{K}$. So $A_S = A_1 \cap A_2 = A_K$, which implies $(A_S, B_S) = (A_K, B_K)$ since $(A_S, B_S)$ is by definition a concept in $\mathbb{K}$. The dual argument shows that $\mathbb{S}$ is closed under suprema. So $\underline{\mathfrak{B}}(\mathbb{S})$ is a sublattice of $\underline{\mathfrak{B}}(\mathbb{K})$. □

Note that the closed-subsets of a formal context do not form a closure system since the intersection of two closed-subcontexts, in general, is not a closed-subcontext, even though the sublattices of formal concept do so. In the construction of $\mathbb{K}_{\underline{S}}$, $\bigcup_{(A,B)\in\underline{S}} A$ is the concept extent of the top element of the sublattice and $\bigcup_{(A,B)\in\underline{S}} B$ is the concept intent of its bottom element.

**Lemma 1.** *Let* $\mathbb{K} = (G, M, I)$ *be a formal context and* $\mathbb{S} = (H, N, J)$ *a closed-subcontext of* $\mathbb{K}$. *Then* $H = G$ *or* $m \in N$ *with* $m' = H$ *exists. And* $N = M$ *or* $g \in H$ *with* $g' = N$ *exists.*

*Proof.* Due to Definition 3, every concept of $\mathbb{S}$ is a concept of $\mathbb{K}$ as well. In particular, this has to hold for the concepts $(\emptyset'', \emptyset')$ and $(H'', H')$ of $\mathbb{S}$. □

We provide next some basic statements about closed-subcontexts. Since the following lemmas are based on the work of Wille [14] and lifted to our approach, the proofs are similar to the ones in [7, Section 3.3].

**Lemma 2.** *For every set $T \subseteq \mathfrak{B}(G, M, I)$ there is a smallest closed-subcontext $\mathbb{S}$ of $\mathbb{K}$, that contains all $(A \times B)$ for $(A, B) \in T$. $\mathfrak{B}(\mathbb{S})$ is the sublattice of $\underline{\mathfrak{B}}(\mathbb{K})$ generated by $T$.*

*Proof.* The proof follows the structure of the proof of Proposition 45 in [7].  □

**Lemma 3.** *$\mathbb{S} = (H, N, J)$ is a closed-subcontext of the formal context $\mathbb{K} = (G, M, I)$ iff $X^{JJ} \supseteq X^{JI}$ holds for each $X \subseteq H$ and for each $X \subseteq N$.*

*Proof.* The proof follows the structure of the proof of Proposition 46 in [7].  □

**Lemma 4.** *The closed-subcontexts $(H, N, J)$ of $(G, M, I)$ are exactly the sub-contexts that satisfy the condition: (C) If $(g, m) \in (H \times N)$ and $(g, m) \in I \setminus J$ then $(h, m) \notin I$ for $h \in H$ with $g^J \subseteq h^J$ and $(g, n) \notin I$ for $n \in N$ with $m^J \subseteq n^J$.*

*Proof.* The proof follows the structure of the proof of Proposition 47 in [7].  □

**Lemma 5.** *Let $\mathbb{K} = (G, M, I)$ be a formal context. A clarified formal context $\mathbb{S} = (H, N, J)$ is a closed-subcontext of $\mathbb{K}$ if and only if $H \subseteq G$, $N \subseteq M$ and $J \subseteq I \cap (H \times N) \subseteq H \times N \setminus (\nearrow^J \cup \swarrow^J)$.*

*Proof.* The proof follows the structure of the proof of Proposition 49 in [7].  □

**Lemma 6.** *Let $\mathbb{K} = (G, M, I)$ be a formal context and $(A, B)$ and $(C, D)$ con-cepts of $\mathbb{K}$. Then $(A, B, A \times B)$, $(A, M, I \cap (A \times M))$ and $(G, B, I \cap (G \times B))$ are closed-subcontexts. If $(A, B) \leq (C, D)$ also $(C, B, (A \times B \cup C \times D))$ and $(C, B, I \cap (C \times B))$ are closed-subcontexts. The corresponding concept lattices are given through $\underline{\mathfrak{B}}(A, B, A \times B) = \{(A, B)\}$, $\underline{\mathfrak{B}}(A, M, I \cap (A \times M)) = ((A, B)]$, $\underline{\mathfrak{B}}(G, B, I \cap (G \times B)) = [(A, B))$, $\underline{\mathfrak{B}}(C, B, (A \times B \cup C \times D)) = \{(A, B), (C, D)\}$, and $\underline{\mathfrak{B}}(C, B, I \cap (C \times B)) = [(A, B), (C, D)]$.*

*Proof.* The proof follows the structure of the proof of Proposition 50 in [7].  □

Also, the set of the arrow relations of a closed-subcontext $\mathbb{S}$ is a subset of the set of the arrow relations of the original context $\mathbb{K}$.

**Lemma 7.** *Let $\mathbb{K} = (G, M, I)$ be a formal context and $\mathbb{S} = (H, N, J)$ a closed-subcontext. Then $\nearrow^J \subseteq \nearrow^I$ and $\swarrow^J \subseteq \swarrow^I$ holds.*

*Proof.* Let $g \in H, m \in N$ and $g \swarrow^J m$. Assumed $g \not\swarrow^I m$. Then there exists $h \in G$ with $g^I \subseteq h^I$ and $(h, m) \notin I$. It follows $g^J \subseteq g^{I \cap (G \times H)} \subseteq h^{I \cap (G \times H)} \Rightarrow h \in h^{I \cap (G \times H)} \subseteq g^{JI} = g^{JJ} \subseteq H \Rightarrow g^J \subseteq h^J$. This is a conflict to $g \swarrow^J m$.  □

Now we transfer our approach to the field of Boolean substructures. To find all Boolean sublattices (of dimension $k$) in a lattice $\underline{\mathfrak{B}}(\mathbb{K})$ the closed-subcontexts of $\mathbb{K}$ that are Boolean subcontexts as well have to be found. Hence, Theorem 1 can be restricted in the following way:

**Lemma 8.** *Let $\mathbb{K}$ be a formal context. $\underline{\mathbb{S}} \in \mathcal{SLB}_k(\underline{\mathfrak{B}}(\mathbb{K}))$ iff $\underline{\mathfrak{B}}(\mathbb{K}_{\underline{S}}) \cong \underline{\mathfrak{B}}(k)$ for $\mathbb{K}_{\underline{S}} = (\bigcup_{(A,B) \in \underline{S}} A, \bigcup_{(A,B) \in \underline{S}} B, \bigcup_{(A,B) \in \underline{S}} A \times B)$.*

To directly identify the Boolean closed-subcontexts in a formal context $\mathbb{K}$, the properties of closed-subcontexts can be utilized. Since every concept in $\mathbb{K}$ is either retained or erased but not altered in a closed-subcontext $\mathbb{S}$, the Boolean structure of $\mathbb{S}$ has to be preserved from $\mathbb{K}$. Every Boolean subcontext $\mathbb{T} = (H, N, J) \in \mathcal{SRB}(\mathbb{K})$ provides the Boolean structure. Lifting each concept $(A_{\mathbb{T}}, B_{\mathbb{T}}) \in \mathfrak{B}(\mathbb{T})$ to a concept $(A_{\mathbb{K}}, B_{\mathbb{K}}) \in \mathfrak{B}(\mathbb{K})$ with $A_{\mathbb{T}} \subseteq A_{\mathbb{K}}$ and $B_{\mathbb{T}} \subseteq B_{\mathbb{K}}$, generates an extention of the sets $H, N$ and $J$ that provides a Boolean closed-subcontext $\mathbb{S} = (\widetilde{H}, \widetilde{N}, \widetilde{J}) \in \mathcal{SC}(\mathbb{K})$ as follows: $\widetilde{H} := H \cup \bigcup_{(A_{\mathbb{T}}, B_{\mathbb{T}}) \in \mathfrak{B}(\mathbb{T})} A_{\mathbb{K}}$, $\widetilde{N} := H \cup \bigcup_{(A_{\mathbb{T}}, B_{\mathbb{T}}) \in \mathfrak{B}(\mathbb{T})} B_{\mathbb{K}}$ and $\widetilde{J} := \bigcup_{(A_{\mathbb{T}}, B_{\mathbb{T}}) \in \mathfrak{B}(\mathbb{T})} (A_{\mathbb{K}} \times B_{\mathbb{K}})$. This approach is represented through the dotted lines in Fig. 1.

# 6   Connecting Boolean Suborders and Boolean Subcontexts

In this section we investigate the relationship between Boolean subcontexts and Boolean suborders. For this purpose, we use the embeddings $\varphi_1$ and $\varphi_2$ and expand them to the set of Boolean subcontexts. Further, we present a construction to get from a Boolean suborder to a corresponding Boolean subcontext. Both approaches are analyzed with focus on the structural information they transfer and their interplay.

## 6.1   Embeddings of Boolean Substructures

To investigate the connection between Boolean subcontexts $\mathbb{S}$ of a formal context $\mathbb{K}$ and Boolean suborders of $\mathfrak{B}(\mathbb{K})$ we consider embeddings of $\mathfrak{B}(\mathbb{S})$ in $\mathfrak{B}(\mathbb{K})$. Therefore we lift the embeddings $\varphi_1$ and $\varphi_2$ introduced in Sect. 2 to the level of subcontexts and suborders:

$$\varphi_1 : \mathcal{S}(\mathbb{K}) \to \mathcal{SO}(\mathfrak{B}(\mathbb{K})), \ \mathbb{S} \mapsto (\{\varphi_1(C) \mid C \in \mathfrak{B}(\mathbb{S})\}, \leq) \text{ and}$$
$$\varphi_2 : \mathcal{S}(\mathbb{K}) \to \mathcal{SO}(\mathfrak{B}(\mathbb{K})), \ \mathbb{S} \mapsto (\{\varphi_2(C) \mid C \in \mathfrak{B}(\mathbb{S})\}, \leq).$$

From the input (concept or context), it is clear whether the original or the lifted versions of the embeddings $\varphi_1$ and $\varphi_2$ are used in the following. We will, in particular, study these mappings for Boolean subcontexts. In this case, an additional structural benefit arises: The images of reduced Boolean subcontexts are sub-$\vee$-semilattice and sub-$\wedge$-semilattices of the original concept lattice:

**Lemma 9.** *Let $\mathbb{K}$ be a formal context, $\mathbb{S} = [H, N] \in \mathcal{SRB}_k(\mathbb{K})$. Then $\varphi_1(\mathfrak{B}(\mathbb{S}))$ is a sub-$\vee$-semilattice of $\mathfrak{B}(\mathbb{K})$ and $\varphi_2(\mathfrak{B}(\mathbb{S}))$ is a sub-$\wedge$-semilattice of $\mathfrak{B}(\mathbb{K})$.*

*Proof.* Consider $\varphi_1$: Let $J := I \cap (H \times N)$ and $(A, B)$ and $(C, D)$ be two concepts of $\mathfrak{B}(\mathbb{S})$. Then $\varphi_1(A, B) \vee \varphi_1(C, D) = (A'', A') \vee (C'', C') = ((A'' \cup C'')'', (A' \cap C')) = ((A' \cap C')', (A \cup C)') = ((A \cup C)'', (A \cup C)') $ and in addition $((A \cup C)'', (A \cup C)') = \varphi_1((A \cup C), (B \cap D)) = \varphi_1((A, B) \vee (C, D))$. Since $\mathbb{S}$ is a reduced Boolean context, it includes all possible object combinations as extents so that $E = E^{JJ}$ holds for every $E \subseteq H$. Therefore, in $\mathfrak{B}(\mathbb{S})$ holds $(A, B) \vee (c; D) = ((A \cup C)^{JJ}, B \cap D) = (A \cup C, B \cap D)$. The procedure for $\varphi_2$ is analogous.   $\square$

Note that this conclusion does not hold for Boolean reducible subcontexts, e.g., the formal context given in Fig. 2 and its subcontext $\mathbb{S} = [\{1237\}, \{abce\}]$.

The images of the two maps of a reduced Boolean context are in general just a sub-$\vee$-semilattice and a sub-$\wedge$-semilattice, respectively. Hence, the images of $\varphi_1$ and $\varphi_2$ have to be identical for $\mathbb{S} \in \mathcal{SRB}_k(\mathbb{K})$ to generate a lattice. This means $\varphi_1(A, B) = (A'', A) = (B', B'') = \varphi_2(A, B)$ has to hold for all $(A, B) \in \mathfrak{B}(\mathbb{S})$.

For every subcontext $\mathbb{S} = (H, N, J) \leq \mathbb{K}$ we can differ between the four cases: Case 1 with $A' = A^J = B$, $B' = B^J = A$, case 2 with $A' = A^J = B$, $A = B^J \subset B'$, case 3 with $B = A^J \subset A'$, $B' = B^J = A$ and case 4 with $B = A^J \subset A'$, $A = B^J \subset B'$. The condition under which $\varphi_1(A, B) = \varphi_2(A, B)$ holds is the following:

**Lemma 10.** *Let $\mathbb{K} = (G, M, I)$ be a formal context and $\mathbb{S} \leq \mathbb{K}$. $\varphi_1(\mathbb{S}) = \varphi_2(\mathbb{S})$ holds if and only if for all $(A, B) \in \mathfrak{B}(\mathbb{S})$ $(A' \setminus B) \times (B' \setminus A) \subseteq I$ holds. If case 1, 2 or 3 holds for all $(A, B) \in \mathfrak{B}(\mathbb{S})$, then $\varphi_1(\mathbb{S}) = \varphi_2(\mathbb{S})$ holds directly.*

*Proof.* For a concept $(A, B) \in \mathfrak{B}(\mathbb{S})$ the identity of both embeddings leads to $\varphi_1(A, B) = \varphi_2(A, B) \Leftrightarrow (A'', A') = (B', B'') = (B', A') \Leftrightarrow (B' \times A') \subseteq I$. This set can be written as $B' \times A' = A \times B \cup (B' \setminus A) \times B \cup A \times (A' \setminus B) \cup (B' \setminus A) \times (A' \setminus B)$. We know $A \times B \subseteq I$ since $(A, B) \in \mathfrak{B}(\mathbb{S})$ and $A \times A' \subseteq I$ and $B' \times B \subseteq I$ by definition of the $\cdot'$ operator. The remaining part equals $(A' \setminus B) \times (B' \setminus A)$. In cases 1 to 3 $(A'', A') = (B', B'')$ holds by construction.                               $\square$

**Proposition 1.** *Let $\mathbb{K} = (G, M, I)$ be a formal context and $\mathbb{S} = [H, N] \in \mathcal{SB}_k(\mathbb{K})$. If $H = G$ or $N = M$, then $\varphi_1(\mathbb{S}) = \varphi_2(\mathbb{S})$ holds.*

However, the relationship between the images of both mappings $\varphi_1$ and $\varphi_2$ of a specific concept is always (not only in the Boolean case) the same, namely:

**Proposition 2.** *Let $\mathbb{K}$ be a formal context and $\mathbb{S} \leq \mathbb{K}$. Then $\varphi_1(A, B) \leq \varphi_2(A, B)$ for all $(A, B) \in \mathfrak{B}(\mathbb{S})$.*

In particular, an interval containing exactly the concepts $(C, D) \in \mathfrak{B}(\mathbb{K})$ with $A \subseteq C$ and $B \subseteq D$ exists between $\varphi_1(A, B)$ and $\varphi_2(A, B)$ with $\varphi_1(A, B)$ as its bottom element and $\varphi_2(A, B)$ as its top element. In the extreme case, this interval can comprise all of $\mathfrak{B}(\mathbb{K})$, as the following example shows.

*Example 2.* Let $\mathbb{K}$ be the formal context in Fig. 3 and $\mathbb{S} = [\{1, 2\}, \{a, b\}] \leq \mathbb{K}$. For the concept $(A, B) = (\{1, 2\}, \{a, b\})$ of $\mathbb{S}$, $\varphi_1(A, B) = (\{1, 2\}, \{a, b, c, d\})$ and $\varphi_2(A, B) = (\{1, 2, 3, 4\}, \{a, b\})$ hold. These are the bottom and the top element of the whole concept lattice of $\mathbb{K}$.

This raises the question whether there is a concept lattice where a Boolean suborder exists that can not be obtained by embedding. This is indeed the case also in Fig. 2; see, e.g., the Boolean order marked with filled red circles.

An approach to make any Boolean suborder of a (concept) lattice reachable is to expand $\mathbb{K}$ by additional objects and attributes so that every formal concept $C \in \mathfrak{B}(\mathbb{K})$ can be generated by one object and by one attribute. For a (concept)

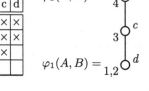

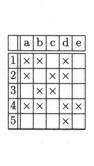

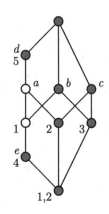

**Fig. 3.** An example of a formal context $\mathbb{K}$ and its subcontext $\mathbb{S} = [\{1,2\}, \{a,b\}] = [A,B]$ with $[\varphi_1(A,B), \varphi_2(A,B)] = \mathfrak{B}(\mathbb{K})$.

**Fig. 4.** Example of a formal context $\mathbb{K}$ with $|\mathcal{SRB}_3(\mathbb{K})| = |\mathcal{SOB}_3(\mathfrak{B}(\mathbb{K}))| = 4$.

lattice $\underline{L}$ this is the case with the context $\mathbb{K} = (L, L, \leq)$. Here $\underline{S} \in \mathcal{SOB}_k(\underline{L})$ is the image of both $\varphi_1(\mathbb{S})$ and $\varphi_2(\mathbb{S})$ for the Boolean subcontext $\mathbb{S} = (S, S, \leq)$.

Since we are interested in the connections between the existence of Boolean subcontexts on the one hand and the existence of Boolean suborders on the other hand, we observe a first relationship between these sets.

**Lemma 11.** *Let $\mathbb{K}$ be a formal context, $\mathcal{SB}_k(\mathbb{K}) \neq \emptyset$. Then $\mathcal{SOB}_k(\mathfrak{B}(\mathbb{K})) \neq \emptyset$.*

*Proof.* Let $\mathbb{S} \in \mathcal{SB}_k(\mathbb{K})$. By definition $\mathfrak{B}(\mathbb{S}) \cong \mathfrak{B}(k)$. Since $\varphi_1 : \mathfrak{B}(\mathbb{S}) \mapsto \mathfrak{B}(\mathbb{K})$ is an order embedding $\varphi_1(\mathfrak{B}(\mathbb{S}))$ is a Boolean suborder of dimension $k$ in $\mathfrak{B}(\mathbb{K})$.□

In general the images of $\varphi_1(\mathbb{S})$ and $\varphi_2(\mathbb{S})$ are neither lattices nor semilattices. However, we know from Lemma 9 that if $\mathbb{S}$ is a reduced Boolean subcontext and $\varphi_1(\mathfrak{B}(\mathbb{S})) = \varphi_2(\mathfrak{B}(\mathbb{S}))$ holds, there exists a Boolean sublattice $\underline{S}$ of the same dimension in $\mathfrak{B}(\mathbb{K})$. We can generalize the previous statement as follows:

**Lemma 12.** *Let $\mathbb{K}$ be a clarified formal context and $\mathbb{S}_1, \mathbb{S}_2 \in \mathcal{SRB}_k(\mathbb{K})$ with $\mathbb{S}_1 = [H_1, N_1], \mathbb{S}_2 = [H_2, N_2]$ and $\mathbb{S}_1 \neq \mathbb{S}_2$. If $H_1 \neq H_2$, then $\varphi_1(\mathbb{S}_1) \neq \varphi_1(\mathbb{S}_2)$ holds. If $N_1 \neq N_2$, then $\varphi_2(\mathbb{S}_1) \neq \varphi_2(\mathbb{S}_2)$ holds.*

*Proof.* Since $\mathbb{S}_1, \mathbb{S}_2 \in \mathcal{SRB}_k(\mathbb{K})$, $|H_1| = |H_2|$ holds. If $H_1 \neq H_2$ holds, $g_1 \in H_1$ with $g_1 \notin H_2$ and $g_2 \in H_2$ with $g_2 \notin H_1$ exist. Since $\mathbb{S}_1$ and $\mathbb{S}_2$ are reduced and Boolean there is a concept $C_1 = (g_1, g_1'') \in \mathfrak{B}(\mathbb{S}_1)$ and a concept $C_2 = (g_2, g_2'') \in \mathfrak{B}(\mathbb{S}_2)$. Hence $\mathbb{K}$ is clarified, $\varphi_1(C_1) = (g_1'', g_1') \neq (g_2'', g_2') = \varphi_1(C_2)$. If $N_1 \neq N_2$ holds, the analogous procedure can be executed using $\varphi_2$. □

Based on this statement, we can assume that the total number of reduced Boolean subcontexts of a formal context $\mathbb{K}$ is a lower bound of the total number of Boolean suborders of $\mathfrak{B}(\mathbb{K})$:

*Conjecture 1.* Let $\mathbb{K}$ be a clarified formal context with $|\mathcal{SRB}_k(\mathbb{K})| = n$. Then $|\mathcal{SOB}_k(\mathfrak{B}(\mathbb{K}))| \geq n$ holds.

This conjecture can not be proved as straight forward as Lemma 12 since $\varphi_1$ and $\varphi_2$ can be identical for some $\mathbb{S} \in \mathcal{SRB}_k(\mathbb{K})$. In addition not every Boolean suborder is the image of $\varphi_1(\mathbb{S})$ or $\varphi_2(\mathbb{S})$ for a $\mathbb{S} \in \mathcal{SRB}_k(\mathbb{K})$. Both phenomena occur in the example given in Fig. 4, where the marked Boolean suborder is not the image of the embedding by $\varphi_1$ or $\varphi_2$ of any Boolean subcontext contained in the given formal context, although in this case the number of Boolean subcontexts of dimension 3 and Boolean suborders of dimension 3 is identical.

## 6.2   Subconcepts Associated to Suborders

After investigating mappings of Boolean subcontexts to Boolean suborders, we now analyze the connection between those substructures the other way around. As presented by Albano and Chornomaz [3, Prop. 1] every formal context $\mathbb{K}$ contains a Boolean subcontext $\mathbb{S} \in \mathcal{SB}_k(\mathbb{K})$ if $\mathfrak{B}(\mathbb{K})$ contains a Boolean suborder $\underline{S} \in \mathcal{SOB}_k(\mathfrak{B}(\mathbb{K}))$. Based on this statement, we introduce a construction to generate a (not necessarily reduced) Boolean subcontext of a formal context based on a Boolean suborder of the corresponding concept lattice.

**Definition 4.** *Let $\mathbb{K}$ be a formal context and $\underline{S} \in \mathcal{SOB}_k(\mathfrak{B}(\mathbb{K}))$. We call $\psi(\underline{S}) := [H, N]$ with $H := \bigcup_{C \in At(\underline{S})} minG_{obj}(C)$ and $N := \bigcup_{C \in CoAt(\underline{S})} minG_{att}(C)$ the subcontext of $\mathbb{K}$ associated to $\underline{S}$.*

Indeed the structure arising from the construction given in Definition 4 is a Boolean subcontext of the same dimension as $\underline{S}$:

**Lemma 13.** *Let $\mathbb{K}$ be a formal context, $\underline{S} \in \mathcal{SOB}_k(\mathfrak{B}(\mathbb{K}))$ and $\mathbb{S} = [H, N] := \psi(\underline{S})$ the subcontext of $\mathbb{K}$ associated to $\underline{S}$. Then $\mathbb{S} \in \mathcal{SB}_k(\mathbb{K})$.*

*Proof.* Let $At(\underline{S}) = \{A_1, A_2, ..., A_k\}$ and $CoAt(\underline{S}) = \{C_1, C_2, ..., C_k\}$. Due to the Boolean structure of $\underline{S}$ the atoms can be ordered holding the following condition: $A_i$ is a lower bound for the set $CoAt(\underline{S}) \setminus C_i$ for all $1 \leq i \leq k$ and analogous $C_i$ is an upper bound for the set $At(\underline{S}) \setminus A_i$ for all $1 \leq i \leq k$. It follows $gIm$ for all $g \in min_{obj}G(A_i)$, $m \in N \setminus minG_{att}(C_i)$ and $g \not{I} m$ else. So $\mathbb{S} \cong \mathbb{N}^c(k)$.    □

In the following, we study the interplay of the mapping $\psi$ from suborders to subcontexts with the mappings $\varphi_1$ and $\varphi_2$ from subcontexts to suborders.

**Lemma 14.** *Let $\mathbb{K}$ be a formal context and $\mathbb{S} = [H, N] \in \mathcal{SRB}_k(\mathbb{K})$. Then $\mathbb{S} = \psi(\varphi_1(\mathbb{S}))$ iff for all $n \in N$ $(n', n'') \in CoAt(\varphi_1(\mathbb{S}))$ holds and $\mathbb{S} = \psi(\varphi_2(\mathbb{S}))$ holds iff for all $h \in H$ $(h'', h') \in At(\varphi_2(\mathbb{S}))$ holds.*

*Proof.* Consider $\varphi_1$: Let $\psi(\varphi_1(\mathbb{S})) = [\tilde{H}, \tilde{N}]$, $H = \{h_1, h_2, ..., h_k\}$ and $N = \{n_1, n_2, ..., n_k\}$. Due to the construction of $\varphi_1$ $At(\varphi_1(\mathbb{S})) = \{A_1, A_2, ..., A_k\}$ with $A_i = (h_i'', h_i')$. Since every $h_i$ is a minimal object generator of an atom of $\varphi_1(\mathbb{S})$ $\tilde{H} = H$ holds. Let $CoAt(\varphi_1(\mathbb{S})) = \{C_1, C_2, ..., C_k\}$. $\tilde{N}$ consists of the minimal attribute generators of the coatoms of $\varphi_1(\mathbb{S})$. Following, $\widetilde{N} = N$ if and only if a renumbering of the coatoms exists so that $C_i = (n'_i, n''_i)$ for all $i \in \{1, 2, ..., k\}$. The procedure for $\varphi_2$ is analogous. $\qquad \square$

*Example 3.* Let $\mathbb{K}$ be the formal context in Fig. 4 and $\mathbb{S}_1 = [\{1, 2, 3\}, \{a, b, c\}]$, $\mathbb{S}_2 = [\{2, 3, 4\}, \{a, b, c\}]$, $\mathbb{S}_3 = [\{1, 2, 3\}, \{b, c, d\}]$ and $\mathbb{S}_4 = [\{2, 3, 4\}, \{b, c, d\}])$ its reduced Boolean subcontexts of dimension 3. Then $\mathbb{S}_1 = \psi(\varphi_1(\mathbb{S}_1)) = \psi(\varphi_2(\mathbb{S}_1))$, $\mathbb{S}_2 = \psi(\varphi_2(\mathbb{S}_2))$ and $\mathbb{S}_3 = \psi(\varphi_1(\mathbb{S}_3))$ hold.

**Lemma 15.** *Let $\mathbb{K}$ be a formal context, $\underline{S} \in \mathcal{SOB}_k(\underline{\mathfrak{B}}(\mathbb{K}))$, $\mathbb{S} := \psi(\underline{S})$. Let $C \in \underline{S} \setminus \{0_{\underline{S}}, 1_{\underline{S}}\}$ with either $C$ not being the supremum (in $\underline{\mathfrak{B}}(\mathbb{K})$) of a subset of $At(\underline{S})$ or $C$ not being the infimum (in $\underline{\mathfrak{B}}(\mathbb{K})$) of a subset of $CoAt(\underline{S})$. Then $(A, B)$ with $A = \bigcup \{minG_{obj}(X) \mid X \in At(\underline{S}), X \leq C\}$ and $B = \bigcup \{minG_{att}(X) \mid X \in CoAt(\underline{S}), X \geq C\}$ is a concept of $\mathbb{S}$ with $\varphi_1(A, B) \neq \varphi_2(A, B)$.*

*Proof.* According to the construction of $\mathbb{S}$ there is a concept $(A, B) \in \mathfrak{B}(\mathbb{S})$ as stated. If $C$ is not the supremum of a subset of $At(\underline{S})$, especially $A$ does not generate $C$. Therefore $\varphi_1(A, B) = (A'', A') < C$, due to the construction of A. Also $\varphi_2(A, B) = (B', B'') \geq C$ and consequently $\varphi_1(A, B) < \varphi_2(A, B)$. Similarly, if $C$ is not the infimum of a subset of $CoAt(\underline{S})$, $\varphi_1(A, B) = (A'', A') \leq C$, $\varphi_2(A, B) = (B', B'') > C$ and $\varphi_1(A, B) < \varphi_2(A, B)$. $\qquad \square$

**Lemma 16.** *Let $\mathbb{K}$ be a formal context, $\underline{S} \in \mathcal{SOB}(\underline{\mathfrak{B}}(\mathbb{K}))$. Then $\varphi_1(\psi(\underline{S}))$ is a sub-$\vee$-semilattice and $\varphi_2(\psi(\underline{S}))$ is a sub-$\wedge$-semilattice of $\underline{\mathfrak{B}}(\mathbb{K})$.*

*Proof.* Let $\mathbb{S} = [H, N] := \psi(\underline{S})$. $H$ is the set of all minimal generators of the atoms of $\underline{S}$. Due to the Boolean structure, all concepts in $\mathbb{K}$ that are generated by a subset of $H$ are exactly the supremum of a subset of $At(\mathbb{S})$. Since this generation corresponds to mapping the concepts $C \in \mathfrak{B}(\mathbb{S})$ with $\varphi_1$, $\varphi_1(\mathbb{S})$ is a sub-$\vee$-semilattice. The second part of the statement is proved similarly. $\qquad \square$

**Definition 5.** *Let $\mathbb{K}$ be a formal context, $\underline{S} \in \mathcal{SOB}_k(\underline{\mathfrak{B}}(\mathbb{K}))$. We call $\varphi_1(\psi(\underline{S}))$ the sub-$\vee$-sublattice of $\underline{\mathfrak{B}}(\mathbb{K})$ associated to $\underline{S}$ and $\varphi_2(\psi(\underline{S}))$ the sub-$\wedge$-sublattice of $\underline{\mathfrak{B}}(\mathbb{K})$ associated to $\underline{S}$.*

The statement in Lemma 16 holds especially for a $\underline{S}$ being a Boolean sub-semilattice or a Boolean sublattice of $\underline{\mathfrak{B}}(\mathbb{K})$ and provides $\varphi_1(\psi(\underline{S})) = \underline{S}$ and $\varphi_2(\psi(\underline{S})) = \underline{S}$, respectively, as follows.

**Lemma 17.** *Let $\mathbb{K}$ be a formal context and $\underline{S} \in \mathcal{SOB}_k(\underline{\mathfrak{B}}(\mathbb{K}))$. If $\underline{S}$ is a sub-$\vee$-semilattice, $\varphi_1(\psi(\underline{S})) = \underline{S}$. If $\underline{S}$ is a sub-$\wedge$-semilattice, $\varphi_2(\psi(\underline{S})) = \underline{S}$.*

*Proof.* Let $\underline{S}$ be a sub-$\vee$-semilattice and $\mathbb{S} = [H, N] := \psi(\underline{S})$. $H$ is the set of minimal generators of the atoms of $\underline{S}$. Due to the Boolean structure all concepts in $\underline{\mathfrak{B}}(\mathbb{K})$ that are generated by a subset of $H$ are exactly the supremums of a subset of the atoms of $\underline{S}$. Since this generation corresponds to mapping the concepts $C \in \mathfrak{B}(\mathbb{S})$ with $\varphi_1$, every image of $\varphi_1(C)$ is contained in $\underline{S}$. The second statement is proved similarly. $\qquad \square$

|   | a | b | c | d | e | f |
|---|---|---|---|---|---|---|
| 1 | × | × |   |   |   |   |
| 2 | × |   | × |   |   |   |
| 3 |   | × | × | × |   |   |
| 4 |   | × | × |   | × |   |
| 5 |   | × | × |   |   | × |
| 6 |   | × | × | × | × | × |

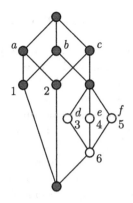

**Fig. 5.** Example of a formal context that shows that neither $\varphi_1$ and $\psi$ nor $\varphi_2$ and $\psi$ are (dually) adjoint mappings.

**Proposition 3.** *Let $\mathbb{K}$ be a formal context and $\underline{S} \in \mathcal{SLB}_k(\underline{\mathfrak{B}}(\mathbb{K}))$ a sublattice. Then $\varphi_1(\psi(\underline{S})) = \varphi_2(\psi(\underline{S})) = \underline{S}$.*

Our research can be concluded in the following theorems. They give an insight into the interplay of $\varphi_1, \varphi_2$ and $\psi$ and the structural properties they transfer.

**Theorem 2.** *Let $\mathbb{K}$ be a formal context and $\mathbb{S} \in \mathcal{SB}(\mathbb{K})$. Then:*

  *i) $\psi(\varphi_1(\mathbb{S})) = \mathbb{S}$ iff a sub-$\vee$-semilattice $\underline{S} \in \mathcal{SOB}(\underline{\mathfrak{B}}(\mathbb{K}))$ exists with $\psi(\underline{S}) = \mathbb{S}$.*
  *ii) $\psi(\varphi_2(\mathbb{S})) = \mathbb{S}$ iff a sub-$\wedge$-semilattice $\underline{S} \in \mathcal{SOB}(\underline{\mathfrak{B}}(\mathbb{K}))$ exists with $\psi(\underline{S}) = \mathbb{S}$.*
  *iii) $\psi(\varphi_1(\mathbb{S})) = \psi(\varphi_2(\mathbb{S})) = \mathbb{S}$ iff a $\underline{S} \in \mathcal{SLB}(\underline{\mathfrak{B}}(\mathbb{K}))$ exists with $\psi(\underline{S}) = \mathbb{S}$.*

*Furthermore, if $\mathbb{S}$ is reduced, $\varphi_1(\mathbb{S}) = \varphi_1(\psi(\varphi_1(\mathbb{S})))$ and $\varphi_2(\mathbb{S}) = \varphi_2(\psi(\varphi_2(\mathbb{S})))$.*

*Proof.* Consider i): ($\Rightarrow$) follows directly from Lemma 16 since $\mathbb{S}$ is the subcontext corresponding to the suborder $\varphi_1(\mathbb{S})$. ($\Leftarrow$) is presented in Lemma 17. ii) is proved similarly and iii) follows from the combination of i) and ii). The last statement follows from the combination of Lemma 9 and Lemma 15.         □

**Theorem 3.** *Let $\mathbb{K}$ be a formal context and $\underline{S} \in \mathcal{SOB}(\underline{\mathfrak{B}}(\mathbb{K}))$.*

  *i) Then $\varphi_1(\psi(\underline{S})) = \underline{S}$ iff $\underline{S}$ is a sub-$\vee$-semilattice.*
  *ii) Then $\varphi_2(\psi(\underline{S})) = \underline{S}$ iff $\underline{S}$ is a sub-$\wedge$-semilattice.*
  *iii) Then $\varphi_1(\psi(\underline{S})) = \varphi_2(\psi(\underline{S})) = \underline{S}$ iff $\underline{S}$ is a sublattice.*

*Proof.* Consider i): ($\Rightarrow$) follows directly from Lemma 16. ($\Leftarrow$) is presented in Lemma 17, ii) is proved similarly, iii) follows from combining i) and ii).         □

Altough $\varphi_1$ and $\psi$ (or $\varphi_2$ and $\psi$) seem to be (dually) adjoint mappings, they are not. E.g., in Fig. 5 consider the subcontexts $\mathbb{S}_1 = [\{1,2,3,4\}, \{a,b,c\}]$, $\mathbb{S} = [\{1,2,3,4,5\}, \{a,b,c\}]$, and $\mathbb{S}_2 = [\{1,2,3,4,5,6\}, \{a,b,c\}]$. It holds $\varphi_1(\mathbb{S}_1) = \varphi_1(\mathbb{S}_2) = \varphi_1(\mathbb{S}) = \varphi_2(\mathbb{S}_2) = \varphi_2(\mathbb{S}_1)$ – the image is highlighted in the line diagram, and its associated context is $\mathbb{S}$. This shows that $\psi \circ \varphi_1$ is neither monotonic nor anti-monotonic, and the same holds for $\psi \circ \varphi_2$.

## 7  Interplay of Both Approaches

In the previous sections, two approaches to relate Boolean substructures of a formal context $\mathbb{K}$ with those of the corresponding concept lattice $\mathfrak{B}(\mathbb{K})$ were introduced. In this section, we set both of them in relation.

In Sect. 5 a one-to-one correspondence between the closed-subcontexts of a formal context $\mathbb{K}$ and the sublattices of $\underline{\mathfrak{B}}(\mathbb{K})$ is presented. However, subsemilattices and suborders are not addressed. In addition, the closed-subcontexts restrict not only the object set and the attribute set of a formal context but also its incidence relation, whereby they could be understood as a more substantial altering of $\mathbb{K}$ compared to the approach presented in Sect. 6. It provides different maps to associate specific Boolean suborders on the one side with Boolean subcontexts on the other side while transferring some structural information.

The intersection of both approaches is localised in the Boolean subcontexts that are closed-subcontexts as well and in general the subcontexts $\mathbb{S} \leq \mathbb{K}$ with $C \in \mathfrak{B}(\mathbb{K})$ for all $C \in \mathfrak{B}(\mathbb{S})$.

**Lemma 18.** *Let $\mathbb{K}$ be a formal context. $\mathbb{S} \leq \mathbb{K}$ is a closed-subcontext of $\mathbb{K}$ iff $\varphi_1(C) = \varphi_2(C) = C$ for all $C \in \mathfrak{B}(\mathbb{S})$.*

This statement can be restricted to Boolean subcontexts. E.g., the Boolean subcontext $\mathbb{S} = [G, \{a, b, c\}]$ in Fig. 2 fulfils the requirement. In general, the set of the Boolean subcontexts of $\mathbb{K}$ that are closed-subcontexts is smaller than the set of all Boolean sublattices of $\underline{\mathfrak{B}}(\mathbb{K})$. So not every Boolean sublattice of $\underline{\mathfrak{B}}(\mathbb{K})$ can be reached by an embedding of a subcontext of such a structure. Refering to those structures we expand the statement of Lemma 11 as follows:

**Lemma 19.** *Let $\mathbb{K}$ be a formal context and $\mathbb{S} \in \mathcal{SB}_k(\mathbb{K})$ with $\mathbb{S}$ a closed-subcontext of $\mathbb{K}$. Then $\underline{S} := \varphi_1(\mathbb{S}) = \varphi_2(\mathbb{S}) \in \mathcal{SLB}_k(\underline{\mathfrak{B}}(\mathbb{K}))$.*

However, in general the subcontext $\tilde{\mathbb{S}}$ associated to $\underline{S}$ is not equal to $\mathbb{S}$. E.g. in Fig. 2 the subcontext $\mathbb{S} = [G, \{a, b, c\}]$ is embedded to a Boolean sublattice $\underline{S}$ but the sublattice, that is associated to $\underline{S}$ is $\tilde{\mathbb{S}} = [\{1, 2, 3, 4\}, \{a, b, c\}]$.

## 8  Conclusion

This work relates Boolean substructures in a formal context $\mathbb{K}$ with those in its concept lattice $\underline{\mathfrak{B}}(\mathbb{K})$. The notion of closed-subcontexts of $\mathbb{K}$ is presented to generalize closed relations and provide a one-to-one correspondence to the set of all sublattices of $\underline{\mathfrak{B}}(\mathbb{K})$ using a direct construction. In particular, this relationship can be restricted to the set of all Boolean closed-subcontexts of $\mathbb{K}$, that can be generated based on the set of all reduced Boolean subcontexts of $\mathbb{K}$, and all Boolean sublattices of $\underline{\mathfrak{B}}(\mathbb{K})$. Moreover, we investigated two embeddings of Boolean subcontexts of $\mathbb{K}$ into $\underline{\mathfrak{B}}(\mathbb{K})$. The images of those embeddings are, in general, not sub(semi)lattices but only Boolean suborders and do not cover $\mathcal{SOB}(\mathbb{K})$ completely. Through the introduction of the subcontext $\mathbb{S}$ associated to a Boolean suborder $\underline{S}$ of $\underline{\mathfrak{B}}(\mathbb{K})$, the investigated connection is investigated the other way around. The combination of both approaches give an insight of their interplay and the structural information they transfer. Through this every subsemilattice $\underline{S}$ can be associated with a concrete subcontext, that can be mapped to $\underline{S}$ by one of the two embeddings.

We conclude this work with two open questions. First, we are curious to which amount the presented findings can be transferred to general substructures of (not necessarily finite) formal contexts and their corresponding concept lattices. Secondly, we are interested in consideration of other special substructures, e.g., the subcontexts of a concept lattice isomorphic to a nominal scale, as those scales also contain nearly identical objects that differ only in one attribute.

# References

1. Albano, A.: Polynomial growth of concept lattices, canonical bases and generators: extremal set theory in formal concept analysis. Ph.D. thesis, SLUB Dresden (2017)
2. Albano, A.: Rich subcontexts. arXiv preprint arXiv:1701.03478 (2017)
3. Albano, A., Chornomaz, B.: Why concept lattices are large - extremal theory for the number of minimal generators and formal concepts. In: International Conference on Concept Lattices and Their Applications. CEUR Workshop Proceedings, vol. 1466, pp. 73–86. CEUR-WS.org (2015)
4. Dias, S.M., Vieira, N.: Reducing the size of concept lattices: the JBOS approach. In: International Conference on Concept Lattices and Their Applications. CEUR Workshop Proceedings, vol. 672, pp. 80–91. CEUR-WS.org (2010)
5. Felde, M., Hanika, T.: Formal context generation using Dirichlet distributions. In: Endres, D., Alam, M., Şotropa, D. (eds.) ICCS 2019. LNCS (LNAI), vol. 11530, pp. 57–71. Springer, Cham (2019). https://doi.org/10.1007/978-3-030-23182-8_5
6. Ganter, B., Kuznetsov, S.O.: Scale coarsening as feature selection. In: Medina, R., Obiedkov, S. (eds.) ICFCA 2008. LNCS (LNAI), vol. 4933, pp. 217–228. Springer, Heidelberg (2008). https://doi.org/10.1007/978-3-540-78137-0_16
7. Ganter, B., Wille, R.: Formal Concept Analysis - Mathematical Foundations. Springer, Heidelberg (1999). https://doi.org/10.1007/978-3-642-59830-2
8. Hanika, T., Koyda, M., Stumme, G.: Relevant attributes in formal contexts. In: Endres, D., Alam, M., Şotropa, D. (eds.) ICCS 2019. LNCS (LNAI), vol. 11530, pp. 102–116. Springer, Cham (2019). https://doi.org/10.1007/978-3-030-23182-8_8
9. Kauer, M., Krupka, M.: Generating complete sublattices by methods of formal concept analysis. Int. J. Gen Syst **46**(5), 475–489 (2017)
10. Kuznetsov, S.: Stability as an estimate of the degree of substantiation of hypotheses derived on the basis of operational similarity. Autom. Doc. Math. Linguist. **24** (1990)
11. Liu, J., Paulsen, S., Sun, X., Wang, W., Nobel, A.B., Prins, J.F.: Mining approximate frequent itemsets in the presence of noise: algorithm and analysis. In: International Conference on Data Mining, pp. 407–418 (2006)
12. Qi, J., Wei, L., Wan, Q.: Multi-level granularity in formal concept analysis. Granular Comput. **4**(3), 351–362 (2018). https://doi.org/10.1007/s41066-018-0112-7
13. Stumme, G., Taouil, R., Bastide, Y., Pasquier, N., Lakhal, L.: Computing iceberg concept lattices with titanic. Data Knowl. Eng. **42**(2), 189–222 (2002)
14. Wille, R.: Bedeutungen von Begriffsverbänden. In: Beiträge zur Begriffsanalyse, pp. 161–211. B.I.-Wissenschaftsverlag, Mannheim (1987)
15. Zadeh, L.A.: Toward a theory of fuzzy information granulation and its centrality in human reasoning and fuzzy logic. Fuzzy Sets Syst. **90**(2), 111–127 (1997)

# Rules

# Enumerating Maximal Consistent Closed Sets in Closure Systems

Lhouari Nourine and Simon Vilmin[(✉)]

LIMOS, Université Clermont Auvergne, Aubière, France
lhouari.nourine@uca.fr, simon.vilmin@ext.uca.fr

**Abstract.** Given an implicational base and an inconsistency binary relation over a finite set, we are interested in the problem of enumerating all maximal consistent closed sets (denoted by MCCEnum for short). We show that MCCEnum cannot be solved in output-polynomial time unless P = NP, even for lower bounded lattices. We give an incremental-polynomial time algorithm to solve MCCEnum for closure systems with constant Carathéodory number. Finally we prove that in biatomic atomistic closure systems MCCEnum can be solved in output-quasipolynomial time if minimal generators obey an independence condition, which holds in atomistic modular lattices. For closure systems closed under union (i.e. distributive), MCCEnum is solved by a polynomial delay algorithm [23,26].

**Keywords:** Closure systems · Implicational base · Inconsistency relation · Enumeration algorithm

## 1 Introduction

In this paper, we consider binary inconsistency relations (i.e. graphs) and implicational bases [8,33] over a same groundset. More precisely, we are interested in the enumeration of maximal closed sets of a closure system given by an implicational base that are consistent with respect to an inconsistency relation. We call this problem MAXIMAL CONSISTENT CLOSED SETS ENUMERATION, or MCCEnum for short.

This problem finds applications for instance in minimization of sub-modular functions [23] or argumentation frameworks [13]. More generally, inconsistency relations combined with posets appear also in event structures [29], representations of median-semilattices [4] or cubical complexes [2] in which the term *"inconsistency"* is used. Recently in [22,23], the authors derive a representation for modular semi-lattices based on inconsistency and projective ordered spaces [21]. Furthermore, they characterize the cases where given an implicational base and an inconsistency relation, maximal consistent closed sets coincide with maximal independent sets of the inconsistency relation, seen as a graph.

S. Vilmin—The second author is funded by the CNRS, France, ProFan project. This research is also supported by the French government IDEXISITE initiative 16-IDEX-0001 (CAP 20-25).

© Springer Nature Switzerland AG 2021
A. Braud et al. (Eds.): ICFCA 2021, LNAI 12733, pp. 57–73, 2021.
https://doi.org/10.1007/978-3-030-77867-5_4

The problem MCCENUM is also a particular case of dualization in closure systems given by an implicational bases, ubiquitous in computer science [8,12,16]. This latter problem however cannot be solved in output-polynomial time unless P = NP [3] even when the input implicational base has premises of size at most two [11]. When restricted to graphs and implicational bases with premises of size one, or posets, the problem can be solved in polynomial delay [23,26].

In this paper, we show first that enumerating maximal consistent closed sets cannot be solved in output-polynomial time unless P = NP, even for the well-known class of lower bounded lattices [1,10,17]. This surprising result further emphasizes the hardness of dualization in lattices given by implicational bases [3,11]. On the positive side, we show that when the maximal size of minimal generators is bounded by a constant, the problem can be solved in incremental-polynomial time. As a direct corollary, we obtain that MCCENUM can be solved efficiently in several classes of convex geometries where this parameter, also known as the Carathéodory number, is constant [27]. Finally, we focus on biatomic atomistic closure systems [6,9]. We show that under an independence condition, the size of a minimal generator is logarithmic in the size of the groundset. As a consequence, we get a quasi-polynomial time algorithm for enumerating maximal consistent closed sets which can be applied to the well-known class of atomistic modular lattices [19,21,30,32].

The rest of the paper is organized as follows. Section 2 gives necessary definitions about closure systems and implicational bases. In Sect. 3 we show that MCCENUM cannot be solved in output-polynomial time, in particular for lower bounded closure systems. In Sect. 4, we show that if the size of a minimal generator is bounded by a constant, MCCENUM can be solved efficiently. Section 5 is devoted to the class of biatomic atomistic closure systems. We conclude with open questions and problems in Sect. 6.

## 2   Preliminaries

All the objects considered in this paper are finite. Let $X$ be a set. We denote by $2^X$ its powerset. For any $n \in \mathbb{N}$, we write $[n]$ for the set $\{1, \ldots, n\}$. We will sometimes use the notation $x_1 \ldots x_n$ as a shortcut for $\{x_1, \ldots, x_n\}$. The size of a subset $A$ of $X$ is denoted by $|A|$. If $\mathcal{H} = (X, \mathcal{E})$ is a hypergraph, we denote by $\mathsf{IS}(\mathcal{H})$ its independent sets (or stable sets). We write $\mathsf{MIS}(\mathcal{H})$ for its maximal independent sets. Similarly, if $G = (X, E)$ is a graph, its independent sets (resp. maximal independent sets) are written $\mathsf{IS}(G)$ (resp. $\mathsf{MIS}(G)$).

We recall principal notions on lattices and closure systems [19]. A mapping $\phi \colon 2^X \to 2^X$ is a *closure operator* if for any $Y, Z \subseteq X$, $Y \subseteq \phi(Y)$ (extensive), $Y \subseteq Z$ implies $\phi(Y) \subseteq \phi(Z)$ (isotone), and $\phi(\phi(Y)) = \phi(Y)$ (idempotent). We call $\phi(Y)$ the *closure* of $Y$. The family $\mathcal{F} = \{\phi(Y) \mid Y \subseteq X\}$ ordered by set-inclusion forms a *closure system* or *lattice*. A closure system $\mathcal{F} \subseteq 2^X$ is a set system such that $X \in \mathcal{F}$ and for any $F_1, F_2 \in \mathcal{F}$, $F_1 \cap F_2$ also belongs to $\mathcal{F}$. Elements of $\mathcal{F}$ are *closed sets* and we say that $F$ is *closed* if $F \in \mathcal{F}$. Each closure system $\mathcal{F}$ induces a unique closure operator $\phi$ such that $\phi(Y) = \bigcap\{F \in \mathcal{F} \mid Y \subseteq F\}$, for any $Y \subseteq X$. Thus, there is a one-to-one correspondence between closure systems and operators. In this paper, We assume that $\phi$ and $\mathcal{F}$ are *standard*: $\phi(\emptyset) = \emptyset$ and for any $x \in X$, $\phi(x)\backslash\{x\}$ is closed. Note that $\emptyset$ is thus the minimum element of $\mathcal{F}$, called the *bottom*. Similarly, $X$ is the *top* of $\mathcal{F}$.

Let $\phi$ be a closure operator with corresponding closure system $\mathcal{F}$. Let $F_1, F_2 \in \mathcal{F}$. We say that $F_1$ and $F_2$ are *comparable* if $F_1 \subseteq F_2$ or $F_2 \subseteq F_1$. They are *incomparable* otherwise. A subset $\mathcal{S}$ of $\mathcal{F}$ is an *antichain* if its elements are pairwise incomparable. If for any $F \in \mathcal{F}$, $F_1 \subset F \subseteq F_2$ implies $F = F_2$, we say that $F_2$ *covers* $F_1$, and denote it $F_1 \prec F_2$. An *atom* is a closed set covering the bottom $\emptyset$ of $\mathcal{F}$. Dually, a *co-atom* is a closed set covered by the top $X$ of $\mathcal{F}$. We denote by $\mathcal{C}(\mathcal{F})$ the set of co-atoms of $\mathcal{F}$. Let $M \in \mathcal{F}$. We say that $M$ is *meet-irreducible* in $\mathcal{F}$ if for any $F_1, F_2 \in \mathcal{F}$, $M = F_1 \cap F_2$ entails either $F_1 = M$ or $F_2 = M$. In this case, $M$ has a unique cover $M^*$ in $\mathcal{F}$. The set of meet-irreducible elements of $\mathcal{F}$ is denoted by $\mathcal{M}(\mathcal{F})$. Dually, $J \in \mathcal{F}$ is a *join-irreducible* element of $\mathcal{F}$ if for any $F_1, F_2 \in \mathcal{F}$, $J = \phi(F_1 \cup F_2)$ implies $J = F_1$ or $J = F_2$. Then, $J$ covers a unique element $J_*$ in $\mathcal{F}$. We denote by $\mathcal{J}(\mathcal{F})$ the join-irreducible elements of $\mathcal{F}$. When $\mathcal{F}$ and $\phi$ are standard, there is a one-to-one correspondence between $X$ and $\mathcal{J}(\mathcal{F})$ given by $\mathcal{J}(\mathcal{F}) = \{\phi(x) \mid x \in X\}$. Furthermore, $x_* = \phi(x)_* = \phi(x)\backslash x$. Consequently, we will identify $X$ with $\mathcal{J}(\mathcal{F})$.

Let $x \in X$. A *minimal generator* of $x$ is an inclusion-wise minimal subset $A_x$ of $X$ such that $x \in \phi(A_x)$. We consider $\{x\}$ as a trivial minimal generator of $x$. Following [27], the *Carathéodory number* $c(\mathcal{F})$ of $\mathcal{F}$ is the least integer $k$ such that for any $A \subseteq X$ and any $x \in X$, $x \in \phi(A)$ implies the existence of some $A' \subseteq A$ with $|A'| \leq k$ such that $x \in \phi(A')$. At first, this notion was used for convex geometries, but its definition applies to any closure system. Moreover, the Carathéodory number of $\mathcal{F}$ is the maximal possible size of a minimal generator (see Proposition 4.1 in [27], which can be applied to any closure system). A *key* of $\mathcal{F}$ is a minimal subset $K \subseteq X$ such that $\phi(K) = X$. We denote by $\mathcal{K}$ the set of keys of $\mathcal{F}$. The number of keys in $\mathcal{K}$ is denoted by $|\mathcal{K}|$. It is well-known (see for instance [12]) that maximal independent sets $\mathsf{MIS}(\mathcal{K})$ of $\mathcal{K}$, viewed as a hypergraph over $X$, are exactly co-atoms of $\mathcal{F}$. We define arrow relations from [18]. Let $x \in X$ and $M \in \mathcal{M}(\mathcal{F})$. We write $x \uparrow M$ if $x \notin M$ but $x \in M^*$. Dually, we write $M \downarrow x$ if $x \notin M$ but $x_* \subseteq M$.

We move to implicational bases [8,33]. An *implication* is an expression of the form $A \to B$ with $A, B \subseteq X$. We call $A$ the *premise* and $B$ the *conclusion*. A set $\Sigma$ of implications over $X$ is an *implicational base* over $X$. We denote by $|\Sigma|$ the number of implications in $\Sigma$. A subset $F \subseteq X$ *satisfies* or *models* $\Sigma$ if for any $A \to B \in \Sigma$, $A \subseteq F$ implies $B \subseteq F$. The family $\mathcal{F} = \{F \subseteq X \mid F \text{ satisfies } \Sigma\}$ is a closure system whose induced closure operator $\phi$ can be computed by the *forward chaining algorithm*. This procedure starts from any subset $Y$ of $X$ and constructs a sequence $Y = Y_0 \subseteq \cdots \subseteq Y_k = \phi(Y)$ of subsets of $X$ such that for any $i \in [k]$, $Y_i = Y_{i-1} \cup \bigcup\{B \mid \exists A \to B \in \Sigma \text{ s.t. } A \subseteq Y_{i-1}\}$. The algorithm stops when $Y_{i-1} = Y_i$. Given an implicational base $\Sigma$ and an implication $A \to B$, we say that $A \to B$ *holds* in $\Sigma$ if for any $F \subseteq X$, $F$ satisfies $A \to B$ if it satisfies $\Sigma$. We say that $\Sigma$ is *standard* if the associated closure system $\mathcal{F}$ is standard.

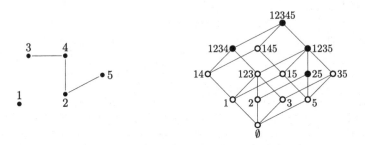

**Fig. 1.** On the left, a consistency-graph $G_c$ over $X = \{1, 2, 3, 4, 5\}$ with inconsistent pairs 34, 24 and 25. On the right, the closure system associated to $\Sigma = \{13 \to 2, 12 \to 3, 23 \to 1, 4 \to 1\}$. Black and white dots stand for inconsistent and consistent closed sets respectively. We have $\mathsf{maxCC}(\Sigma, G_c) = \{145, 123, 35\}$.

We now introduce the main problem. Following [2,22,23] we call an *inconsistency relation* any symmetric and irreflexive relation over $X$. Such a relation is sometimes called a *site* [4] or a *conflict relation* [29]. Usually, inconsistency relations need to satisfy more conditions in order to capture median or modular-semilattices [4,22]. As we do not need further restrictions here, we can choose to model inconsistency as a graph $G_c = (X, E_c)$, and call it a *consistency-graph*. An edge $uv$ of $E_c$ represents an *inconsistent pair* of elements in $X$. A subset $Y$ which does not contain any inconsistent pair (i.e. an independent set of $G_c$) is called *consistent*. Let $\Sigma$ be an implicational base over $X$ and a $G_c = (X, E_c)$ consistency-graph. We denote by $\mathsf{maxCC}(\Sigma, G_c)$ the set of maximal consistent closed sets of $\mathcal{F}$, that is $\mathsf{maxCC}(\Sigma, G_c) = \mathsf{max}_{\subseteq}(\mathcal{F} \cap \mathsf{IS}(G_c))$. If $G_c$ is empty, $X$ is the unique element of $\mathsf{maxCC}(\Sigma, G_c)$. Hence, we will assume without loss of generality that $G_c$ is not empty. An example of implicational base along with a consistency-graph is given in Fig. 1. Our problem is the following.

MAXIMAL CONSISTENT CLOSED-SETS ENUMERATION (MCCENUM)
**Input:** A standard implicational base $\Sigma$ over $X$, a non-empty consistency-graph $G_c = (X, E_c)$.
**Output:** The set $\mathsf{maxCC}(\Sigma, G_c)$ of maximal consistent closed sets of $\mathcal{F}$ with respect to $G_c$.

The standard property for $\Sigma$ is crucial. If $\Sigma$ is non-standard, MCCENUM becomes equivalent to the more general problem of dualization in lattices given by implicational bases, where a hypergraph $\mathcal{H} = (X, \mathcal{E})$ is given instead of a graph $G_c$. This problem cannot be solved in output-polynomial time unless P = NP [3,11]. Let $\Sigma$ be an implicational base over some $X$ and $\mathcal{H} = (X, \mathcal{E})$ a hypergraph. To each $E_i \in \mathcal{E}$, we create a new pair of vertices $x_i, y_i$ such that

$\phi(y_i) = \phi(x_i) = \phi(E_i)$. This amounts to add the implications $E_i \rightarrow x_i$, $x_i \rightarrow y_i$ and $y_i \rightarrow E_i$ to $\Sigma$. The edges of $G_c$ are the pairs $x_i y_i$ for any $E_i \in \mathcal{E}$. Then, a closed set in the resulting closure system will be consistent with respect to $G_c$ if and only if it is an independent set of $\mathcal{H}$.

If $\Sigma$ is empty, then MCCEnum is equivalent to the enumeration of maximal independent sets of a graph which can be efficiently solved [24]. If premises of $\Sigma$ have size 1, the problem also reduces to maximal independent sets enumeration [23,26]. In [22] the authors identify, for a fixed $\Sigma$, the consistency-graphs $G_c$ such that $\mathsf{MIS}(G_c) = \mathsf{maxCC}(\Sigma, G_c)$.

We conclude with a recall on enumeration algorithms [24]. Let $\mathcal{A}$ be an algorithm with input $x$ and output a set of solutions $R(x)$. We denote by $|R(x)|$ the number of solutions in $R(x)$. We assume that each solution in $R(x)$ has size $\mathsf{poly}(|x|)$. The algorithm $\mathcal{A}$ is running in *output-polynomial* time if its execution time is bounded by $\mathsf{poly}(|x| + |R(x)|)$. It is *incremental-polynomial* if for any $1 \leq i \leq |R(x)|$, the time spent between the $i$-th and $i+1$-th output is bounded by $\mathsf{poly}(|x|+i)$, and the algorithm stops in time $\mathsf{poly}(|x|)$ after the last output. If the delay between two solutions output and after the last one is $\mathsf{poly}(|x|)$, $\mathcal{A}$ has *polynomial-delay*. Note that if $\mathcal{A}$ is running in incremental-polynomial time, it is also output-polynomial. Finally, we say that $\mathcal{A}$ runs in *output-quasipolynomial* time if is execution time is bounded by $N^{\mathsf{polylog}(N)}$ where $N = |x| + |R(x)|$.

## 3    Closure Systems Given by Implicational Bases

Let $\Sigma$ be an implicational base over $X$ and $G_c$ a non-empty consistency-graph. Observe that $\mathsf{IS}(G_c) \cup \{X\}$ is a closure system where a set $F \subseteq X$ is closed if and only if $F = X$ or it is an independent set of $G_c$. From this point of view, elements of $\mathsf{maxCC}(\Sigma, G_c)$ are those maximal proper subsets of $X$ that are both closed in $\mathcal{F}$ and $\mathsf{IS}(G_c) \cup \{X\}$. In other words, the maximal consistent closed sets $\mathsf{maxCC}(\Sigma, G_c)$ of $\mathcal{F}$ with respect to $G_c$ are exactly the co-atoms of $\mathcal{F} \cap (\mathsf{IS}(G_c) \cup \{X\})$. An implicational base for this closure system is $\Sigma \cup \{uv \rightarrow X \mid uv \in E_c\}$. Consequently, MCCEnum is a restricted version of the following problem

Co-atoms Enumeration (CE)
**Input:** An implicational base $\Sigma$ over a set $X$.
**Output:** The co-atoms $\mathcal{C}(\mathcal{F})$ of the closure system $\mathcal{F}$ associated to $\Sigma$.

where there exists at least one implication of the form $uv \rightarrow X$ in $\Sigma$ (or holding in $\Sigma$). In [26], the authors prove that CE cannot be solved in output-polynomial time unless $\mathsf{P} = \mathsf{NP}$. To show that even MCCEnum is intractable, in particular in lower bounded closure systems, we use a generalization of CE:

Dualization of a set (DualS)
**Input:** An implicational base $\Sigma$ over $X$, and $B \subseteq X$.
**Output:** The family of maximal closed sets $F$ of $\mathcal{F}$ such that $B \nsubseteq F$.

Note that CE is the particular case of DUALS where $B = X$. For a given $\Sigma$ and $B$, let us denote by $\mathsf{max}(\Sigma, B)$ the family $\mathsf{max}_{\subseteq}\{F \in \mathcal{F} \mid B \not\subseteq F\}$. We first need the following preparatory lemma.

**Lemma 1.** *Let $\Sigma_Y$ be an implicational base over $Y$, and $B \subseteq Y$. Let $X = Y \cup \{u, v\}$, $\Sigma = \Sigma_Y \cup \{B \to uv\}$ and let $G_c = (X, E_c = \{uv\})$ be a consistency-graph. The following equality holds:*

$$\mathsf{maxCC}(\Sigma, G_c) = \bigcup_{C \in \mathsf{max}(\Sigma_Y, B)} \{C \cup \{u\}, C \cup \{v\}\} \tag{1}$$

*Proof.* Let $C$ be an element in $\mathsf{max}(\Sigma_Y, B)$. We show that $C \cup \{u\}$ and $C \cup \{v\}$ are in $\mathsf{maxCC}(\Sigma, G_c)$. As no implication of $\Sigma$ has $u$ or $v$ in its premise, we have that $C \cup \{u\}$ and $C \cup \{v\}$ are consistent and closed with respect to $\Sigma$. Let $y \in Y \setminus C$. As $C \in \mathsf{max}(\Sigma_Y, B)$, it must be that $B \subseteq \phi_Y(C \cup \{y\})$. Since $B \to uv$ is an implication of $\Sigma$, it follows that $uv \subseteq \phi(C \cup \{u, y\})$. Thus, for any $x \in X \setminus (C \cup \{u\})$, $\phi(C \cup \{u, x\})$ is inconsistent. We conclude that $C \cup \{u\} \in \mathsf{maxCC}(\Sigma, G_c)$. Similarly we obtain $C \cup \{v\} \in \mathsf{maxCC}(\Sigma, G_c)$.

Let $S \in \mathsf{maxCC}(\Sigma, G_c)$. We show that $S$ can be written as $C \cup \{u\}$ or $C \cup \{v\}$ for some $C$ in $\mathsf{max}(\Sigma_Y, B)$. First, let $F$ be a consistent closed set in $\mathcal{F}$ such that $u \notin F$ and $v \notin F$. As $\Sigma$ has no implication with $u$ or $v$ in its premise, it follows that both $F \cup \{u\}$ and $F \cup \{v\}$ are closed and consistent. Hence, either $u \in S$ or $v \in S$. Without loss of generality, let us assume $u \in S$. Let $C = S \setminus \{u\}$. As $S \in \mathsf{maxCC}(\Sigma, G_c)$, it is closed with respect to $\Sigma_Y$ and does not contain $B$. Thus, $C \in \mathcal{F}_Y$ and $B \not\subseteq C$. Let $y \in Y \setminus C$. As $S \in \mathsf{maxCC}(\Sigma, G_c)$, it must be that $\phi(S \cup \{y\})$ contains the inconsistent pair $uv$ of $G_c$. Hence, $B \subseteq \phi(S \cup \{y\})$ by construction of $\Sigma$. Consequently, we have that $B \subseteq \phi_Y(C \cup \{y\})$ for any $y \in Y \setminus C$. Hence, we conclude that $C \in \mathsf{max}(\Sigma_Y, B)$ as expected. $\square$

Therefore, if there is an algorithm solving MCCENUM in output-polynomial time, it can be used to solve DUALS within the same running time using the reduction of Lemma 1. Consequently, we have the following theorem.

**Theorem 1.** *The problem MCCENUM cannot be solved in output-polynomial time unless $\mathsf{P} = \mathsf{NP}$.*

In fact, we can strengthen the preceding theorem by a careful analysis of the closure system used in the reduction in [11,26]. More precisely, we show that the problem remains untractable for lower bounded closure systems. These have been introduced with the doubling construction in [10] and then studied in [1,7,17]. More precisely, a lattice is lower bounded if it is obtained from a boolean lattice by repeated duplications of lower pseudo-intervals.

A characterization of lower bounded lattices is given in [17] in terms of the $D$-relation. This relation relies on $\mathcal{J}(\mathcal{F})$ and we say that $x$ depends on $y$, denoted by $xDy$ if there exists a meet-irreducible element $M \in \mathcal{M}(\mathcal{F})$ such that $x \uparrow M \downarrow y$, see for instance Lemma 11.10 in [17]. A $D$-cycle is a sequence $x_1, \ldots, x_k$ of elements of $X$ such that $x_1 D x_2 D \ldots D x_k D x_1$.

**Theorem 2.** *(Reformulated from Corollary 2.39, [17]) A closure system $\mathcal{F}$ is lower bounded if and only if it contains no D-cycle.*

**Corollary 1.** *The problem MCCENUM cannot be solved in output-polynomial time unless $\mathsf{P} = \mathsf{NP}$, even in lower bounded closure systems.*

*Proof.* We start from the implicational base of [11], where the authors show that DUALS cannot be solved in output-polynomial time unless $\mathsf{P} = \mathsf{NP}$ even if $\Sigma$ has premises of size at most two. Then, we use Lemma 1 and show that the resulting closure system is lower bounded.

Following [11], consider a positive 3-CNF over $n$ variables and $m$ clauses

$$\psi(x_1, \ldots, x_n) = \bigwedge_{i=1}^{m} C_i = \bigwedge_{i=1}^{m} (x_{i,1} \vee x_{i,2} \vee x_{i,3})$$

Let $Y = \{x_1, \ldots x_n, y_1, \ldots, y_m, z\}$ and consider the following sets of implications:

- $\Sigma_1 = \{x_{i,k} x_{i,\ell} \to z \mid i \in [m] \text{ and } k, \ell \in [3], k \neq \ell\}$,
- $\Sigma_2 = \{y_i \to z \mid i \in [m]\}$,
- $\Sigma_3 = \{x_{i,k} z \to y_i \mid i \in [m], k \in [3]\}$.

And let $\Sigma_Y = \Sigma_1 \cup \Sigma_2 \cup \Sigma_3$. In [11] the authors show that DUALS is already intractable for these instances with $B = \{y_1 \ldots y_m, z\}$. Observe that in general, no implication of the form $uv \to Y$ holds in $\Sigma_Y$, so that a straightforward identification with MCCENUM is not possible.

Therefore, applying Lemma 1, we obtain that MCCENUM cannot be solved in output-polynomial time in the following case: $X = Y \cup \{u, v\}$, $\Sigma = \Sigma_Y \cup \{B \to uv\}$, $G_c = (X, E_c = \{uv\})$.

Let us show that $\mathcal{F}$, the closure system associated to $\Sigma$, is indeed lower bounded. We proceed by analysing the $D$-relation. Observe first that $\mathcal{F}$ is standard. We begin with $u, v$. Let $t \in X \setminus \{u\}$ and $M \in \mathcal{M}(\mathcal{F})$ such that $t \uparrow M$. As no premise of $\Sigma$ contains $u$, it must be that for any $F \in \mathcal{F}$, $F \cup \{u\} \in \mathcal{F}$. In particular, we deduce that $u \in M$. Hence for any $t \in X \setminus \{u\}$, $t$ does not depend on $u$. Applying the same reasoning on $v$, we obtain that no $D$-cycle can contain $u$ or $v$. Let $x_i \in X$, $i \in [n]$. As $x_i$ is the conclusion of no implication in $\Sigma$, we have that the unique meet-irreducible element $M_i$ satisfying $x_i \uparrow M_i$ is $X \setminus x_i$. Therefore, there is no element in $X \setminus \{x_i\}$ on which $x_i$ depends, so that no $D$-cycle can contain $x_i$, for any $i \in [n]$. Let us move to $z$. As $y_j \to z \in \Sigma$ for any $j \in [m]$, we have $y_{j_*} = \phi(y_j)_* = \{z\}$. Hence, $zDy_j$ cannot hold since $M \downarrow y_j$ implies $z \in M$, for any $M \in \mathcal{M}(\mathcal{F})$. Thus, $z$ only depends on some of the $x_i$'s, $i \in [n]$, and no $D$-cycle can contain $z$ either.

Henceforth, the only possible $D$-cycles must be contained in $\{y_1, \ldots, y_m\}$. We show that for any $i, k \in [m]$, $y_i Dy_k$ does not hold. For any $y_i$, $i \in [m]$, we have $y_{i_*} = \{z\}$ as $y_i \to z \in \Sigma$. Hence, a meet-irreducible element $M_i$ satisfying $y_i \uparrow M_i \downarrow y_k$ must contain $z$. Let $F \in \mathcal{F}$ be any closed set satisfying $y_i \notin F$ but $z \in F$. Assume there exists some $y_k$ such that $y_k \notin F$. Then $F \cup \{y_k\} \in \mathcal{F}$, as $y_k \to z$ is the only implication having $y_k$ in its premise, and $z \in F$. Therefore, it must be that for any $M_i \in \mathcal{M}(\mathcal{F})$ such that $z \in M_i$ and $y_i \notin M_i$, $\{y_1, \ldots, y_m\} \setminus \{y_i\} \subseteq M_i$ is verified, so that $y_i \uparrow M_i \downarrow y_k$ is not possible. As a consequence $y_i D y_k$ cannot hold, for any $i, k \in [m]$. We conclude that $\mathcal{F}$ has no $D$-cycles and that it is lower bounded by Theorem 2. □

Therefore, there is no algorithm solving MCCENUM in output-polynomial time unless $\mathsf{P} = \mathsf{NP}$ even when restricted to lower bounded closure systems. In the next section, we consider classes of closure systems where MCCENUM can be solved in incremental-polynomial time.

## 4    Minimal Generators with Bounded Size

In this section, we identify classes of closure systems for which MCCENUM can be solved in output-polynomial time. Let $\Sigma$ be an implicational base over $X$ and $G_c$ a non-empty consistency-graph. Recall that elements of $\mathsf{maxCC}(\Sigma, G_c)$ are exactly the co-atoms of $\mathcal{F} \cap (\mathsf{IS}(G_c) \cup \{X\})$. Therefore, $\mathsf{maxCC}(\Sigma, G_c)$ can be obtained as $\mathsf{MIS}(\mathcal{K})$, where $\mathcal{K}$ is the set of keys of $\mathcal{F} \cap (\mathsf{IS}(G_c) \cup \{X\})$.

*Remark 1.* In general, $\mathcal{F} \cap (\mathsf{IS}(G_c) \cup \{X\})$ does not belong to the same class as $\mathcal{F}$ (distributive, modular, lower bounded, ...). Hence, for a given class of closure system, MCCENUM differ from CE. For instance if $\mathcal{F} = 2^X$, CE is trivial while MCCENUM requires to enumerate maximal independent sets of $G_c$.

Now, if we can guarantee that any element of $\mathcal{K}$ has constant size, then $\mathcal{K}$ has polynomial size with respect to $G_c, \Sigma$ and $X$. Hence, we can derive an incremental-polynomial time algorithm computing $\mathsf{maxCC}(\Sigma, G_c)$ in two steps:

1. Compute the set of keys $\mathcal{K}$,
2. Compute $\mathsf{MIS}(\mathcal{K}) = \mathsf{maxCC}(\Sigma, G_c)$.

To identify such cases, we have to characterize elements of $\mathcal{K}$. To do so, we have to guarantee that a set $Y \subset X$ contains a key of $\mathcal{K}$ whenever $Y$ or $\phi(Y)$ is inconsistent with respect to $G_c$. Looking at $G_c$ is sufficient to distinguish between consistent and inconsistent closed sets of $\mathcal{F}$. However, there may be consistent (non-closed) sets $Y$ such that $\phi(Y)$ contains an edge of $G_c$. These will not be seen by just considering $G_c$. Thus, if $uv$ is the edge of $G_c$ contained in $\phi(Y)$, we deduce that there must be a minimal generator $A_u$ of $u$ contained in $Y$, possibly $A_u = \{u\}$. Similarly, $Y$ contains a minimal generator $A_v$ of $v$. The fact that $G_c$ is a graph plays an important role here, as it guarantees that $Y$ can be identified by combining only two minimal generators, one for each vertices of some edge in $G_c$. In particular, keys in $\mathcal{K}$ will share the following property.

**Proposition 1.** *Let $K \in \mathcal{K}$. Then there exists $uv \in E_c$, a minimal generator $A_u$ of $u$, and a minimal generator $A_v$ of $v$ such that $K = A_u \cup A_v$.*

*Proof.* Let $K \in \mathcal{K}$. As $\phi(K) = X$ in $\mathcal{F} \cap (\mathsf{IS}(G_c) \cup \{X\})$ and $E_c$ is non-empty by assumption, there exists an edge $uv$ of $G_c$ such that $uv$ is in the closure of $K$ with respect to $\Sigma$. Thus, there exist minimal generators $A_u$ of $u$ and $A_v$ of $v$

such that $A_u \cup A_v \subseteq K$. Assume that $A_u \cup A_v \subset K$ and let $x \in K \setminus (A_u \cup A_v)$. As $u \in \phi(A_u)$ and $v \in \phi(A_v)$, we get $uv \in \phi(K \setminus \{x\})$, a contradiction with the minimality of $K$. $\qquad\square$

*Example 1.* We consider $\Sigma$, $X$ and $G_c$ of Fig. 1. We have that $\phi(135) = 1235$ is inconsistent as it contains 25. However 135 is consistent with respect to $G_c$. For this example, we will have $\mathcal{K} = \{135, 34, 24, 25\}$. Note that 135 can be decomposed following Proposition 1 as the minimal generator 13 of 2, and 5 as a trivial minimal generator for itself.

Remark that $E_c \not\subseteq \mathcal{K}$ in the general case, as there may be an implication $u \to v$ in $\Sigma$ for some inconsistent pair $uv \in E_c$. Thus $u$ is a key which satisfies Proposition 1 with $A_u = A_v = \{u\}$. It also follows from Proposition 1 that $c(\mathcal{F})$, the maximum size of a minimal generator or Carathéodory number of $\mathcal{F}$, plays an important role for MCCEnum. When no restriction on $c(\mathcal{F})$ holds, $\mathcal{K}$ can have exponential size with respect to $\Sigma$ and $G_c$. The next example drawn from [26] illustrates this exponential growth.

*Example 2.* Let $X = \{x_1, \ldots, x_n, y_1, \ldots, y_n, u, v\}$ and $\Sigma = \{x_i \to y_i \mid i \in [n]\} \cup \{y_1 \ldots y_n \to uv\}$. The consistency-graph is $G_c = (X, \{uv\})$. The set of non-trivial minimal generators of $u$ and $v$ is $\{z_1 \ldots z_n \mid z_i \in \{x_i, y_i\}, i \in [n]\}$. Moreover, minimal generators of $u$ and $v$ are also the keys of $\mathcal{F} \cap (\mathsf{IS}(G_c) \cup \{X\})$. Thus, $|\mathcal{K}| = 2^n$, which is exponential with respect to $\Sigma$ and $G$. Observe that $\Sigma$ is acyclic [20,33]: for any $x, y \in X$ if $y$ belongs to some minimal generator of $x$, then $x$ is never contained in a minimal generator of $y$.

Hence, computing $\mathsf{maxCC}(\Sigma, G_c)$ through the intermediary of $\mathcal{K}$ is in general impossible in output-polynomial time. In fact, this exponential blow up occurs even for small classes of closure systems where the Carathéodory number $c(\mathcal{F})$ is unbounded. In Example 2 for instance, the closure system induced by $\Sigma$ is acyclic [20,33], a particular case of lower boundedness [1].

On the other hand, let us assume now that $c(\mathcal{F})$ is bounded by some constant $k \in \mathbb{N}$. Then, by Proposition 1, every key in $\mathcal{K}$ has at most $2 \times k$ elements. As a consequence we show in the next theorem that the two-steps algorithm we described can be conducted in incremental-polynomial time.

**Theorem 3.** *Let $\Sigma$ be an implicational base over $X$ with induced $\mathcal{F}$, and $G_c = (X, E_c)$ a consistency-graph. If $c(\mathcal{F}) \leq k$ for some constant $k \in \mathbb{N}$, the problem MCCEnum can be solved in incremental-polynomial time.*

*Proof.* The set of keys $\mathcal{K}$ can be computed in incremental-polynomial time with respect to $\mathcal{K}$, $\Sigma$, $X$ and $G_c$ using the algorithm of Lucchesi and Osborn [28] with input $\Sigma' = \Sigma \cup \{uv \to X \mid uv \in E_c\}$. Observe that the closure system associated to $\Sigma'$ is exactly $\mathcal{F} \cap \{\mathsf{IS}(G_c) \cup \{X\}\}$. Indeed, a consistent closed set of $\mathcal{F}$ models $\Sigma'$ and a subset $F \subseteq X$ which satisfies $\Sigma'$ must also satisfy $\Sigma$ and being an independent set of $G_c$ if $F \subset X$. Note that $\mathcal{K}$ is then computed in time $\mathsf{poly}(|\Sigma| + |X| + |G_c| + |\mathcal{K}|)$. As the total size of $\mathcal{K}$ is bounded by $|X|^{2k}$ by Proposition 1, we get that $\mathcal{K}$ is computed in time $\mathsf{poly}(|\Sigma| + |X| + |G_c|)$.

Then, we apply the algorithm of Eiter and Gottlob [14] to compute $\mathsf{MIS}(\mathcal{K}) = \mathsf{maxCC}(\Sigma, G_c)$ which runs in incremental polynomial time. Since $\mathcal{K}$ has polynomial size with respect to $|X|$, the delay between the $i$-th and $(i+1)$-th solution of $\mathsf{maxCC}(\Sigma, G_c)$ output is bounded by $\mathsf{poly}(|X|^{2k}+i)$, that is $\mathsf{poly}(|X|+i)$. Furthermore, the delay after the last output is also bounded by $\mathsf{poly}(|X|^{2k}) = \mathsf{poly}(|X|)$. As the time spent before the first solution output is $\mathsf{poly}(|\Sigma| + |X| + |G_c|)$, the whole algorithm has incremental delay as expected.    □

To conclude this section, we show that Theorem 3 applies to various classes of closure systems known in the theory of convex geometries [27].

A closure system $\mathcal{F}$ is *distributive* if for any $F_1, F_2 \in \mathcal{F}$, $F_1 \cup F_2 \in \mathcal{F}$. Implicational bases of distributive closure systems have premises of size one [19].

Let $P = (X, \leq)$ be a partially ordered set, or poset. A subset $Y \subseteq X$ is *convex* in $P$ if for any triple $x \leq y \leq z$, $x, z \in Y$ implies $y \in Y$. The family $\{Y \subseteq X \mid Y \text{ is convex in } P\}$ is known to be closure system over $X$ [9,25].

Let $G = (X, E)$ be a graph. We say that $G$ is *chordal* if it has no induced cycle of size $\geq 4$. A *chord* in a path from $x$ to $y$ is an edge connecting to non-adjacent vertices of the path. A subset $Y$ of $X$ is *monophonically convex* in $G$ if for every pair $x, y$ of elements in $Y$, every $z \in X$ which lies on a chordless path from $x$ to $y$ is in $Y$. The family $\{Y \subseteq X \mid Y \text{ is monophonically convex in } G\}$ is a closure system [15,27].

Finally, let $X \subseteq \mathbb{R}^n$, $n \in \mathbb{N}$, be a finite set of points, and denote by $\mathsf{ch}(Y)$ the *convex hull* of $Y$. The set system $\{\mathsf{ch}(Y) \mid Y \subseteq X\}$ forms a closure system [27] usually known as an *affine convex geometry*.

**Corollary 2.** *Let $\Sigma$ be an implicational base over $X$ and $G_c = (X, E_c)$. MCCEnum can be solved in incremental-polynomial time in the following cases:*

- *$\mathcal{F}$ is distributive,*
- *$\mathcal{F}$ is the family of convex subsets of a poset,*
- *$\mathcal{F}$ is The family of monophonically convex subsets of a chordal graph has Carathéodory number at most 2,*
- *$\mathcal{F}$ is an affine convex geometry in $\mathbb{R}^k$ for a fixed constant $k$.*

*Proof.* Distributive lattices have Carathéodory number 1 as they can be represented by implicational bases with singleton premises. The family of convex subsets of a poset has Carathéodory number 2 [25] (Corollary 13). The family of monophonically convex subsets of a chordal graphs have Carathéodory number at most 2 [15] (Corollary 3.4). The Carathéodory number of an affine convex geometry in $\mathbb{R}^k$ is $k - 1$ (see for instance [27], p. 32).    □

In the distributive case, the algorithm can perform in polynomial delay using the algorithm of [24] since $\mathcal{K}$ will be a graph by Proposition 1. This connects with previous results on distributive closure systems by Kavvadias et al. [26].

## 5 Biatomic Atomistic Closure Systems

In this section, we are interested in biatomic atomistic closure systems. Namely, we show that when minimal generators obey an independence condition, the size of $X$ is exponential with respect to $c(\mathcal{F})$. To do so, we show that in biatomic atomistic closure systems, each subset of a minimal generator is itself a minimal generator. This result applies to atomistic modular closure systems, which can be represented by implications with premises of size at most two [32]. This suggests that MCCENUM becomes more difficult when implications have binary premises.

First, we need to define atomistic biatomic closure systems. Let $\mathcal{F}$ be a closure system over $X$ with associated closure operator $\phi$. We say that $\mathcal{F}$ is *atomistic* if for any $x \in X$, $\phi(x) = \{x\}$. Equivalently, $\mathcal{F}$ is atomistic if its join-irreducible elements equal its atoms. Note that in a standard closure system, an atom is a singleton element. Biatomic closure systems have been studied by Birkhoff and Bennett in [6,9]. We reformulate their definition in terms of closure systems. A closure system $\mathcal{F}$ is *biatomic* if for every closed sets $F_1, F_2 \in \mathcal{F}$ and any atom $\{x\} \in \mathcal{F}$, $x \in \phi(F_1 \cup F_2)$ implies the existence of atoms $\{x_1\} \subseteq F_1$, $\{x_2\} \subseteq F_2$ such that $x \in \phi(x_1 x_2)$. In atomistic closure systems in particular, the biatomic condition applies to every element of $X$. Hence the next property of biatomic atomistic closure systems.

**Proposition 2.** *Let $\mathcal{F}$ be a biatomic atomistic closure system. Let $F \in \mathcal{F}$ and $x, y \in X$ with $x, y \notin F$. If $y \in \phi(F \cup \{x\})$, then there exists an element $z \in F$ such that $y \in \phi(xz)$.*

*Proof.* In atomistic closure systems, every element of $X$ is closed, therefore we apply the definition to the closed sets $F$ and $\{x\}$. □

We will also make use of the following folklore result about minimal generators. We give a proof for self-containment.

**Proposition 3.** *If $A_x$ is a minimal generator of $x \in X$, then $\phi(A) \cap A_x = A$ for any $A \subseteq A_x$.*

*Proof.* First, we have that $A \subseteq \phi(A) \cap A_x$ as $A \subseteq \phi(A)$ and $A \subseteq A_x$. Now suppose that there exists $a \in \phi(A) \cap A_x$ such that $a \notin A$. Then, $a \in \phi(A_x \setminus \{a\})$ as $A \subseteq A_x \setminus \{a\}$. Hence, $\phi(A_x) = \phi(A_x \setminus \{a\})$ and $x \in \phi(A_x \setminus \{a\})$, a contradiction with $A_x$ being a minimal generator of $x$. □

Our first step is to show that in a biatomic atomistic closure system, if $A_x$ is a minimal generator for some $x \in X$, then every non-empty subset $A$ of $A_x$ is itself a minimal generator for some $y \in X$. We prove this statement in Lemmas 2 and 3. Recall that an element $x \in X$ is a (trivial) minimal generator of itself.

**Lemma 2.** *Let $x \in X$ and let $A_x$ be a minimal generator of $x$ with size $k \geq 2$. Then for any $a_i \in A_x$, $i \in [k]$, there exists $y_i \in X$ such that $A_x \setminus \{a_i\}$ is a minimal generator of $y_i$.*

*Proof.* Let $A_x = \{a_1, \dots, a_k\}$ be a minimal generator of $x$ such that $k \geq 2$. Then, for any $a_i \in A_x$, $i \in [k]$, we have $a_i \notin \phi(A_x \setminus \{a_i\})$ by Proposition 3. However, we have $x \in \phi(\{a_i\} \cup \phi(A_x \setminus \{a_i\})) = \phi(A_x)$. Thus, by Proposition 2, there must exists $y_i \in \phi(A_x \setminus \{a_i\})$ such that $x \in \phi(a_i y_i)$.

Let us show that $A_x \setminus \{a_i\}$ is a minimal generator of $y_i$. Assume for contradiction this is not the case. As $y_i \in \phi(A_x \setminus \{a_i\})$, there must be a proper subset $A$ of $A_x \setminus \{a_i\}$ which is a minimal generator for $y_i$. Note that since $A_x$ has at least 2 elements, at least one proper subset of $A_x \setminus \{a_i\}$ exists. As $A \subset A_x \setminus \{a_i\}$, there exists $a_j \in A_x$, $a_j \neq a_i$, such that $a_j \notin A$. Therefore, $A \subseteq A_x \setminus \{a_j\}$ and $\phi(A) \subseteq \phi(A_x \setminus \{a_j\})$. More precisely, $y_i \in \phi(A)$ and hence $y_i \in \phi(A_x \setminus \{a_j\})$. However, we also have that $a_i \in \phi(A_x \setminus \{a_j\})$ as $a_i \in A_x$, $a_i \neq a_j$, and since $x \in \phi(a_i y_i)$, we must have $x \in \phi(A_x \setminus \{a_j\})$, a contradiction with $A_x$ being a minimal generator of $x$. Thus, we deduce that $A_x \setminus \{a_i\}$ is a minimal generator for $y_i$, concluding the proof.                                    □

In the particular case where $A_x$ has only two elements, say $a_1$ and $a_2$, then $A_x \setminus \{a_1\} = \{a_2\}$ and the element $a_2$ is a trivial minimal generator of itself. By using inductively Lemma 2 on the size of $A_x$, one can derive the next straightforward lemma.

**Lemma 3.** *Let $\mathcal{F}$ be a biatomic atomistic closure system. Let $A_x$ be a minimal generator of some $x \in X$. Then, for any $A \subseteq A_x$ with $A \neq \emptyset$, there exists $y \in X$ such that $A$ is a minimal generator of $y$.*

Thus, for a given minimal generator $A_x$ of $x$, any non-empty subset $A$ of $A_x$ is associated to some $y \in X$. We show next than when $A_x$ also satisfies an independence condition, $A$ will be the unique subset of $A_x$ associated to $y$. Following [19], we reformulate the definition of independence in an atomistic closure system $\mathcal{F}$, but restricted to its atoms. A subset $Y$ of $X$ is *independent* in $\mathcal{F}$ if for any $Y_1, Y_2 \subseteq Y$, $\phi(Y_1 \cap Y_2) = \phi(Y_1) \cap \phi(Y_2)$.

**Lemma 4.** *Let $\mathcal{F}$ be a biatomic atomistic closure system. Let $A_x$ be an independent minimal generator of $x \in X$, and let $A$ be a non-empty subset of $A_x$. Then, there exists $y \in X$ such that $A$ is the unique minimum subset of $A_x$ satisfying $y \in \phi(A)$.*

*Proof.* Let $A_x$ be an independent minimal generator of $x \in X$, and let $A$ be a non-empty subset of $A_x$. By Lemma 3, there exists $y \in X$ such that $A$ is a minimal generator for $y$, which implies $y \in \phi(A)$.

To prove that $A$ is the unique minimum subset of $A_x$ such that $y \in \phi(A)$, we show that for any $B \subseteq A_x$ such that $A \nsubseteq B$, $y \in \phi(B)$ cannot hold. Consider $B \subseteq A_x$ with $A \nsubseteq B$ and suppose that $y \in \phi(B)$. Note that $B$ must exist as the empty set is always a possible choice. Since $y \in \phi(A)$, we have $y \in \phi(A) \cap \phi(B)$. Furthermore, $\phi(A \cap B) \subset \phi(A)$ as $A \cap B \subset A$ and $\phi(A \cap B) \cap A_x = A \cap B$ by Proposition 3. Moreover, $A_x$ is independent, so that $\phi(A) \cap \phi(B) = \phi(A \cap B)$. Hence, $y \in \phi(A \cap B) \subset \phi(A)$, a contradiction with $A$ being a minimal generator of $y$.                                    □

Hence, when $A_x$ is independent, each non-empty subset $A$ of $A_x$ is the unique minimal generator of some $y$ being included in $A_x$. As a consequence, we obtain the following theorem.

**Theorem 4.** *Let $\mathcal{F}$ be a biatomic atomistic closure system. If for any $x \in X$ and any minimal generator $A_x$ of $x$, $A_x$ is independent, then $c(\mathcal{F}) \leq \lceil \log_2(|X|+1) \rceil$.*

*Proof.* Let $A_x$ be a minimal generator of $x$, $x \in X$ such that $c(\mathcal{F}) = |A_x|$. As $A_x$ is a minimal generator, $\phi(A) \neq \phi(A')$ for any distinct $A, A' \subseteq A_x$, due to Proposition 3. Furthermore $A_x$ is independent by assumption. Thus, by Lemma 4, for each non-empty subset of $A$, there exists $y \in X$ such that $A$ is the unique minimum subset of $A_x$ with $y \in \phi(A)$. Consequently, $X$ must contain at least $2^{|A_x|} - 1$ elements in order to cover each non-empty subset of $A_x$, that is $2^{|A_x|} - 1 \leq |X|$, which can be rewritten as $|A_x| = c(\mathcal{F}) \leq \lceil \log_2(|X|+1) \rceil$ as required. $\square$

Now let $\mathcal{F}$ be a biatomic atomistic closure system on $X$ given by some implicational base $\Sigma$ and let $G_c = (X, E_c)$ be a consistency-graph. Assume that every minimal generator is independent. By Theorem 4, we have that $|X|$ has exponential size with respect to $c(\mathcal{F})$, and by Proposition 1, it must be that the size of a key in $\mathcal{K}$ cannot exceed $2 \times \lceil \log_2(|X|+1) \rceil$. Thus, with respect to $\Sigma$, $G_c$ and $X$, $\mathcal{K}$ will have size quasi-polynomial in the worst case. Using the same algorithm as in Sect. 4, we obtain the next theorem.

**Theorem 5.** *Let $\Sigma$ be an implicational base of a biatomic atomistic closure system $\mathcal{F}$ over $X$ and $G_c$ a consistency-graph. If for any $x \in X$ and any minimal generator $A_x$ of $x$, $A_x$ is independent, then MCCENum can be solved in output-quasipolynomial time.*

*Proof.* For clarity, we put $n = |X|$ and $k$ as the total size of the output $\mathsf{MIS}(\mathcal{K})$. $\mathcal{K}$ can be computed in incremental-polynomial time with the algorithm in [28]. Furthermore, by Theorem 4, the total size of $\mathcal{K}$ is bounded by $n^{\log(n)}$. Thus, this first step runs in time $\mathsf{poly}(|\Sigma| + |G_c| + n + n^{\log(n)})$, which is bounded by $\mathsf{poly}(|\Sigma| + |G_c| + n)^{\log(n)}$ being quasipolynomial in the size of $\Sigma$, $G_c$, $\mathcal{K}$ and $X$. To compute $\mathsf{MIS}(\mathcal{K}) = \mathsf{maxCC}(\Sigma, G_c)$ we use the algorithm of Fredman and Khachiyan [16] whose running time is bounded by $(n^{\log(n)} + k)^{o(\log(n^{\log(n)} + k))}$. In our case, we can derive the following upper bounds:

$$(n^{\log(n)} + k)^{o(\log(n^{\log(n)} + k))} \leq (k + n)^{\log(n) \times o(\log(k+n)^{\log(n)})}$$
$$\leq (k + n)^{O(\log^3(k+n))}$$

Thus, the time needed to compute $\mathsf{MIS}(\mathcal{K})$ from $\mathcal{K}$ is output-quasipolynomial in the size of $X$ and $\mathsf{maxCC}(\Sigma, G_c)$. Consequently, the running time of the whole algorithm is bounded by

$$\mathsf{poly}(|\Sigma| + |G_c| + n)^{\log(n)} + (k + n)^{O(\log^3(k+n))}$$

which is indeed quasipolynomial in the size of the input $\Sigma$, $X$, $G_c$ and the output $\mathsf{MIS}(\mathcal{K}) = \mathsf{maxCC}(\Sigma, G_c)$. $\square$

To conclude this section, we show that atomistic modular closure systems [19, 30] satisfy conditions of Theorem 5. Recall that a closure system $\mathcal{F}$ is modular if for any $F_1, F_2, F_3 \in \mathcal{F}$, $F_1 \subseteq F_2$ implies $\phi(F_1 \cup (F_2 \cap F_3)) = \phi(F_1 \cup F_3) \cap F_2$. It was proved for instance in [6] (Theorem 7) that atomistic modular closure systems are biatomic. To show that any minimal generator is independent, we make use of the following result.

**Theorem 6.** *(Reformulated from [19], Theorem 360) Let $\mathcal{F}$ be a modular closure system. A subset $A = \{a_1, \ldots, a_k\} \subseteq X$ is independent if and only if $\phi(a_1) \cap \phi(a_2) = \phi(a_1 a_2) \cap \phi(a_3) = \cdots = \phi(a_1 \ldots a_{k-1}) \cap \phi(a_k) = \emptyset$.*

**Proposition 4.** *Let $\mathcal{F}$ be an atomistic modular closure system. Let $A_x$ be a minimal generator of some $x \in X$. Then $A_x$ is independent.*

*Proof.* Let $A_x = \{a_1, \ldots, a_k\}$ be a minimal generator for some $x \in X$. Then, by Proposition 3, $\phi(a_1 \ldots a_i) \cap A_x = a_1 \ldots a_i$ for any $i \in [k]$. Furthermore, $\phi(a) = \{a\}$ for any $a \in X$ since $\mathcal{F}$ is atomistic. Thus we conclude that $\phi(a_1 \ldots a_i) \cap \phi(a_{i+1}) = \emptyset$ for any $i \in [k-1]$ as $a_{i+1} \notin a_1 \ldots a_i$. It follows by Theorem 6 that $A_x$ is indeed independent. □

**Corollary 3.** *Let $\Sigma$ be an implicational base over $X$ and $G_c = (X, E_c)$. Then MCCENUM can be solved in output-quasipolynomial time if:*

- *$\mathcal{F}$ is biatomic atomistic and has Carathéodory number 2 (including convex subsets of a poset and monophonically convex sets of a chordal graph),*
- *$\mathcal{F}$ is atomistic modular.*

*Proof.* For the first statement, note that in an atomistic closure system with Carathéodory number 2, any minimal generator $A_x$ contains exactly two elements $a_1, a_2$. Since $\mathcal{F}$ is atomistic, $a_1$ and $a_2$ are closed and the independence of $A_x$ follows. If $\mathcal{F}$ is atomistic modular, biatomicity follows from [6] (Theorem 7), and independence from Proposition 4. □

*Remark 2.* For atomistic modular closure systems, the connection between the size of $X$ and the Carathéodory number may also be derived from counting arguments on subspaces of vector spaces [31].

## 6    Conclusions

In this paper we proved that given a consistency-graph over and an implicational base over a finite set, the enumeration of maximal consistent closed sets is impossible in output-polynomial time unless P = NP. Moreover, we showed that this problem, called MCCENUM, is already intractable for the well-known class of lower bounded closure systems. On the positive side, we proved that when the size of a minimal generator is bounded by a constant, the enumeration of maximal consistent closed sets can be conducted in incremental polynomial time. This result covers various classes of convex geometries. Finally, we proved

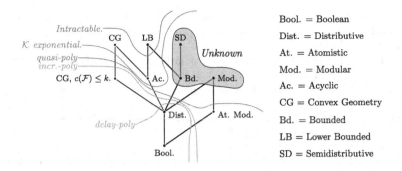

**Fig. 2.** The complexity of MCCEnum in the hierarchy of closure systems

that in biatomic atomistic closure systems, MCCEnum can be solved in output-quasipolynomial time provided minimal generators obey an independence condition. This applies in particular to atomistic modular closure systems. In Fig. 2, we summarize our results in the hierarchy of closure systems.

For future research, we would like to understand which properties or parameters of closure systems affect the tractability of the problem. We have seen that a bounded Carathéodory number gives an incremental-polynomial time algorithm, while lower boundedness makes the problem intractable. Another question is the following: is the problem still hard if the closure system is given by a context (equivalently, its meet-irreducible elements)? The question is particularly interesting for classes such as semidistributive lattices where we can compute the context in polynomial time in the size of an implicational base [5].

**Acknowledgment.** Authors would like to thank reviewers for their useful corrections and suggestions, in particular on standard closure systems.

# References

1. Adaricheva, K.: Optimum basis of finite convex geometry. Discrete Appl. Math. **230**, 11–20 (2017)
2. Ardila, F., Owen, M., Sullivant, S.: Geodesics in CAT(0) cubical complexes. Adv. Appl. Math. **48**(1), 142–163 (2012)
3. Babin, M.A., Kuznetsov, S.O.: Dualization in lattices given by ordered sets of irreducibles. Theoret. Comput. Sci. **658**, 316–326 (2017)
4. Barthélemy, J.-P., Constantin, J.: Median graphs, parallelism and posets. Discrete Math. **111**(1–3), 49–63 (1993)
5. Beaudou, L., Mary, A., Nourine, L.: Algorithms for k-meet-semidistributive lattices. Theoret. Comput. Sci. **658**, 391–398 (2017)
6. Bennett, M.K.: Biatomic lattices. Algebra Univers. **24**(1–2), 60–73 (1987)
7. Bertet, K., Caspard, N.: Doubling convex sets in lattices: characterizations and recognition algorithms. Order **19**(2), 181–207 (2002). https://doi.org/10.1023/A: 1016524118566

8. Bertet, K., Demko, C., Viaud, J.-F., Guérin, C.: Lattices, closures systems and implication bases: a survey of structural aspects and algorithms. Theoret. Comput. Sci. **743**, 93–109 (2018)

9. Birkhoff, G., Bennett, M.K.: The convexity lattice of a poset. Order **2**(3), 223–242 (1985). https://doi.org/10.1007/BF00333128

10. Day, A.: A simple solution to the word problem for lattices. Can. Math. Bull. **13**(2), 253–254 (1970)

11. Defrain, O., Nourine, L.: Dualization in lattices given by implicational bases. Theoret. Comput. Sci. **814**, 169–176 (2020)

12. Demetrovics, J., Libkin, L., Muchnik, I.B.: Functional dependencies in relational databases: a lattice point of view. Discrete Appl. Math. **40**(2), 155–185 (1992)

13. Dung, P.M.: On the acceptability of arguments and its fundamental role in nonmonotonic reasoning, logic programming and n-person games. Artif. Intell. **77**(2), 321–357 (1995)

14. Eiter, T., Gottlob, G.: Identifying the minimal transversals of a hypergraph and related problems. SIAM J. Comput. **24**(6), 1278–1304 (1995)

15. Farber, M., Jamison, R.E.: Convexity in graphs and hypergraphs. SIAM J. Algebraic Discrete Methods **7**(3), 433–444 (1986)

16. Fredman, M.L., Khachiyan, L.: On the complexity of dualization of monotone disjunctive normal forms. J. Algorithms **21**(3), 618–628 (1996)

17. Freese, R., Ježek, J., Nation, J.B.: Free Lattices, vol. 42. American Mathematical Society, Providence (1995)

18. Ganter, B., Wille, R.: Formal Concept Analysis: Mathematical Foundations. Springer, Heidelberg (2012)

19. Grätzer, G.: Lattice Theory: Foundation. Springer, Heidelberg (2011)

20. Hammer, P.L., Kogan, A.: Quasi-acyclic propositional horn knowledge bases: optimal compression. IEEE Trans. Knowl. Data Eng. **7**(5), 751–762 (1995)

21. Herrmann, C., Pickering, D., Roddy, M.: A geometric description of modular lattices. Algebra Univers. **31**(3), 365–396 (1994)

22. Hirai, H., Nakashima, S.: A compact representation for modular semilattices and its applications. Order **37**(3), 479–507 (2020). https://doi.org/10.1007/s11083-019-09516-0

23. Hirai, H., Oki, T.: A compact representation for minimizers of k-submodular functions. J. Comb. Optim. **36**(3), 709–741 (2018). https://doi.org/10.1007/s10878-017-0142-0

24. Johnson, D.S., Yannakakis, M., Papadimitriou, C.H.: On generating all maximal independent sets. Inf. Process. Lett. **27**(3), 119–123 (1988)

25. Kashiwabara, K., Nakamura, M.: Characterizations of the convex geometries arising from the double shellings of posets. Discrete Math. **310**(15–16), 2100–2112 (2010)

26. Kavvadias, D.J., Sideri, M., Stavropoulos, E.C.: Generating all maximal models of a Boolean expression. Inf. Process. Lett. **74**(3–4), 157–162 (2000)

27. Korte, B., Lovász, L., Schrader, R.: Greedoids, vol. 4. Springer, Heidelberg (2012)

28. Lucchesi, C.L., Osborn, S.L.: Candidate keys for relations. J. Comput. Syst. Sci. **17**(2), 270–279 (1978)

29. Nielsen, M., Plotkin, G., Winskel, G.: Petri nets, event structures and domains, part I. Theoret. Comput. Sci. **13**(1), 85–108 (1981)

30. Stern, M.: Semimodular Lattices: Theory and Applications, vol. 73. Cambridge University Press, Cambridge (1999)

31. Wild, M.: The minimal number of join irreducibles of a finite modular lattice. Algebra Univers. **35**(1), 113–123 (1996)

32. Wild, M.: Optimal implicational bases for finite modular lattices. Quaest. Math. **23**(2), 153–161 (2000)
33. Wild, M.: The joy of implications, aka pure horn formulas: mainly a survey. Theoret. Comput. Sci. **658**, 264–292 (2017)

# A New Kind of Implication to Reason with Unknown Information

Francisco Pérez-Gámez$^{(\boxtimes)}$ ⓘ, Pablo Cordero ⓘ, Manuel Enciso ⓘ, and Angel Mora ⓘ

University of Malaga, Malaga, Spain
{franciscoperezgamez,pcordero,enciso,amora}@uma.es

**Abstract.** Formal Concept Analysis (FCA) extracts knowledge from an object-attribute relation. In the classical case, it focuses on positive information, i.e. attributes that are satisfied by objects. Several papers have recently been published extending FCA to manage negative information, i.e. attributes that are not satisfied by objects. However, the study of unknown information –being unknown, whether it is positive or negative value– is an issue to be explored. In this paper, we approach this problem by using a 4-valued logic. Specifically, given a context with partial information that corresponds to a 3-valued relation, we define a 4-valued Galois connection from where we extend the notions of concept and implication. Also, we present Amstrong's axioms in this new framework, and we prove that this inference system is sound and complete.

**Keywords:** Implications · Unknown information · Galois connection · Logic

## 1 Introduction

Since Wille's introduction of Formal Concept Analysis (FCA) in the early 80s [22], it has established itself as a very successful mathematical approach to knowledge management, with a rich theory as well as numerous practical applications. Essentially, FCA provides methods for data analysis, knowledge representation, information management and reasoning [9]. From a data-set (called *formal context*) describing a set of objects and their attributes, FCA extracts knowledge that can mainly be represented in two different ways: by *formal concepts* ordered in a hierarchy or by *attribute implications* [10]. Formal concepts represent sets of objects with common attributes, whereas implications can be seen as Horn formulas describing relations between sets of attributes. Implications should naturally lead to the use of logic to automate reasoning. The success of FCA is based on its solid mathematical foundations, which lie in the notions of Galois connection and closure operator in the Boolean algebras of attribute/object sets.

Supported by Grants TIN2017-89023-P and PRE2018-085199 of the Science and Innovation Ministry of Spain and UMA2018-FEDERJA-001 of the Junta de Andalucia, and European Social Fund.

ⓒ Springer Nature Switzerland AG 2021
A. Braud et al. (Eds.): ICFCA 2021, LNAI 12733, pp. 74–90, 2021.
https://doi.org/10.1007/978-3-030-77867-5_5

Classical FCA focuses on positive information, i.e. on the attributes that objects have. On the other side, the information provided by the attributes that the objects do not fulfil (negative information) is also relevant in a wide variety of applications. See, for instance, the [14] first approach to this issue, which can already be found in the early works of Ganter and Wille [10], consists of duplicating the context by the apposition of the context and its negation. This representation tends to be inefficient and redundant, as observed by Missaoui et al. [17]. The reason is that this approach does not take advantage of the relationship between positive and negative information. Thus, for instance, the attributes "switched_on" and "switched_off" can be conceived as opposite attributes and related with the implication "switched_on implies not switched_off". Considering this kind of implications as baseline knowledge would allow FCA methods to improve their performance.

In this research line, the foundations of FCA have been extended to deal with mixed (positive and negative) information by studying mixed concept lattices in-depth [20]. Methods to compute the mixed concept lattice are explored in [21] and a logic for reasoning about mixed implications is introduced in [3]. In [11], Ganter and Kwuida define a negation, not at the level of attributes, but at the level of concepts by using pseudocomplemented lattices, also known as p-algebras.

Information (positive or negative) is usually assumed to be known. Nevertheless, there are many cases where it is necessary to work and reason with partial information, i.e. lack of information or unknowledge. The reasons why there may be unknown information are varied. For example:

- In some cases, speaking about some attributes has no sense. For instance, in a dataset about patients, the attribute "regular period" has no sense for the males.
- Sometimes, we cannot determine whether an attribute is positive or negative because we have not got this information yet. We can not affirm if a student has passed a subject if the final exam has not already held.
- In other situations, the positive or negative nature of the information may be uncertain. For example, if we collect information on opinions from users about touristic resources, we may not be able to conclude, from a particular opinion, that it is positive or negative.

Fuzzy set theory can provide tools to address this problem. Extensions of FCA to the fuzzy framework that considers positive and negative information can be found in [13]. Nevertheless, other, not necessarily fuzzy, approaches are possible. Thus, [2] provides a survey of several approaches to the treatment of incomplete knowledge in FCA.

We tackle this problem with a different strategy. We consider three-valued sets of attributes, i.e. sets of attributes that can be positive, negative or unknown. In the same way, partial formal context will be three-valued relations between objects and attributes. Several authors have explored this line. As far as we know, the pioneer work is due to S. Kuznetsov [15] where he introduced the notion of positive, negative and undefined *examples* (objects). These three kinds of examples are used, by the so-called JSM-Method [5], to automatically generate *hypothesis*, providing a forecast for the undefined information. The work established a proper introduction of the undefined information and a solid extension of FCA to deal with it. However, it is focused on the Concept Lattice, but no further study is introduced for implications. In a later paper, B. Ganter and S. Kuznetsov [8] used this framework to built a machine learning method, showing the benefits of the use of FCA to approach problems in other areas. In the work, the authors also introduced the patter structures notion to take advantage of the notion of *example*, already introduced in the previous article. [16] also illustrates how other popular methods and techniques are closer to FCA than they are shown in the literature. In that work, S. Kuznetsov detailed how classification methods can be approached by using FCA, making a proper translation of the notions of Taxonomy, Meronomy, Tolerance and Similarity Relations, which are strongly connected to set of objects, set of attributes, a particular representation of formal contexts and a meet operation in semilattices, respectively. In that work, the author introduces a four value logic, namely "empirically false," "empirically true," "empirical contradiction", and "empirically undeterminate". However, the author introduced this set of truth values to be used in a second-order logic, which is different from the proper framework in FCA. Finally, in [19], S. Obiedkov used a set of three values to build an extension of the original FCA. It can be considered a very solid approach to introduce the management of incomplete information and can be considered a paper close to our work. However, we are interested in the further development of efficient methods to manage implications and formal context. In this line, we balanced the expressive power of the logic and the efficiency of the methods, remaining on a propositional logic with no disjunction operator, neither in the syntax nor in the implicit semantics. At this point, our approach differs from Obiedkov's paper, which introduces a modal logic in his extension.

Then, we introduce an algebraic structure to be the truthfulness values in which the formal framework is generalized. The key point is to establish this algebraic structure in such way that the classical framework could be naturally adaptable. We propose a new structure being an algebra of four-valued sets in which the fourth value captures the inconsistency of some attribute sets. The origin of this structure is the FCA conjunctive interpretation of attribute sets.

The paper is structured as follows: Sect. 2 presents some preliminary notions and results, Sect. 3 introduces the algebra that we use as underlying structure of truthfulness values, Sect. 4 studies the generalization of the concept lattice, Sect. 5 presents the notion of weak implications and its also provides a version of Armstrong's axioms proving its soundness and completeness. Finally, conclusions and further works are presented in Sect. 6.

## 2   Preliminaries

In this chapter, we are going to introduce the main notions that we are going to use in the paper. For more details, we refer to [1,4]. The basic structure that we are going to use is the complete lattice $\mathbb{L} = (L, \leq)$, i.e. an ordered set in which any subset $X \subseteq L$ has supremum and infimum (denoted by $\bigvee X$ and $\bigwedge X$, respectively). As a consequence, $\mathbb{L}$ has a maximum ($\top$) and a minimum ($\bot$). A subset $H \subseteq L$ is said to be $\vee$-dense (respectively $\wedge$-dense) if, for all $\ell \in L$, there exists a subset $T \subseteq H$ such that $\ell = \bigvee T$ (respectively $\ell = \bigwedge T$).

A *closure operator* $\varphi$ on $\mathbb{L}$ is a mapping being monotone ($x \leq y$ implies $\varphi(x) \leq \varphi(y)$, for all $x, y \in L$), extensive ($x \leq \varphi(x)$ for all $x \in L$) and idempotent ($\varphi(\varphi(x)) = \varphi(x)$ for all $x \in L$). An element $x \in L$ is said to be *closed* (w.r.t $\varphi$) if $\varphi(x) = x$. The notion of closure operator is closely related to the concept of closure system [10, Chapter 0]. A subset $S \subseteq L$ is said to be a *closure system* –also known as Moore family– if $\top \in S$ and $X \subseteq S$ implies $\bigwedge X \in S$.

We have seen that closure systems and closure operators are two faces of the same phenomenon; there is also a third face in terms of Galois connections. Given two complete lattices $\mathbb{L}_1 = (L_1, \leq)$ and $\mathbb{L}_2 = (L_2, \leq)$, a *Galois connection* between $\mathbb{L}_1$ and $\mathbb{L}_2$ is pair of mappings $\phi : L_1 \to L_2$ and $\psi : L_2 \to L_1$ such that both of them are antitone[1] and both compositions, $\phi \circ \psi$ and $\psi \circ \phi$, are extensive. It is well-known that the pair $(\phi, \psi)$ is a Galois connection if and only if, for all $\ell_1 \in L_1$ and $\ell_2 \in L_2$,

$$\ell_1 \leq \psi(\ell_2) \quad \text{if and only if} \quad \ell_2 \leq \phi(\ell_1) \tag{1}$$

Galois connections and closure operators are strongly related. This relationship is partially described in the following theorem.

**Theorem 1.** *Given a Galois connection $(\phi, \psi)$ between $\mathbb{L}_1$ and $\mathbb{L}_2$, the map $\psi \circ \phi$ is a closure operator on $\mathbb{L}_1$ and the map $\phi \circ \psi$ is a closure operator on $\mathbb{L}_2$. The maps $\phi$ and $\psi$, respectively, define dual isomorphisms between the corresponding closure systems. Specifically, the set $\mathfrak{B} = \{(x, y) \mid \phi(x) = y, \psi(y) = x\}$ with the order $\leq$ defined as*

$$(x_1, y_1) \leq (x_2, y_2) \quad \text{iff} \quad x_1 \leq x_2 \quad \text{or, equivalently, iff} \quad y_2 \leq y_1$$

*form a complete lattice such that, for any family $\{(x_j, y_j) \in \mathfrak{B} : j \in J\}$, the supremum and the infimum are given by:*

$$\sup_{j \in J}(x_j, y_j) = \left(\psi\phi\left(\bigvee_{j \in J} x_j\right), \bigwedge_{j \in J} y_j\right) \qquad \inf_{j \in J}(x_j, y_j) = \left(\bigwedge_{j \in J} x_j, \phi\psi\left(\bigvee_{j \in J} y_j\right)\right)$$

For more details, see [4, Chapters 3 & 7].

## 2.1 FCA Preliminaries

Now, we introduce the basic notions of FCA. For more information, we refer to [9,10].

First, FCA considers, as the starting point, a formal context $\mathbb{K} = (G, M, I)$, which consists in two not empty sets $G$ (whose elements are called objects) and

---

[1] A mapping $\varphi$ in $(L, \leq)$ is antitone if $x \leq y$ implies $\varphi(x) \geq \varphi(y)$, for all $x, y \in L$.

$M$ (whose elements are called attributes) and a relation $I$ between $G$ and $M$. The meaning of $(g, m) \in I$ is that the object $g$ has the attribute $m$. From this context, a Galois connection is defined by the pair $\uparrow: 2^G \to 2^M$ and $\downarrow: 2^M \to 2^G$ such that, for a set $A \subseteq G$ of objects, we have that $A^\uparrow = \{m \in M \mid (g, m) \in I \quad \forall g \in A\}$ and, for a set $B \in M$ of attributes, we have that $B^\downarrow = \{g \in G \mid (g, m) \in I \quad \forall m \in B\}$. Namely, $A^\uparrow$ is the set of properties that are shared by all the objects in $A$, and $B^\downarrow$ is the set of objects having all the attributes in $B$.

This Galois connection, as established in Theorem 1, allows to consider the *formal concepts*, which are the fixed points (closed sets), namely the pairs $(A, B)$ with $A \subseteq G$ and $B \subseteq M$ such that $A^\uparrow = B$ and $B^\downarrow = A$. These "are formal abstractions of *concepts* of human thought allowing meaningful and comprehensible interpretation". The prefix *formal* emphasizes that "they are mathematical entities and must not be identified with concepts of the mind" [12].

From Theorem 1, the formal concepts with the order $\leq$ defined as

$$(A_1, B_1) \leq (A_2, B_2) \quad \text{iff} \quad A_1 \leq A_2 \quad \text{or, equivalently, iff} \quad B_2 \leq B_1$$

form a complete lattice, which is called the *concept lattice* and denoted by $\mathbb{B}(G, M, I)$, whose supremum and infimum are described in the following theorem.

**Theorem 2.** *The concept lattice $\mathbb{B}(G, M, I)$ is a complete lattice in which infimum and sumpremum are given by:*

$$\bigwedge_{t \in T} (A_t, B_t) = \left( \bigcap_{t \in T} A_t (\bigcup_{t \in T} B_t)^{\downarrow\uparrow} \right)$$

$$\bigvee_{t \in T} (A_t, B_t) = \left( (\bigcup_{t \in T} A_t)^{\uparrow\downarrow}, \bigcap_{t \in T} B_t \right)$$

*A complete lattice $\mathbb{L}$ is isomorphic to $\mathbb{B}(G, M, I)$ if there are mappings $\overline{\gamma}: G \to L$ and $\overline{\mu}: M \to L$ such that $\overline{\gamma}(G)$ is a $\vee$-dense in $\mathbb{L}$, $\overline{\mu}(M)$ is $\wedge$-dense in $\mathbb{L}$ and $(g, m) \in I$ is equivalent to $\overline{\gamma}(g) \leq \overline{\mu}(m)$ for all $g \in G$ and all $m \in M$. In particular, $\mathbb{L} \cong \mathbb{B}(L, L, \leq)$.*

This theorem is known as *the Basic Theorem of Concept Lattices* [10, Theorem 3, Chapter 1].

## 2.2 Attribute Implications

Given a formal context $(G, M, I)$, an *implication between attributes* in $M$ is a pair of subsets of $M$, denoted by $A \to B$. We say that a subset $T \subseteq M$ is a model of $A \to B$ if $A \not\subseteq T$ or $B \subseteq T$. In that case, we denote it by $T \models A \to B$. $T$ is model of a set $\Sigma$ of implications ($T \models \Sigma$) if $T$ is model of every single implication in $\Sigma$. We say that $A \to B$ holds in a context $(G, M, I)$ if $\{g\}^\uparrow$ is model of $A \to B$ for all $g \in G$, that is, if each object that has all the attributes from $A$ has all the attributes from $B$ as well. In that case, we say that $A \to B$ is a (valid) implication of $(G, M, I)$. The following proposition characterizes the validity of implications.

**Proposition 1.** *An implication* $A \to B$ *holds in* $(G, M, I)$ *if and only if* $B \subseteq A^{\downarrow\uparrow}$, *which is equivalent to* $A^{\downarrow} \subseteq B^{\downarrow}$.

We are going to establish when an implication $A \to B$ follows semantically from $\Sigma$, notated $\Sigma \models A \to B$: it holds if each subset of $M$ that is model of $\Sigma$ is also model of $A \to B$. A set of implications $\Sigma$ is complete with respect to $(G, M, I)$ if every implication that holds in $(G, M, I)$ follows from $\Sigma$.

The above semantic definition of inference has a syntactic counterpart. There are different axiomatic systems, being the most popular one the so-called *Armstrong's Axioms* that consider a scheme of an axiom and two inference rules: Let $A, B, C \subseteq M$,

[Inc] Inclusion: $\vdash_A A \cup B \to A$
[Aum] Augmentation: $A \to B \vdash_A A \cup C \to B \cup C$
[Trans] Transitivity $A \to B, B \to C \vdash_A A \to C$.

An implication $A \to B$ is said to be derived from a set $\Sigma$, denoted by $\Sigma \vdash A \to B$, if there exists a sequence of implications $A_i \to B_i$ with $1 \leq i \leq n$ such that $A_n = A$, $B_n = B$, and each $A_i \to B_i$ is either an axiom or $A_i \to B_i \in \Sigma$ or it is obtained from the formulas in $\{A_j \to B_j \mid j < i\}$ by using one of the inference rules.

It is well-known that this axiomatic system is sound and complete working with finite sets, i.e. $\Sigma \models A \to B$ iff $\Sigma \vdash A \to B$.

## 3   An Algebraic Framework for Unknown Information

In this section, we present the algebra that is defined to be the underlying structure of our formal framework. We begin by extending the algebra of two elements $\mathbf{2} = \{0, 1\}$ to consider unknown information.

We use the set of three elements $\{+, -, \circ\}$, denoted by $\mathbf{3}$. The element $+$ represents the information that we know to be true (we call it positive information), the element $-$ represents the information that we know to be untrue (we call it negative information) and the element $\circ$ represents the information that we do not know whether it is true or false (we call it unknown information). Finally, we endow this set with a $\wedge$-semilattice structure by considering the reflexive clousure of $\{(\circ, +), (\circ, -)\}$ (see Fig. 1a). This $\wedge$-semilattice will be denoted by $\underline{\mathbf{3}} = (\mathbf{3}, \leq)$.

In the following section, we study sets valued in $\mathbf{3}$, to which, as we will see, we will give a conjunctive interpretation. This interpretation leads us to introduce a fourth element, denoted $\iota$. This element is called oxymoron and represents inconsistent or contradictory information when a conjunction plays this role. This element will be the maximum of the completion of $\mathbf{3}$ to be a lattice.

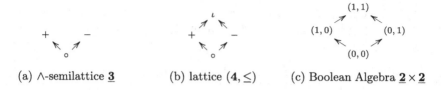

(a) ∧-semilattice **3**          (b) lattice (**4**, ≤)          (c) Boolean Algebra **2** × **2**

**Fig. 1.** Truthfulness's values

This lattice, denoted by $\underline{4} = (\mathbf{4}, \leq)$, is shown in Fig. 1b and is isomorphic to the Boolean algebra $\mathbf{2} \times \mathbf{2}$ (see Fig. 1c).

The lattice $\underline{4}$ can be considered as an example of the bilattices [6,7], which are doubly-ordered sets $(A, \leq_k, \leq_t)$ where $(A, \leq_k)$ and $(A, \leq_t)$ are complete lattices. In $\leq_k$ the key to order is the amount of information that we have, and in $\leq_t$ the key is the veracity of the information. In $\underline{4}$, the knowledge order is given when we read it from the bottom to up, and the truthfulness order is obtained by reading it from right to left. We specially focus on the first view.

### 3.1   The ∧-semilattice of 3-sets

Given a universal set, $U$, we denote by $\underline{\mathbf{3}}^U$ the ∧-semilattice of the **3**-sets, that is, the set of the functions $X : U \to \mathbf{3}$ with the structure of ∧-semilattice considering the point-wise extension of the order $\leq$:

$$X \sqsubseteq Y \text{ iff } X(u) \leq Y(u) \text{ for all } u \in U.$$

A **3**-set $X$ provides information about the knowledge that we have from each element in $U$. These elements usually correspond to attributes or properties that can hold. We call *support* of a **3**-set $X$ to the set $\mathrm{Spp}(X) = \{u \in U \mid X(u) \neq \circ\}$. The support collects those elements that we have the absolute knowledge about them.

To simplify the notation, when $\mathrm{Spp}(X) = \{u_1, \dots, u_n\}$ is finite, we write $\{u_1/X(u_1), \dots, u_n/X(u_n)\}$. In particular, the unique set having empty support is denoted by $\varnothing$. In addition, when no confusion arises, the support is expressed as a sequence of elements, where $u$ means that its image is $+$ and $\bar{u}$ denotes that $u$ has the value $-$. The elements that are not in the support of $X$, that is, with the value $\circ$, do not appear in the sequence.

*Example 1.* Given the universe $U = \{a, b\}$, the **3**-sets $X = \{a/+, b/-\}$ and $Y = \{a/+\}$ are denoted by $X = a\bar{b}$ and $Y = a$ respectively.

We also consider the functions $\mathrm{Pos}, \mathrm{Neg}, \mathrm{Unk} : \mathbf{3}^U \to \mathbf{2}^U$ defined as follows:

$$\mathrm{Pos}(X) = X^{-1}(+) = \{u \in U \mid X(u) = +\}$$
$$\mathrm{Neg}(X) = X^{-1}(-) = \{u \in U \mid X(u) = -\}$$
$$\mathrm{Unk}(X) = X^{-1}(\circ) = \{u \in U \mid X(u) = \circ\}$$

for each $X \in \mathbf{3}^U$.

We can see that Pos and Neg are isotone functions between $\mathbf{3}^U$ and $\mathbf{2}^U = (2^U, \subseteq)$, while that Unk is antitone. We have that $\mathrm{Spp}(X) = \mathrm{Pos}(X) \cup \mathrm{Neg}(X) = U \setminus \mathrm{Unk}(X)$, for all $X \in \mathbf{3}^U$.

Finally, we define $\overline{(\ )}: \mathbf{3}^U \to \mathbf{3}^U$ where, for all $X \in \mathbf{3}^U$ and $u \in U$,

$$\overline{X}(u) = \begin{cases} - \ \text{if } X(u) = + \\ \circ \ \text{if } X(u) = \circ \\ + \ \text{if } X(u) = - \end{cases}$$

Given $X \in \mathbf{3}^U$, $\overline{X}$ is named the *opposite* of $X$. Obviously, $\mathrm{Pos}(\overline{X}) = \mathrm{Neg}(X)$, $\mathrm{Neg}(\overline{X}) = \mathrm{Pos}(X)$, and $\mathrm{Unk}(\overline{X}) = \mathrm{Unk}(X)$.

## 3.2   The Lattices of 4-sets and 3̇-sets

As we have mentioned, we conceive the **3**-sets as properties and their knowledge about them, considering a conjunctive interpretation. When we join two different **3**-sets, we can find inconsistencies: a property can be positive in one of the sets and negative in the other set. So in the final set, we have an inconsistent element. This makes us introduce a new kind of set that is valued in **4**.

Given a universe $U$, we denote by $\mathbf{4}^U$ the set of the **4**-sets, that is, of the functions $X: U \to \mathbf{4}$. We can assume that $\mathbf{3}^U \subseteq \mathbf{4}^U$. We can point-wise extend the order of $\mathbf{4}$ to the set of the **4**-sets as usual:

$$X \sqsubseteq Y \qquad \text{iff} \qquad X(u) \leq Y(u) \text{ for all } u \in U$$

obtaining a complete lattice where

$$\left( \bigvee_{i \in I} X_i \right)(u) = \bigvee_{i \in I} X_i(u) \qquad \left( \bigwedge_{i \in I} X_i \right)(u) = \bigwedge_{i \in I} X_i(u) \qquad \text{for all } u \in U$$

The infimum of this complete (bi)lattice is $\varnothing$ and the supremum, denoted by $\dot{\imath}$, is the **4**-set such that $\dot{\imath}(u) = \iota$ for all $u \in U$.

*Example 2.* In $\mathbf{4}^{\{a,b\}}$, we have that:

$$\{a/+\} \sqsubseteq \{a/+, b/-\} \sqsubseteq \{a/\iota, b/-\} \sqsubseteq \dot{\imath}$$

$\{a/+\} \vee \{b/-\} = \{a/+, b/-\}$, $\{a/+\} \wedge \{a/-, b/-\} = \varnothing$ and $\{a/+\} \vee \{a/-, b/-\} = \{a/\iota, b/-\}$.

Following with the conjunctive interpretation of the sets, we consider all the **4**-sets having some $X(u) = \iota$ as equivalent sets. These sets are called inconsistent sets or oxymorons. To formalize it, we define the function:

$$\mathcal{O}: \mathbf{4}^U \to \mathbf{4}^U \text{ where } \mathcal{O}(X) = \begin{cases} X \ \text{if } X \in \mathbf{3}^U, \\ \dot{\imath} \ \text{otherwise.} \end{cases}$$

This function is a closure operator in $\mathbf{4}^U$, so it is an $\wedge$-homomorphism. The set of their images, $\mathbf{3}^U \cup \{\dot{\imath}\}$, denoted by $\dot{\mathbf{3}}^U$, is a closure system and, as a consequence, it is also a complete lattice.

Specifically, if we consider the relation $\sqsubseteq$ of $\mathbf{4}^U$ restricted to $\dot{\mathbf{3}}^U$ we have that $(\dot{\mathbf{3}}^U, \sqsubseteq)$ is a $\wedge$-subsemilattice of $(\mathbf{4}^U, \sqsubseteq)$, but not a sublattice. Moreover, $(\dot{\mathbf{3}}^U, \sqsubseteq)$ is a complete lattice. In order to differentiate the join of $(\dot{\mathbf{3}}^U, \sqsubseteq)$ from the join of $\mathbf{4}^U$ we use the symbol $\sqcup$ for the first one, that is, given $\{X_j : j \in J\} \subseteq \dot{\mathbf{3}}^U$,

$$\bigsqcup_{j \in J} X_j = \mathcal{O}(\bigvee_{j \in J} X_j)$$

We can check that $(\dot{\mathbf{3}}^U, \sqsubseteq)$ is the completion to lattice from $\wedge$-semilattice $\mathbf{3}^U$, that is, $(\dot{\mathbf{3}}^U, \sqsubseteq)$ is obtained by adding the element $i$ as supremum to the $\wedge$-semilattice $\mathbf{3}^U$ (see Fig. 2). The complete lattice $(\dot{\mathbf{3}}^U, \sqcup, \wedge, \varnothing, i)$ will be denoted by $\dot{\underline{\mathbf{3}}}^U$, being our target algebraic structure.

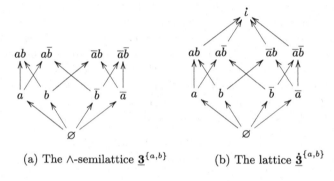

(a) The $\wedge$-semilattice $\underline{\mathbf{3}}^{\{a,b\}}$     (b) The lattice $\dot{\underline{\mathbf{3}}}^{\{a,b\}}$

**Fig. 2.** The caption would be semi-lattice and lattice with three values of $U = a, b$

The set $i$ is named the *inconsistent set*, and the sets of $\mathbf{3}^U$ are named *consistent sets*. Moreover, the maximal sets of $\underline{\mathbf{3}}^U$, i.e. those that have support $U$, are the super-atoms of $\dot{\underline{\mathbf{3}}}^U$. These sets are named *full sets*, and the set of all of them will be denoted by $\mathfrak{Full}(U)$.

*Example 3.* In $(\dot{\mathbf{3}}^{\{a,b\}}, \sqsubseteq)$, we have that $a \sqsubseteq a\bar{b} \sqsubseteq i$, $a \sqcup \bar{b} = a\bar{b}$, $a \wedge \bar{a}\bar{b} = \varnothing$ and $a \sqcup \bar{a}\bar{b} = i$.

Now, we extend the functions Pos, Neg and Unk considering

$$\mathrm{Pos}(i) = \mathrm{Neg}(i) = U, \qquad \mathrm{Unk}(i) = \varnothing$$

**Proposition 2.** *Let $U$ be a non-empty set and $X, Y \in \dot{\mathbf{3}}^U$. Then,*

1. $X \sqsubseteq Y$ *iff* $\mathrm{Pos}(X) \subseteq \mathrm{Pos}(Y)$ *and* $\mathrm{Neg}(X) \subseteq \mathrm{Neg}(Y)$.
2. $X \in \mathfrak{Full}(U)$ *iff* $\mathrm{Unk}(X) = \varnothing$ *iff* $\mathrm{Pos}(X) \cup \mathrm{Neg}(X) = U$.
3. *The restriction of the functions* Pos *and* Neg *to* $\mathfrak{Full}(U)$ *are bijections in* $\mathbf{2}^U$.

The proof of this proposition is straightforward from the definitions.

## 4   Extending the Concept of Lattice

In this section, we extend FCA's basic results to consider unknown information using the algebraic framework introduced in the previous section. Due to space limitations, some proofs of the theoretical results cannot be introduced.

We begin by defining a *partial formal context* as a triple $\mathbb{P} = (G, M, I)$ where $G$ and $M$ are non-empty sets, whose elements are called *objects* and *attributes* respectively, and $I \colon G \times M \to \mathbf{3}$ is called the incidence relation.

$I(g, m) = +$ means that the attribute $m$ is present in the object $g$; $I(g, m) = -$ means that the attribute $m$ is not present in the object $g$; and $I(g, m) = \circ$ means that we do not know whether the attribute $m$ is present in the object $g$ or not. We represent these contexts as tables (see Fig. 3 for instance).

Given a partial formal context $\mathbb{P} = (G, M, I)$, for each $g \in G$ and for each $m \in M$ we consider the **3**-sets $I(g, \ ) \in \mathbf{3}^M$ and $I( \ , m) \in \mathbf{3}^G$ defined as:

$$I(g, \ )(x) = I(g, x) \text{ for all } x \in M; \quad I( \ , m)(x) = I(x, m) \text{ for all } x \in G$$

If a partial formal context $\mathbb{P} = (G, M, I)$ satisfies that $I(g, \ )$ is a full set for all $g \in G$ we say that it is a total formal context. Moreover, any (classic) formal context $\mathbb{K} = (G, M, I)$ can be seen as a partial formal context $\mathbb{P} = (G, M, I')$ where $I'(g, m) = +$ iff $g \ I \ m$, and $I'(g, m) = \circ$ otherwise. In addition, a partial formal context $\mathbb{P} = (G, M, I)$ can induce the following formal contexts:

- $\mathbb{K}_{\mathbb{P}}^+ = (G, M, I^+)$ where $I^+ = I^{-1}(+)$, that is $gI^+m$ iff $I(g, m) = +$.
  Their derivation operators are denoted by the symbol $+$, that is, for all $X \subseteq G$ and $Y \subseteq M$

$$X^+ = \bigcap_{g \in X} gI^+( \ ) = \{m \in M \mid gI^+m, \forall g \in X\}$$

$$Y^+ = \bigcap_{m \in Y} ( \ )I^+m = \{g \in G \mid gI^+m, \forall m \in Y\}$$

- $\mathbb{K}_{\mathbb{P}}^- = (G, M, I^-)$ where $I^- = I^{-1}(-)$ and their derivation operators are denoted by the symbol $-$ and defined in a similar way.

We use these formal contexts to define the derivation operators in the partial formal context as follows.

**Theorem 3.** *Given a partial formal context* $\mathbb{P} = (G, M, I)$, *the derivation operators* $( \ )^{\uparrow} \colon \mathbf{2}^G \to \dot{\mathbf{3}}^M$ *and* $( \ )^{\downarrow} \colon \dot{\mathbf{3}}^M \to \mathbf{2}^G$ *defined as*

$$X^{\uparrow} = \bigwedge_{g \in X} I(g, \ ), \quad \text{and} \quad Y^{\downarrow} = \mathrm{Pos}(Y)^+ \cap \mathrm{Neg}(Y)^-$$

*form a Galois connection between the lattices* $\underline{\mathbf{2}}^G$ *and* $\underline{\dot{\mathbf{3}}}^M$.

*Proof.* We prove Condition (1), i.e. for all the subsets $X \subseteq G$ and $Y \in \dot{\mathbf{3}}^M$ we have that

$$X \subseteq Y^{\downarrow} \quad \text{iff} \quad Y \sqsubseteq X^{\uparrow}$$

Let's suppose that $X \subseteq Y^{\downarrow} = \text{Pos}(Y)^+ \cap \text{Neg}(Y)^-$, i.e. $X \subseteq \text{Pos}(Y)^+$ and $X \subseteq \text{Neg}(Y)^-$. Since $\mathbb{K}_{\mathbb{P}}^+$ and $\mathbb{K}_{\mathbb{P}}^-$ are (classical) formal contexts, $X \subseteq Y^{\downarrow}$ holds if and only if $\text{Pos}(Y) \subseteq X^+$ and $\text{Neg}(Y) \subseteq X^-$. By Proposition 2, we can ensure that it is equivalent to $Y \sqsubseteq X^{\uparrow}$ because it is straightforwardly proved that $X^+ \subseteq \text{Pos}(X^{\uparrow})$ and $X^- \subseteq \text{Neg}(X^{\uparrow})$. ∎

Following Theorem 1, we center on the fixed-points of this Galois connection.

**Definition 1.** *Given a partial formal context* $\mathbb{P} = (G, M, I)$, *a concept is a pair* $(A, B) \in 2^G \times \dot{3}^M$ *such that* $A^{\uparrow} = B$ *and* $B^{\downarrow} = A$. *The set of concepts will be denoted by* $\mathfrak{B}_*(\mathbb{P})$.

The next corollary is a consequence of the previous theorem and Theorem 1.

**Corollary 1.** *Given a partial formal context* $\mathbb{P} = (G, M, I)$, *the set* $\mathfrak{B}_*(\mathbb{P})$ *with the order defined as*

$$(A_1, B_1) \leq (A_2, B_2) \quad \text{iff} \quad A_1 \subseteq A_2 \quad \text{iff} \quad B_2 \sqsubseteq B_1$$

*is a complete lattice, denoted by* $\underline{\mathfrak{B}}_*(\mathbb{P})$, *such that, for all the families of concepts* $\{(A_j, B_j) \in \mathfrak{B}_*(\mathbb{P}) : j \in J\}$, *the join and the meet are given by:*

$$\sup_{j \in J}(A_j, B_j) = \left(\left(\bigcup_{j \in J} A_j\right)^{\uparrow \downarrow}, \bigwedge_{j \in J} B_j\right) \qquad \inf_{j \in J}(A_j, B_j) = \left(\bigcap_{j \in J} A_j, \left(\bigsqcup_{j \in J} B_j\right)^{\downarrow \uparrow}\right)$$

In the following theorem, we can see a connection between the concepts of a formal context and the concepts of a partial formal context:

**Theorem 4.** *Given a partial formal context* $\mathbb{P} = (G, M, I)$, *we have that* $\underline{\mathfrak{B}}_*(\mathbb{P})$ *is isomorphic to* $\underline{\mathfrak{B}}(\mathbb{K}_{\mathbb{P}}^- \mid \mathbb{K}_{\mathbb{P}}^+)$.
*Conversely, for any formal context* $\mathbb{K} = (G, M, I)$ *and* $X \subseteq M$, *one has that* $\underline{\mathfrak{B}}(\mathbb{K})$ *is isomorphic to* $\underline{\mathfrak{B}}_*(\mathbb{P}_X)$ *where* $\mathbb{P}_X = (G, M, I_X)$ *with*

$$I_X(g, m) = \begin{cases} + & \text{if } g \, I \, m \text{ and } m \in X, \\ - & \text{if } g \, I \, m \text{ and } m \notin X, \\ \circ & \text{otherwise.} \end{cases}$$

| $\mathbb{P}$ | $a$ | $b$ | $c$ |
|---|---|---|---|
| 1 | + | ∘ | − |
| 2 | ∘ | + | + |
| 3 | − | − | ∘ |

| $\mathbb{K}_{\mathbb{P}}^-\mid\mathbb{K}_{\mathbb{P}}^+$ | a0 | b0 | c0 | a1 | b1 | c1 |
|---|---|---|---|---|---|---|
| 1 |  |  | × | × |  |  |
| 2 |  |  |  |  | × | × |
| 3 | × | × |  |  |  |  |

**Fig. 3.** Partial formal context $\mathbb{P}$ and formal context $(\mathbb{K}_{\mathbb{P}}^- \mid \mathbb{K}_{\mathbb{P}}^+)$

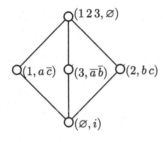

**Fig. 4.** The lattice $\underline{\mathfrak{B}}_\star(\mathbb{P}) \cong \underline{\mathfrak{B}}(\mathbb{K}_\mathbb{P}^- \mid \mathbb{K}_\mathbb{P}^+)$

*Example 4.* Figure 3 shows a partial formal context $\mathbb{P}$, and the classical formal context $(\mathbb{K}_\mathbb{P}^- \mid \mathbb{K}_\mathbb{P}^+)$ built from it. The lattice $\underline{\mathfrak{B}}_\star(\mathbb{P})$ is shown in Fig. 4. As Theorem 4 ensures, it is isomorphic to the lattice $\underline{\mathfrak{B}}(\mathbb{K}_\mathbb{P}^- \mid \mathbb{K}_\mathbb{P}^+)$.

**Corollary 2.** *Let* $\mathbb{L} = (L, \leq)$ *be a complete lattice and $G$ and $M$ be not empty sets. If there exist mappings $\overline{\gamma}\colon G \to L$ and $\overline{\mu}\colon M \to L$ such that $\overline{\gamma}(G)$ is $\vee$-dense and $\overline{\mu}(G)$ is $\wedge$-dense in $\mathbb{L}$, then $\mathbb{L} \cong \underline{\mathfrak{B}}_\star(\mathbb{P})$ where $\mathbb{P} = (G, M, I)$ with $I(g, m) = +$ iff $g \leq m$, and $I(g, m) = \circ$ otherwise.*

Moreover, as consequence of the last theorem, we have immediately that $\mathbb{P}$ is no just one partial formal context that satisfies the conditions given before. As particular case of this corollary, one has that $\mathbb{L} \cong \underline{\mathfrak{B}}(L, L, \leq) \cong \underline{\mathfrak{B}}_\star(\mathbb{P})$, considering $G = M = L$ and $\overline{\gamma} = \overline{\mu} = id$.

## 5   Reasoning with Weak Implications

In this section, we are going to introduce the notion of weak implication and a logic to reason about this kind of implications. The name is intended to reflect that these are implications that, with the information currently available, are true; but which, when we obtain new knowledge, may no longer be true.

Given a not empty set of attributes $M$, we call weak implication (of attributes) to the expression $A \rightsquigarrow B$ where $A, B \in \dot{\mathbf{3}}^M$. The set of weak implications will be denoted by

$$\mathcal{L}_M^w = \{A \rightsquigarrow B \colon A, B \in \dot{\mathbf{3}}^M\}$$

In this set, which we consider to be the language of the logic, we introduce the semantics as follows:

**Definition 2.** *Let $C$ be a $\dot{\mathbf{3}}$-set on $M$. We say that $C$ is model of a weak implication $A \rightsquigarrow B \in \mathcal{L}_M^w$ if satisfies that $A \sqsubseteq C$ implies $B \sqsubseteq C$. The set of the models of $A \rightsquigarrow B$ is denoted by $\mathrm{Mod}(A \rightsquigarrow B)$.*

*We say that $C$ is model of a theory $\Sigma \subseteq \mathcal{L}_M^w$ if it is model of all weak implication $A \rightsquigarrow B \in \Sigma$, that is,*

$$\mathrm{Mod}(\Sigma) = \bigcap_{A \rightsquigarrow B \in \Sigma} \mathrm{Mod}(A \rightsquigarrow B)$$

As usual, we can consider that a partial formal context is a model of a weak implication when the set $\{I(g, \ ) \mid g \in G\}$ is a model.

**Definition 3.** *Let $\mathbb{P} = (G, M, I)$ be a partial formal context and $A \rightsquigarrow B \in \mathcal{L}_M^w$. We say that $\mathbb{P}$ is model of $A \rightsquigarrow B$, or that $A \rightsquigarrow B$ is satisfied in $\mathbb{P}$, and it will be denoted by $\mathbb{P} \models A \rightsquigarrow B$, if $\{g\}^{\uparrow} \in \mathrm{Mod}(A \rightsquigarrow B)$ for all $g \in G$.*

*We say that a partial formal context $\mathbb{P}$ is model of a set $\Sigma \subseteq \mathcal{L}_M^w$, denoted by $\mathbb{P} \models \Sigma$, if $\mathbb{P} \models X \rightsquigarrow Y$ for all $X \rightsquigarrow Y \in \Sigma$.*

As in the classical case, we can easily characterize the implications that are satisfied by a context by using the derivation operators.

**Proposition 3.** *Let $\mathbb{P} = (G, M, I)$ be a partial formal context and $A \rightsquigarrow B \in \mathcal{L}_M^w$.*

$$\mathbb{P} \models A \rightsquigarrow B \quad \text{iff} \quad A^{\downarrow} \subseteq B^{\downarrow} \quad \text{iff} \quad B \sqsubseteq A^{\downarrow\uparrow}.$$

Now, we introduce the notion of semantic derivation.

**Definition 4.** *Let $A \rightsquigarrow B \in \mathcal{L}_M^w$ and $\Sigma \subseteq \mathcal{L}_M^w$. We say that $A \rightsquigarrow B$ is semantically derived from $\Sigma$, denoted by $\Sigma \models A \rightsquigarrow B$, when, for all partial formal context $\mathbb{P}$, we have that $\mathbb{P} \models \Sigma$ implies $\mathbb{P} \models A \rightsquigarrow B$.*

About notation, when there is not any possible confusion, we denote the sets of implications without curly brackets. In the same way, we write $\models A \rightsquigarrow B$ when we have that $\varnothing \models A \rightsquigarrow B$.

**Proposition 4.** *Let $\mathbb{P} = (G, M, I)$ be a partial formal context and $A, B \subseteq M$.*

1. *If $\mathbb{K}_{\mathbb{P}}^{+} \models A \rightarrow B$ then $\mathbb{P} \models A \rightsquigarrow B$.*
2. *If $\mathbb{K}_{\mathbb{P}}^{-} \models A \rightarrow B$ then $\mathbb{P} \models \overline{A} \rightsquigarrow \overline{B}$.*

The third pillar of the logic is the axiomatic system. In this case, we consider Armstrong's axioms that we will prove to be correct and complete.

**Definition 5.** *The axiomatic system $\mathcal{A}$ considers one axiom and two inference rules. They are the following: for all $A, B, C \in \dot{\underline{\mathbf{3}}}^M$,*

[Inc] *Inclusion:* $\vdash_{\mathcal{A}} A \sqcup B \rightsquigarrow A$
[Augm] *Augmentation:* $A \rightsquigarrow B \vdash_{\mathcal{A}} A \sqcup C \rightsquigarrow B \sqcup C$
[Trans] *Transitivity:* $A \rightsquigarrow B, B \rightsquigarrow C \vdash_{\mathcal{A}} A \rightsquigarrow C$

The notion of syntactic derivation is introduced in the standard way.

**Definition 6.** *A weak implication $\phi$ is said to be syntactically derived, or it is inferred, from a set of weak implications $\Sigma$, denoted by $\Sigma \vdash_{\mathcal{A}} \phi$, if there is a sequence of weak implications $\phi_1, ..., \phi_n$ such that $\phi_n = \phi$ and, for all $1 \leq i \leq n$, one of the following conditions is satisfied: $\phi_i \in \Sigma$, $\phi_i$ is an axiom, or $\phi_i$ is obtained from implications belonging to $\{\phi_j \mid 1 \leq j < i\}$ by applying the inferences rules of $\mathcal{A}$. In this case, we say that the sequence $\{\phi_i \mid 1 \leq i \leq n\}$ is a proof for $\Sigma \vdash_{\mathcal{A}} \phi$.*

As usual, we consider some derived rules from the Armstrong's axioms which are easily proved.

**Proposition 5.** *Let $M$ be a finite set of attributes. The following inference rules are derived from the Amstrong's axioms: for all $A, B, C \in \dot{\mathbf{3}}^M$,*

[Frag] *Fragmentation:* $A \rightsquigarrow B \sqcup C \vdash_A A \rightsquigarrow B$
[Un] *Union:* $A \rightsquigarrow B,\ A \rightsquigarrow C \vdash_A A \rightsquigarrow B \sqcup C$
[gTr] *Generalized Transitivity* $A \rightsquigarrow B \sqcup C,\ B \rightsquigarrow D \vdash_A A \rightsquigarrow D.$

From Proposition 3 and the fact that $(\ )^{\downarrow\uparrow}$ is a closure operator in $\dot{\mathbf{3}}^M$, we have that Amstrong's axioms are correct.

**Theorem 5 (Soundness).** *For all weak implication $A \rightsquigarrow B \in \mathcal{L}_M^w$ and all set $\Sigma \subseteq \mathcal{L}_M^w$, we have that $\Sigma \vdash_A A \rightsquigarrow B$ implies $\Sigma \models A \rightsquigarrow B$.*

To prove the completeness of the axiomatic system, we first introduce some necessary results.

**Theorem 6.** *Let $M$ be a finite set and $\Sigma \in \mathcal{L}_M^w$. The mapping $(\ )_\Sigma^\sharp : \dot{\mathbf{3}}^M \to \dot{\mathbf{3}}^M$ defined as*

$$A_\Sigma^\sharp = \bigsqcup \{X \in \dot{\mathbf{3}}^M \mid \Sigma \vdash_A A \rightsquigarrow X\}$$

*is a closure operator in $\dot{\mathbf{3}}^M$ that we name the syntactic closure with respect to $\Sigma$. In addition, $\Sigma \vdash_A A \rightsquigarrow A_\Sigma^\sharp$ for all $A \in \dot{\mathbf{3}}^M$.*

**Corollary 3.** *Let $M$ be a finite set of attributes. For all $\Sigma \subseteq \mathcal{L}_M^w$ and $A, B \in \dot{\mathbf{3}}^M$, we have that*

$$\Sigma \vdash_A A \rightsquigarrow B \quad \text{if and only if} \quad B \sqsubseteq A_\Sigma^\sharp.$$

*Proof.* The direct implication is a consequence of Theorem 6, and the converse result is obtained by using [Inc] and [Trans]. □

**Lemma 1.** *Let $M$ be a finite set of attributes. For all $\Sigma \subseteq \mathcal{L}_M^w$ and $A \in \dot{\mathbf{3}}^M$, we have that*
$$A_\Sigma^\sharp = \min\{X \in \mathrm{Mod}(\Sigma) \mid A \sqsubseteq X\}.$$

*Proof.* Let $X \in \mathrm{Mod}(\Sigma)$ such that $A \sqsubseteq X$. From Theorem 5, $X \in \mathrm{Mod}(\Sigma)$ implies that $X \in \mathrm{Mod}(B \rightsquigarrow C)$ for all $B \rightsquigarrow C \in \mathcal{L}_M^w$ with $\Sigma \vdash_A B \rightsquigarrow C$ and, particularly, by Theorem 6, $X \in \mathrm{Mod}(A \rightsquigarrow A_\Sigma^\sharp)$. Thus, $A \sqsubseteq X$ implies $A_\Sigma^\sharp \sqsubseteq X$.

Moreover, we prove that $A_\Sigma^\sharp \in \mathrm{Mod}(\Sigma)$. For all $B \rightsquigarrow C \in \Sigma$, if $B \sqsubseteq A_\Sigma^\sharp$, by Corollary 3, we have that $\Sigma \vdash_A A \rightsquigarrow B$. Then, by [Trans], $\Sigma \vdash_A A \rightsquigarrow C$ and, again, by Corollary 3, we have that $C \sqsubseteq A_\Sigma^\sharp$. □

Let's see now the completeness of the axiomatic system presented:

**Theorem 7 (Completeness).** *Let $M$ be a finite set of attributes. For all $A \rightsquigarrow B \in \mathcal{L}_M^w$ and $\Sigma \subseteq \mathcal{L}_M^w$, we have that $\Sigma \models A \rightsquigarrow B$ implies $\Sigma \vdash_{\mathcal{A}} A \rightsquigarrow B$.*

*Proof.* Let's prove that $\Sigma \nvdash_{\mathcal{A}} A \rightsquigarrow B$ implies $\Sigma \nmodels A \rightsquigarrow B$. Using the Corollary 3, we have that $\Sigma \nvdash_{\mathcal{A}} A \rightsquigarrow B$ implies that $B \not\sqsubseteq A_\Sigma^\sharp$, and, therefore, $A_\Sigma^\sharp \neq i$.

Let us consider the partial formal context $\mathbb{P} = (G, M, I)$ being $G = \mathrm{Mod}(\Sigma) \setminus \{i\}$ and $I \colon G \times M \to \mathbf{3}$ where $I(g, \ ) = g$ for each $g \in G$.

It is straightforward that $\mathbb{P} \models \Sigma$ because $\{g\}^\uparrow = I(g, \ ) = g \in \mathrm{Mod}(\Sigma)$. However, $\mathbb{P} \nmodels A \rightsquigarrow B$ because, by Lemma 1, $A_\Sigma^\sharp \in G$, $A \sqsubseteq A_\Sigma^\sharp$ and $B \not\sqsubseteq A_\Sigma^\sharp$. $\square$

# 6  Conclusion and Further Work

In this paper, we have extended FCA to consider not only positive but also negative and unknown information in a natural way. The key to this has been the selection of the algebraic structure on which to define the semantics and the derivation operators by ensuring that they still form a Galois connection. The starting point is a three-valued relationship between objects and attributes, which we call partial formal context. Since the interpretation of attribute sets is conjunctive, when we join different sets together, contradictions may arise. Therefore, we need to enrich the structure of truthfulness values to take this into account. The essence of this approach is to replace the attribute powerset with a new structure, $(\dot{\mathbf{3}}^M, \sqsubseteq)$, which is very close to bilattices. Considering this structure, we present the Galois connection formed by the derivation operators and establish the existent relationship between the concept lattices obtained with the classical ones.

Furthermore, we have presented a new kind of attribute implication, which we name weak implications because they can change when new information is added. For these implications, we consider Armstrong's axioms, which have the same appearance as the classical axioms but, being defined on the new structure, incorporate in the union the semantics that allows dealing with positive/negative information, as well as with unknown information and even contradiction or inconsistency. Finally, we prove the soundness and completeness of this axiomatic system.

As further work, we are working on a new axiomatic system in the framework of Simplification Logic [18], that is closer to applications in the sense that, like the previous ones, it can be considered as an executable logic. Moreover, we are also working on another definition of attribute implication that we name strong implications. The idea is that they would remain unchanged when new information is provided. The ultimate goal is to establish a logic that allows reasoning simultaneously with both types of implications.

# References

1. Birkhoff, G.: Lattice Theory, 1st edn. American Mathematical Society Colloquium Publications, Providence (1940)
2. Burmeister, P., Holzer, R.: Treating incomplete knowledge in formal concept analysis. In: Ganter, B., Stumme, G., Wille, R. (eds.) Formal Concept Analysis. LNCS (LNAI), vol. 3626, pp. 114–126. Springer, Heidelberg (2005). https://doi.org/10.1007/11528784_6
3. Cordero, P., Enciso, M., Mora, A., Rodríguez-Jiménez, J.M.: Inference of mixed information in formal concept analysis. Stud. Comput. Intell. **796**, 81–87 (2019)
4. Davey, B., Priestley, H.: Introduction to Lattices and Order, vol. 2. Cambridge University Press, Cambridge (2002)
5. Finn, V.: About machine-oriented formalization of plausible reasonings. F. Beckon-J.S. Mill Style, Semiotika I Informatika **20**, 35–101 (1983)
6. Fitting, M.: Bilattices and the semantics of logic programming. J. Logic Programm. **11**(2), 91–116 (1991)
7. Fitting, M.: Bilattices are nice things. In: Hendricks, V.F., Pedersen, S.A., Bolander, T. (eds.) Self-reference, pp. 53–77. Cambridge University Press, CSLI Publications, Cambridge (2006)
8. Ganter, B., Kuznetsov, S.: Hypotheses and version spaces. ICCS, pp. 83–95 (2003)
9. Ganter, B., Obiedkov, S.: More expressive variants of exploration. In: Conceptual Exploration, pp. 237–292. Springer, Heidelberg (2016). https://doi.org/10.1007/978-3-662-49291-8_6
10. Makhalova, T., Trnecka, M.: A study of boolean matrix factorization under supervised settings. In: Cristea, D., Le Ber, F., Sertkaya, B. (eds.) ICFCA 2019. LNCS (LNAI), vol. 11511, pp. 341–348. Springer, Cham (2019). https://doi.org/10.1007/978-3-030-21462-3_24
11. Ganter, B., Kwuida, L.: Which concept lattices are pseudocomplemented? Lect. Notes Comput. Sci. **3403**, 408–416 (2005)
12. Ganter, B., Wille, R.: Applied lattice theory: formal concept analysis. In: Grätzer, G. (ed.) General Lattice Theory. Birkhäuser. Preprints (1997)
13. Konecny, J.: Attribute implications in L-concept analysis with positive and negative attributes: validity and properties of models. Int. J. Approximate Reason. **120**, 203–215 (2020)
14. Kuznetsov, S.O., Revenko, A.: Interactive error correction in implicative theories. Int. J. Approximate Reason. **63**, 89–100 (2015)
15. Kuztnesov, S.O.: Mathematical aspects of concept analysis. J. Math. Sci. **80**, 1654–1698 (1996)
16. Kuznetsov, S.O.: Galois connections in data analysis: contributions from the soviet era and modern russian research. In: Ganter, B., Stumme, G., Wille, R. (eds.) Formal Concept Analysis. LNCS (LNAI), vol. 3626, pp. 196–225. Springer, Heidelberg (2005). https://doi.org/10.1007/11528784_11
17. Missaoui, R., Nourine, L., Renaud, Y.: Computing implications with negation from a formal context. Fundam. Informaticae **115**(4), 357–375 (2012)
18. Mora, A., Cordero, P., Enciso, M., Fortes, I., Aguilera, G.: Closure via functional dependence simplification. Int. J. Comput. Math. **89**(4), 510–526 (2012)
19. Obiedkov, S.: Modal logic for evaluating formulas in incomplete contexts. In: Priss, U., Corbett, D., Angelova, G. (eds.) ICCS-ConceptStruct 2002. LNCS (LNAI), vol. 2393, pp. 314–325. Springer, Heidelberg (2002). https://doi.org/10.1007/3-540-45483-7_24

20. Rodríguez-Jiménez, J., Cordero, P., Enciso, M., Rudolph, S.: Concept lattices with negative information: a characterization theorem. Inform. Sci. **369**, 51–62 (2016)
21. Rodríguez-Jiménez, J.M., Cordero, P., Enciso, M., Mora, A.: Data mining algorithms to compute mixed concepts with negative attributes: an application to breast cancer data analysis. Math. Methods Appl. Sci. **39**(16), 4829–4845 (2016)
22. Wille, R.: Restructuring lattice theory: an approach based on hierarchies of concepts. Ordered Sets **83**, 445–470 (1982)

# Pruning Techniques in LinCbO for Computation of the Duquenne-Guigues Basis

Radek Janostik, Jan Konecny$^{(\boxtimes)}$, and Petr Krajča

Department of Computer Science, Palacký University Olomouc,
17. listopadu 12, 77146 Olomouc, Czech Republic
{radek.janostik,jan.konecny,petr.krajca}@upol.cz

**Abstract.** We equip our algorithm LinCbO with a pruning technique similar to that of LCM. Our experimental evaluation shows that it significantly improves the performance of the algorithm.

**Keywords:** Duquenne-Guigues basis · Close-By-One · Linclosure · Algorithm

## 1 Introduction

In our recent work [10], we introduced LinCbO – a fast algorithm for computation of the Duquenne-Guigues basis [8] of a formal context. We also showed that LinCbO performs better than other approaches, such as Ganter$^+$[4] and the attribute incremental approach by Obiedkov and Duquenne [18].

The core of LinCbO is Close-by-One (CbO) [15], which computes pseudointents (and intents) using an enhanced LinClosure algorithm. CbO has received a few improvements in the last two decades, like parallel and distributed computation [12,13], partial closures [1], or execution using the map-reduce framework [11,14]. Arguably, the most efficient improvement of CbO is the use of monotony property of closure operators to avoid some unnecessary computation of closures. This is utilized in FCbO [19], InClose-4 [2], InClose-5 [3], and LCM [9,20–23]. We call these methods pruning techniques.

In the case of pseudointents, the computation of closure is much more expensive. Therefore, it seems a good idea to incorporate a pruning technique into LinCbO. In the present paper, we utilize two pruning techniques inspired by the algorithm LCM[1]. This is the first study of application of pruning techniques for computation of the Duquenne-Guigues basis.

The paper is structured as follows: first, we recall basic notions of formal concept analysis used in the rest of the paper (Sects. 2.1 and 2.2) and introduce the LinCbO algorithm (Sect. 2.3). Then, we describe the idea of pruning techniques for the computation of the Duquenne-Guigues basis generally (Sect. 3).

---

[1] The pruning is utilized in the implementation LCM2 (available at http://research. nii.ac.jp/~uno/codes.htm). However it is not described in the related paper. An interested reader can find the description of the LCM's pruning in [9].

© Springer Nature Switzerland AG 2021
A. Braud et al. (Eds.): ICFCA 2021, LNAI 12733, pp. 91–106, 2021.
https://doi.org/10.1007/978-3-030-77867-5_6

Consequently, we turn our attention to two specific techniques we incorporated to LinCbO (Sect. 4). Then we provide results of our experimental evaluation (Sect. 5). Finally, we summarize our conclusions and present our ideas for future research.

## 2   Preliminaries

Here, we recall only the notions which are necessary for the rest of the paper. For more detailed introduction to FCA is needed, see [7], for a thorough description of LinCbO see [10].

### 2.1   Formal Concept Analysis, Theories, Models, and Bases

An input to FCA [7] is a triplet $\langle X, Y, I \rangle$, called a *formal context*, where $X, Y$ are non-empty sets of objects and attributes respectively, and $I$ is a binary relation between $X$ and $Y$. The presence of an object-attribute pair $\langle x, y \rangle$ in the relation $I$ means that the object $x$ has the attribute $y$.

The formal context $\langle X, Y, I \rangle$ induces so-called *concept-forming operators*:

$^\uparrow : 2^X \to 2^Y$ assigns to a set $A$ of objects the set $A^\uparrow$ of all attributes shared by all the objects in $A$.
$^\downarrow : 2^Y \to 2^X$ assigns to a set $B$ of attributes the set $B^\downarrow$ of all objects which share all the attributes in $B$.

Formally, for all $A \subseteq X, B \subseteq Y$ we have

$$A^\uparrow = \{y \in Y \mid \forall x \in A : \langle x, y \rangle \in I\},$$
$$B^\downarrow = \{x \in X \mid \forall y \in B : \langle x, y \rangle \in I\}.$$

A pair $\langle A, B \rangle \in 2^X \times 2^Y$ satisfying $A^\uparrow = B$ and $B^\downarrow = A$ is called *formal concept*. The set $A$ is then called the *extent* of the formal concept and $B$ is called the *intent* of the formal concept. We denote the set of all intents in $\langle X, Y, I \rangle$ by $\mathrm{Int}(X, Y, I)$. We can also characterize the intents as

$$\mathrm{Int}(X, Y, I) = \{B \subseteq Y \mid B^{\downarrow\uparrow} = B\}.$$

An *attribute implication* is an expression of the form $L \Rightarrow R$, where $L, R \subseteq Y$. We say, that $L \Rightarrow R$ is valid in a set $M \subseteq Y$ of attributes if

$$L \subseteq M \text{ implies } R \subseteq M.$$

The fact that $L \Rightarrow R$ is valid in $M$ is written as $\|L \Rightarrow R\|_M = 1$.

A set of attributes $M$ is called a *model of theory* $\mathcal{T}$ if every attribute implication in $\mathcal{T}$ is valid in $M$. The set of all models of $\mathcal{T}$ is denoted $\mathrm{Mod}(\mathcal{T})$, i.e.

$$\mathrm{Mod}(\mathcal{T}) = \{M \mid \forall L \Rightarrow R \in \mathcal{T} : \|L \Rightarrow R\|_M = 1\}.$$

A theory $T$ is called

- *complete* in $\langle X, Y, I \rangle$ if $\mathrm{Mod}(T) = \mathrm{Int}(X, Y, I)$;
- a *basis* of $\langle X, Y, I \rangle$ if no proper subset of $T$ is complete in $\langle X, Y, I \rangle$.

For convenience, we assume $Y = \{1, 2, \ldots, n\}$. Whenever we write about lower attributes or higher attributes, we refer to the natural ordering of the numbers in $Y$.

## 2.2 Duquenne-Guigues Basis

A set $\mathcal{P} \subseteq 2^Y$ is called the *system of pseudointents* of $\langle X, Y, I \rangle$ if for each $P \subseteq Y$, we have

$$P \in \mathcal{P} \quad \text{iff} \quad P \neq P^{\downarrow\uparrow} \text{ and } Q \subset P \text{ implies } Q^{\downarrow\uparrow} \subset P \text{ for each } Q \in \mathcal{P}.$$

If $\mathcal{P}$ is the system of pseudointents of $\langle X, Y, I \rangle$, the theory

$$T = \{P \Rightarrow P^{\downarrow\uparrow} \mid P \in \mathcal{P}\} \tag{1}$$

forms a basis, called the *Duquenne-Guigues basis.*

Having the *Duquenne-Guigues basis* $T$, we define an operator $c_T : 2^Y \to 2^Y$ as follows:

$$c_T(Z) = \bigcup_{n=0}^{\infty} Z^{T_n} \qquad \text{for all } Z \subseteq Y \tag{2}$$

where

1. $Z^T = Z \cup \bigcup\{R \mid L \Rightarrow R \in T, L \subset Z\}$,
2. $Z^{T_0} = Z$,
3. $Z^{T_n} = (Z^{T_{n-1}})^T$.

The operator $c_T$ is a closure operator inducing $\mathcal{P} \cup \mathrm{Int}(X, Y, I)$. That means,

$$\mathcal{P} \cup \mathrm{Int}(X, Y, I) = \{Z \subseteq Y \mid c_T(Z) = Z\}. \tag{3}$$

At first glance, the above represents a cycle: by (3) we need $c_T$ to find $\mathcal{P}$, by (2) we need $T$ to find $c_T$, and by (1) we need $\mathcal{P}$ to find $T$. However, this cycle can be easily broken as we only need a part of $T$ to compute $c_T(Z)$ – specifically, we need only the attribute implications $L \Rightarrow R$ in $T$ satisfying $L \subseteq c_T(Z)$. We need to compute pseudointents in any order which extends subsethood, i.e., $P_1 \subset P_2$ implies that $P_1$ is computed before $P_2$. Whenever we compute a pseudointent $P$, we update the theory $T$ by adding $P \Rightarrow P^{\downarrow\uparrow}$. This way, we keep $T$ ready for the computation of the next pseudointent. An example of an order which extends subsethood is the lectic order [7].

The interested reader can find results on complexity of the enumeration of pseudointents in [16].

## 2.3  LinCbO

In [10], we delivered a fast algorithm, called LinCbO, for computing the Duquenne-Guigues basis. The core of the algorithm is an adapted CbO [15] (Algorithm 2) with the closure operator $c_T$ (2), implemented as LinClosure [5, 17].

**LinClosure.** LinClosure computes the smallest model of a theory $T$ containing a given input set $B$, i.e. $T$-closure of $B$. It works as follows (refer to Algorithm 1[2]):

For each attribute implication $L \Rightarrow R \in T$, it sets an attribute counter $count[L \Rightarrow R]$ to the size $|L|$ of the left side (lines 4, 5). For each attribute $a \in Y$, it forms a list $list[a]$ of attribute implications $L \Rightarrow R \in T$ with $a \in L$ (lines 4, 6, 7). It initializes a set $Z$ of attributes which need to be explored (line 8) and a result set $D$ (line 1). At the beginning, both the sets contain all attributes from the input set $B$.

While there is any attribute $(m)$ in $Z$, this attribute is removed from $Z$ and the attribute counters of all the attribute implications in $m$'s list are decreased (lines 9–13). Whenever an attribute counter $count[L \Rightarrow R]$ reaches zero, the attributes on the right side $R$ which are not yet in the result set $D$ are added to $D$ and $Z$ (lines 14–18). If there is an attribute implication with an empty left side, it must be handled separately (lines 2, 3), as it is not present in any list. When $Z$ is empty, the algorithm terminates and returns the set $D$.

Time complexity of the LinClosure is linear w.r.t. size of $T$.

**CbO (and LinCbO).** Close-by-one is an algorithm for the enumeration of closed sets of a given closure operator. It traverses a prefix tree of all subsets in the depth-first manner. It makes jumps in it using the closure operator, but only ones which preserve the prefix. In other words, it makes only jumps which land in the subtree of the current node.

The test for whether the prefix is kept is called canonicity test. See Fig. 1.

LinCbO, is an adaptation of CbO for LinClosure. It is represented by the recursive procedure, LinCbOStep (refer to Algorithm 2). The procedure accepts two arguments: $B$ – an attribute set to be closed, $y$ – the last attribute added to $B$ by the previous invocations. It also uses a global variable $T$, which is initially empty and represents the computed basis.

First, LinCbOStep finds a closure of $B$ using LinClosure and stores the result as $D$. If $D$ passes the canonicity test (lines 2, 3; ignore the bracketed text for now), then the procedure ends. Next, it computes $B^{\downarrow\uparrow}$ and stores the result as $C$ (line 4). If it is different from $D$, i.e. $D$ is not an intent, we add $D \Rightarrow C$ to $T$. Then, we invoke LinCbOStep with $D \cup \{i\}$ and $i$, for all attributes $i \notin D$ which are higher than $y$. We do so in descending order to ensure the lectic order.

The procedure is initially called with $B = \varnothing$ and $y = 0$ (representing an invalid attribute, which is lower than any valid attribute).

---

[2] For now, ignore the argument $y$ and line 16 as it will be explained later.

---

**Algorithm 1:** LinClosure

---

def LinClosure($B$, $y$):

    input  :  $B$ – set of attributes

                $y$ – last attribute added to $B$

1    $D \leftarrow B$

2    if $\exists \varnothing \Rightarrow R \in T$ *for some* $R$ then

3        $D \leftarrow D \cup R$

4    for all $L \Rightarrow R \in T$ do

5        $count[L \Rightarrow R] \leftarrow |L|$

6        for all $a \in L$ do

7            add $L \Rightarrow R$ to $list[a]$

8    $Z \leftarrow D$

9    while $Z \neq \varnothing$ do

10        $m \leftarrow \min(Z)$

11        $Z \leftarrow Z \backslash \{m\}$

12        for all $L \Rightarrow R \in list[m]$ do

13            $count[L \Rightarrow R] \leftarrow count[L \Rightarrow R] - 1$

14            if $count[L \Rightarrow R] = 0$ then

15                $add \leftarrow R \backslash D$

16                if $\min(add) < y$ then return *fail*

17                $D \leftarrow D \cup add$

18                $Z \leftarrow Z \cup add$

19    return $D$

---

The LinCbO algorithm additionally has the following features:

*Early stop of LinClosure* – The CbO's canonicity test is incorporated into LinClosure: before $D$ is updated (line 16 in Algorithm 1), we check whether it still would keep the prefix, i.e. no attribute lower than CbO's argument $y$ is added. If it fails the canonicity, we stop the computation and return an indication of the fail. LinClosure then uses this indication (bracketed text in line 2 in Algorithm 2).

*Reuse of attribute counters in LinClosure* – instead of filling attribute counters from scratch (lines 4–7) we reuse the counters which we got from the computation of the predecessor in the CbO's tree. As this improvement is not important for the present paper, we skip the details and do not include it in the pseudocodes.

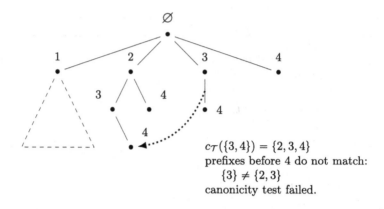

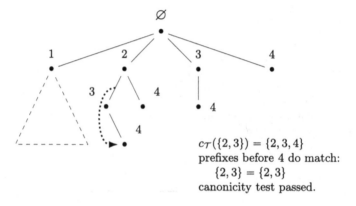

$c_T(\{3,4\}) = \{2,3,4\}$
prefixes before 4 do not match:
$\{3\} \neq \{2,3\}$
canonicity test failed.

$c_T(\{2,3\}) = \{2,3,4\}$
prefixes before 4 do match:
$\{2,3\} = \{2,3\}$
canonicity test passed.

**Fig. 1.** Closure jumps within the prefix tree and canonicity tests

*Exploiting relationships between intents and pseudointents* – Bazhanov and Obiedkov [4] introduced improvements of the NextClosure algorithm (called Ganter in [4]) for computation of the Duquenne-Guigues basis. We directly applied these improvements in LinClosure. Again, these improvements are not important for the present paper, therefore we skip the details and do not include them in the pseudocodes.

CbO, as well as LinCbO, has time delay (the time to generate the first output, the time between two consequent outputs, and the time between the last output and termination) in $O(|Y| \cdot C)$, where $C$ is the time complexity of the utilized closure operator.

---

**Algorithm 2:** LinCbO, simplified

---

def LinCbOStep($B$, $y$):
   **input** : $B$ – a set
          $y$ – last added attribute

1    $D \leftarrow$ LinClosure($B,y$)
2    **if** *prefixes of B and D before y do not match*
     [LinClosure($B,y$) returned fail] **then**
3      **return**

4    $C \leftarrow B^{\downarrow\uparrow}$
5    **if** $C \neq D$ **then**
6      $\mathcal{T} \leftarrow \mathcal{T} \cup (D \Rightarrow C)$;

7    **for** $i \in \{y+1,\dots,n\} \setminus D$ *in decreasing order* **do**
8      LinCbOStep($D \cup \{i\}$, $i$)

---

LinCbOStep($\emptyset$, 0)

---

## 3   Pruning in Pseudointent Computation

The operator $c_{\mathcal{T}}$ (2) is a closure operator; therefore, it satisfies the monotony property, i.e. for any $B, D \subseteq Y$ we have

$$B \subseteq D \text{ implies } c_{\mathcal{T}}(B) \subseteq c_{\mathcal{T}}(D). \tag{4}$$

Furthermore, for any two theories $\mathcal{T}$ and $\mathcal{S}$ with $\mathcal{T} \subseteq \mathcal{S}$, we have $\mathrm{Mod}(\mathcal{S}) \subseteq \mathrm{Mod}(\mathcal{T})$ and, consequently, for all $B \subseteq Y$

$$\mathcal{T} \subseteq \mathcal{S} \text{ implies } c_{\mathcal{T}}(B) \subseteq c_{\mathcal{S}}(B). \tag{5}$$

Putting (4) and (5) together, we get that for any sets $B, D \subseteq Y$ of attributes and theories $\mathcal{T}$ and $\mathcal{S}$, we have that

$$B \subseteq D \text{ and } \mathcal{T} \subseteq \mathcal{S} \text{ imply } c_{\mathcal{T}}(B) \subseteq c_{\mathcal{S}}(D). \tag{6}$$

From (6), we have that for any $i \in Y$:

$$B \subseteq D, \mathcal{T} \subseteq \mathcal{S} \text{ imply } [\text{ if } i \in c_{\mathcal{T}}(B \cup \{y\}) \text{ then } i \in c_{\mathcal{S}}(D \cup \{y\})]. \tag{7}$$

Now, consider $B \cup \{y\}$ being a set to which $y$ is added as the last attribute. Let $i$ be an attribute with $i < y$ and $i \notin B$ and the theories $\mathcal{S}$ and $\mathcal{T}$ be the partially computed Duquenne-Guigues basis at different times. Obviously, $i \in c_{\mathcal{T}}(B \cup \{y\})$ means that the closure of $c_{\mathcal{T}}(B \cup \{y\})$ fails the canonicity test. In words, (7) says that if the canonicity fails for $c_{\mathcal{T}}(B \cup \{y\})$ then it will also fail for $c_{\mathcal{S}}(D \cup \{y\})$.

We can store the information about the failed canonicity test for $c_T(B \cup \{y\})$ and use it later to avoid the computation of $c_S(D \cup \{y\})$. This is what we call a *pruning*, as it effectively prunes branches of the search tree.

Specifically, in our case, we store a rule of form "$y$ adds $i$". This means that when we add the attribute $y$ to the set $B$, the attribute $i$ occurs in the closure $c_T(B \cup \{y\})$ and causes the canonicity test to fail. We use the rule only in subtrees of $B$, as they contain only supersets of $B$.

*Example 1.* In Fig. 2 we illustrate a case for $B = \varnothing$, $y = d$ and $i = a$. In the right-most branch, we observe that the canonicity test fails for $\varnothing \cup \{d\}$, because $a < d$ occurs in the closure $c_T(\varnothing \cup \{d\})$. We store the rule "$d$ adds $a$" and use it in subtrees of $\varnothing$ whenever we add $d$ to a set. This enables us to avoid computation of $c_T(c, d)$, $c_T(b, d)$, and $c_T(b, c, d)$.

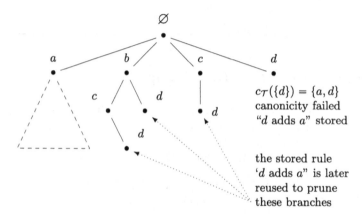

**Fig. 2.** Idea of pruning

## 4   How LinCbO Utilizes the Pruning

We use a global array, *rules*, to store the rules for pruning. A rule "$y$ adds $i$" is stored as $rules[y] = i$. The absence of such a rule is represented by $rules[y] = 0$. Note that it means that a new rule can overwrite an old rule if it has the same attribute on the left side.

We need to modify LinClosure (recall that the canonicity test is incorporated in LinClosure) to provide us information for pruning. The modified LinClosure returns a pair $\langle D, fail \rangle$ where:

- $D$ is the closed set if it passes the canonicity test.
- *fail* is the lowest attribute which violates the prefix if the canonicity test failed; otherwise it is 0.

In Algorithm 1, we only need to accordingly modify lines 16 and 19.

We furthermore modify LinCbO as follows (refer to Algorithm 3):

(p0) Whenever the canonicity test fails, `LinCboStep` returns the attribute *fail*, which LinClosure detects to violate the prefix (line 4). If an invocation of `LinCboStep` returns a non-zero value *fail*, it stores the rule "$y$ adds *fail*" in the array *rules* (lines 10–12).

(p1) At the beginning of `LinCboStep`, i.e. when descending to a subtree, all rules having the last added attribute (argument $y$) on the right side are removed from the stored rules. In the pseudocode, this is performed by a subroutine called `RemoveRulesByRightSide` (line 1).

(p2) At the end of `LinCboStep`, i.e. when backtracking from the current subtree, all rules from this call are removed. In the pseudocode, this is performed by a subroutine called `RemoveAllRulesAddedThisCall` (line 13).

(p3) Before computing a closure $c_T(D \cup \{i\})$ in a subtree of $B$, we check the stored rules to find whether adding $i$ does not add an attribute which causes the canonicity test to fail (line 9).

We use two versions of the pruning:

lcm: does exactly what is described above. Notice that in (p3) it needs only to check existence of a rule with $i$ on the left side; it does not need to check whether the attribute on its right side is in $B$ (the part "$rules[i] \in D$" of the condition in line 9 of Algorithm 3 can be skipped).

lcmx: does what is described above but skips the step (p1) (line 1 of Algorithm 3 is skipped).

*Remark 1.* Due to the early stop utilized in LinClosure, the information for pruning is not as complete as in the case for intents. We do not actually obtain $c_T(D \cup \{y\})$ used in (7) when the canonicity is violated. Instead, we obtain an intermediate set. Still, it is usable to form the pruning rules, as at least one attribute causing the canonicity test to fail is present in the set.

We only need to store the array *rules* and a stack of maximal size $n$ for tracking which rules were added in the present call. Therefore, the memory complexity of the pruning is in $O(n)$.

---

**Algorithm 3:** LinCbO with pruning, simplified

---

```
def LinCbOStep(B, y):
    input : B – a set
            y – last added attribute
```

| | |
|---|---|
| 1 | RemoveRulesByRightSide($y$) |
| 2 | $\langle D, fail \rangle \leftarrow$ LinClosure($B,y$) |
| 3 | if $fail > 0$ then |
| 4 | $\quad$ return $fail$ |
| | |
| 5 | $C \leftarrow B^{\downarrow\uparrow}$ |
| 6 | if $C \neq D$ then |
| 7 | $\quad \mathcal{T} \leftarrow \mathcal{T} \cup (D \Rightarrow C);$ |
| | |
| 8 | for $i \in \{y+1, \ldots, n\} \setminus D$ in decreasing order do |
| 9 | $\quad$ if $rules[i] = 0$ or $rules[i] \in D$ then |
| 10 | $\qquad fail \leftarrow$ LinCbOStep($D \cup \{i\}$, $i$) |
| 11 | $\qquad$ if $fail > 0$ then |
| 12 | $\qquad\quad rules[i] \leftarrow fail$ |
| | |
| 13 | RemoveAllRulesStoredThisCall() |
| 14 | return 0 |

```
LinCbOStep(∅, 0)
```

---

## 5    Experimental Evaluation

We experimentally compare three versions of LinCbO: without pruning and with the two pruning techniques described above. Additionally, we compare them with algorithms available in the framework made by Bazhanov & Obiedkov [4][3], namely Ganter, Ganter$^+$ [4] – each with naïve closure, LinClosure [17], and Wild's closure [24]—and the attribute incremental approach. To achieve maximal fairness, we implemented all the three versions of LinCbO into their framework.

All experiments have been performed on a computer with 64 GB RAM, two Intel Xeon CPU E5-2680 v2 (at 2.80 GHz), Debian Linux 10, and GNU GCC 8.3.0. All measurements have been taken ten times and the mean value is presented.

We used the following datasets from UC Irvine Machine Learning Repository [6]:

- crx – Credit Approval (37 rows containing a missing value were removed),
- shuttle – Shuttle Landing Control,
- magic – MAGIC Gamma Telescope,
- bikesharing_(day|hour) – Bike Sharing Dataset,
- kegg – KEGG Metabolic Reaction Network – Undirected.

---

[3] Available at https://github.com/yazevnul/fcai.

**Table 1.** Properties of the datasets

| Dataset | $|X|$ | $|Y|$ | $|I|$ | # intents | # ps.intents |
|---|---|---|---|---|---|
| inter10crx | 653 | 139 | 40,170 | 10,199,818 | 20,108 |
| inter10shuttle | 43,500 | 178 | 3,567,907 | 38,199,148 | 936 |
| inter3magic | 19,020 | 52 | 399,432 | 1,006,553 | 4181 |
| inter4magic | 19,020 | 72 | 589,638 | 24,826,749 | 21,058 |
| inter5bike_day | 731 | 93 | 24,650 | 3,023,326 | 20,425 |
| inter5crx | 653 | 79 | 20,543 | 348,428 | 3427 |
| inter5shuttle | 43,500 | 88 | 1,609,510 | 333,783 | 346 |
| inter6shuttle | 43,500 | 106 | 2,002,790 | 381,636 | 566 |
| nom10bike_day | 731 | 100 | 9293 | 52,697 | 29,773 |
| nom10crx | 653 | 85 | 8774 | 51,078 | 6240 |
| nom10magic | 19,020 | 102 | 209,220 | 583,386 | 154,090 |
| nom10shuttle | 43,500 | 97 | 435,000 | 2931 | 810 |
| nom15magic | 19,020 | 152 | 209,220 | 1,149,717 | 397,224 |
| nom20magic | 19,020 | 202 | 209,220 | 1,376,212 | 654,028 |
| nom5bike_day | 731 | 65 | 9293 | 61,853 | 16,296 |
| nom5bike_hour | 17,379 | 90 | 238,292 | 1,868,205 | 320,679 |
| nom5crx | 653 | 55 | 8774 | 29,697 | 2162 |
| nom5keg | 65,554 | 144 | 1,834,566 | 13,262,627 | 42,992 |
| nom5shuttle | 43,500 | 52 | 435,000 | 1461 | 319 |
| ord10bike_day | 731 | 93 | 28,333 | 664,713 | 11,795 |
| ord10crx | 653 | 79 | 37,005 | 1,547,971 | 2906 |
| ord10shuttle | 43,500 | 88 | 1,849,216 | 97,357 | 279 |
| ord5bike_day | 731 | 58 | 14,929 | 81,277 | 5202 |
| ord5bike_hour | 17,379 | 83 | 457,578 | 2,174,964 | 99,691 |
| ord5crx | 653 | 49 | 19,440 | 139,752 | 973 |
| ord5magic | 19,020 | 42 | 535,090 | 821,796 | 1267 |
| ord5shuttle | 43,500 | 43 | 868,894 | 4068 | 119 |
| ord6magic | 19,020 | 52 | 662,177 | 2,745,877 | 2735 |

We binarized the datasets using nominal (nom), ordinal (ord), and interordinal (inter) scaling, where each numerical feature was scaled to $k$ attributes with $k-1$ equidistant cutpoints. Categorical features were scaled nominally to a number of attributes corresponding to the number of categories. After the binarization, we removed full columns. Properties of the resulting datasets are shown in Table 1. The naming convention used in Table 1 (and Table 2) is the following: (scaling)$k$(dataset). For example, inter10shuttle is the dataset 'Shuttle Landing Control' interordinally scaled to 10, using 9 equidistant cutpoints.

All datasets and source codes used in this experimental evaluation are available at http://phoenix.inf.upol.cz/~konecnja/fcalad/.

## 6   Our Observations

We made the following observations from the results of our experimental evaluation (Table 2).

**Table 2.** Runtimes in seconds of algorithms generating Duquenne-Guigues basis

| Dataset | LinCbO | LinCbO + lcm | LinCbO + lcmx | Best of the rest | |
|---|---|---|---|---|---|
| inter10crx | 508.551 | 223.38 | **199.115** | 400.292 | AttInc |
| inter10shuttle | 15852.9 | 14967.7 | **14825.4** | 17664.5 | Ganter[+] |
| inter3magic | 26.156 | 24.289 | **24.192** | 106.341 | Ganter |
| inter4magic | 965.353 | **771.084** | 835.315 | 4027.48 | Ganter |
| inter5bike_day | 85.591 | 44.012 | **40.349** | 72.952 | AttInc |
| inter5crx | 3.176 | 1.855 | **1.802** | 5.863 | AttInc |
| inter5shuttle | 120.003 | **112.034** | 112.638 | 137.211 | Ganter |
| inter6shuttle | 133.288 | **126.91** | 126.946 | 164.355 | Ganter |
| nom10bike_day | 7.099 | 1.682 | **1.545** | 4.515 | AttInc |
| nom10crx | 0.944 | 0.332 | **0.328** | 1.227 | AttInc |
| nom10magic | 206.797 | 96.377 | **96.662** | 486.926 | AttInc |
| nom10shuttle | 0.425 | **0.382** | 0.396 | 1.102 | Ganter[+] |
| nom15magic | 1509.86 | 557.051 | **544.459** | 3358.44 | AttInc |
| nom20magic | 4437.05 | 1211.66 | **1210.46** | 7882.15 | AttInc |
| nom5bike_day | 2.219 | 0.833 | **0.804** | 2.580 | AttInc |
| nom5bike_hour | 1410.11 | **476.592** | 481.241 | 1893.33 | AttInc |
| nom5crx | 0.193 | 0.114 | **0.106** | 0.406 | AttInc |
| nom5keg | 1936.7 | **1116.51** | 1139.87 | 7564.710 | Ganter[+] |
| nom5shuttle | 0.309 | 0.297 | **0.292** | 0.481 | Ganter[+] |
| ord10bike_day | 24.997 | 15.947 | **15.108** | 21.884 | AttInc |
| ord10crx | 11.653 | 10.5153 | **10.147** | 28.367 | AttInc |
| ord10shuttle | **34.293** | 36.2858 | 36.2079 | 40.338 | Ganter |
| ord5bike_day | 0.936 | 0.713 | **0.669** | 2.080 | AttInc |
| ord5bike_hour | 321.147 | 273.862 | **258.072** | 1107.570 | AttInc |
| ord5crx | 0.610 | 0.559 | **0.551** | 1.468 | AttInc |
| ord5magic | **46.982** | 48.429 | 48.4259 | 93.845 | Ganter |
| ord5shuttle | **1.319** | 1.345 | 1.349 | 1.380 | Ganter[+] |
| ord6magic | **158.227** | 158.466 | 162.65 | 335.947 | Ganter |

## Comparison of LinCbO with and without Pruning

The pruning techniques seem to have different effect for various types of formal contexts:

- For interordinally scaled data, LinCbO with pruning performed better than without pruning. However, the improvement is significant only for the crx datasets (inter10crx and inter5crx) and for inter5bike_day. For other datasets, the improvement seems insignificant.
- For nominally scaled data, LinCbO with pruning performed significantly better with the exception of shuttle dataset (nom5shuttle and nom10shuttle).
- For ordinally scaled data, LinCbO without pruning performed slightly better than with pruning – namely, for the magic and shuttle datasets (ord5magic, ord6magic, ord10shuttle, and ord5shuttle). LinCbO with pruning performed better in the rest. With the exception of ord5crx and ord10crx, the improvement was significant.

The speed-up factor $\frac{\text{runtime of LinCbO without pruning}}{\text{runtime of LinCbO with pruning}} \cdot 100\%$ of the two pruning methods is shown in Table 3.

## Comparison of the Two Variants of Pruning in LinCbO

The lcmx does not remove pruning rules in (p1) and enables them to be used until overwritten by another rule or removed in (p2). That way, it can avoid more closure computation than lcm at the cost of an inexpensive check of attribute presence (Algorithm 3, line 9).

Indeed, our experimental comparison shows that LinCbO with lcmx performs slightly better than lcm in most cases (Table 2) and avoids more closure computation (Table 3). However, the difference in the performance is not significant.

## Comparison with Other Algorithms

The column 'best of the rest' in Table 2 represents the best algorithm from Bazhanov & Obiedkov's framework. We tested all seven algorithms listed above, however only Ganter, Ganter[+] (both with the naïve closure implementation) and the attribute incremental approach appear in the column, as these performed best in our evaluation. Among these algorithms, the attribute incremental approach was often the fastest one. In some cases, it was even faster than LinCbO without pruning. However, we encountered limits with this algorithm as it runs out of available memory in three cases: inter10shuttle, inter4magic, and nom5keg.

**Table 3.** Numbers of skipped recursive calls and speed-up factors of the two pruning techniques.

| Dataset | lcm | Speed-up factor (%) | lcmx | Speed-up factor (%) |
|---|---|---|---|---|
| inter10crx | 120,851,019 | 227.66 | 126,403,951 | 255.41 |
| inter10shuttle | 1,321,766,518 | 105.91 | 1,326,688,040 | 106.93 |
| inter3magic | 1,538,199 | 107.69 | 1,637,367 | 108.12 |
| inter4magic | 48,536,834 | 125.19 | 52,180,055 | 115.57 |
| inter5bike_day | 18,193,052 | 194.47 | 19,432,953 | 212.13 |
| inter5crx | 2,345,689 | 171.21 | 2,429,752 | 176.25 |
| inter5shuttle | 7,536,887 | 107.11 | 7,603,108 | 106.54 |
| inter6shuttle | 9,922,755 | 105.03 | 10,029,964 | 105 |
| nom10bike_day | 1,195,268 | 422.08 | 1,229,644 | 459.46 |
| nom10crx | 635,844 | 284.38 | 641,138 | 287.87 |
| nom10magic | 2,974,506 | 214.57 | 2,995,995 | 213.94 |
| nom10shuttle | 39,864 | 111.05 | 40,288 | 107.34 |
| nom15magic | 10,129,231 | 271.05 | 10,185,502 | 277.31 |
| nom20magic | 19,659,598 | 366.2 | 19,756,910 | 366.56 |
| nom5bike_day | 502,879 | 266.27 | 533,577 | 276.04 |
| nom5bike_hour | 16,430,989 | 295.87 | 17,011,991 | 293.02 |
| nom5crx | 169,499 | 169.24 | 171,668 | 181.19 |
| nom5keg | 226,578,200 | 173.46 | 227,020,735 | 169.91 |
| nom5shuttle | 12,983 | 103.91 | 13,338 | 105.71 |
| ord10bike_day | 2,468,278 | 156.75 | 2,848,811 | 165.45 |
| ord10crx | 1,621,895 | 110.82 | 2,169,367 | 114.85 |
| ord10shuttle | 1,144,851 | 94.51 | 1,181,005 | 94.71 |
| ord5bike_day | 121,968 | 131.32 | 156,400 | 139.91 |
| ord5bike_hour | 1,122,408 | 117.27 | 1,677,745 | 124.44 |
| ord5crx | 137,169 | 109.12 | 161,173 | 110.74 |
| ord5magic | 491,174 | 97.01 | 493,737 | 97.02 |
| ord5shuttle | 38,877 | 98.02 | 40,987 | 97.75 |
| ord6magic | 1,856,194 | 99.85 | 1,867,038 | 97.28 |

## 7    Conclusion

We enhanced LinCbO with two pruning techniques inspired by LCM and experimentally evaluated the resulting algorithms in comparison with LinCbO without pruning and seven known algorithms. In all tested cases, some version of LinCbO computed the DG-basis faster than other algorithms.

*Future Research*:

– In the paper, we describe our results on application of the LCM-like pruning technique. Besides this, we also experiment with pruning techniques from FCbO and InClose5. We will bring related results in our upcoming papers.

- Furthermore, the presented pruning techniques were not yet tried for enumeration of intents with CbO-based algorithms (except LCM). This represents another direction of our research.
- We observed that the algorithms behave differently for formal contexts obtained by various scaling techniques. We want to study trends of this behavior and provide more general information about it.

**Acknowledgment.** The authors acknowledge support by the grants
  - IGA UP 2020 of Palacký University Olomouc, No. IGA_PrF_2020_019,
  - JG 2019 of Palacký University Olomouc, No. JG_2019_008.

# References

1. Andrews, S.: In-Close, a fast algorithm for computing formal concepts. In: International Conference on Conceptual Structures (ICCS), Moscow (2009)
2. Andrews, S.: Making use of empty intersections to improve the performance of CbO-type algorithms. In: Bertet, K., Borchmann, D., Cellier, P., Ferré, S. (eds.) ICFCA 2017. LNCS (LNAI), vol. 10308, pp. 56–71. Springer, Cham (2017). https://doi.org/10.1007/978-3-319-59271-8_4
3. Andrews, S.: A new method for inheriting canonicity test failures in close-by-one type algorithms (2018)
4. Bazhanov, K., Obiedkov, S.A.: Optimizations in computing the Duquenne-Guigues basis of implications. Ann. Math. Artif. Intell. **70**(1–2), 5–24 (2014)
5. Beeri, C., Bernstein, P.A.: Computational problems related to the design of normal form relational schemas. ACM Trans. Database Syst. (TODS) **4**(1), 30–59 (1979)
6. Dua, D., Graff, C.: UCI machine learning repository (2017)
7. Ganter, B., Wille, R.: Formal Concept Analysis - Mathematical Foundations. Springer, Heidelberg (1999)
8. Guigues, J.-L., Duquenne, V.: Familles minimales d'implications informatives resultant d'un tableau de données binaires. Math. Sci. Humaines **95**, 5–18 (1986)
9. Janostik, R., Konecny, J., Krajča, P.: LCM is well implemented CbO: study of LCM from FCA point of view. CoRR, abs/2010.06980 (2020)
10. Janostik, R., Konecny, J., Krajča, P.: LinCbO: fast algorithm for computation of the Duquenne-Guigues basis. CoRR, abs/2011.04928 (2020)
11. Konecny, J., Krajča, P.: Pruning in map-reduce style CbO algorithms. In: Alam, M., Braun, T., Yun, B. (eds.) ICCS 2020. LNCS (LNAI), vol. 12277, pp. 103–116. Springer, Cham (2020). https://doi.org/10.1007/978-3-030-57855-8_8
12. Krajča, P., Outrata, J., Vychodil, V.: Advances in algorithms based on CbO. CLA **672**, 325–337 (2010)
13. Krajča, P., Outrata, J., Vychodil, V.: Parallel algorithm for computing fixpoints of Galois connections. Ann. Math. Artif. Intell. **59**(2), 257–272 (2010)
14. Krajca, P., Vychodil, V.: Distributed algorithm for computing formal concepts using map-reduce framework. In: Adams, N.M., Robardet, C., Siebes, A., Boulicaut, J.-F. (eds.) IDA 2009. LNCS, vol. 5772, pp. 333–344. Springer, Heidelberg (2009). https://doi.org/10.1007/978-3-642-03915-7_29
15. Kuznetsov, S.O.: A fast algorithm for computing all intersections of objects from an arbitrary semilattice. Nauchno-Tekhnicheskaya Informatsiya Seriya 2- Informatsionnye Protsessy i Sistemy, (1), 17–20 (1993)

16. Kuznetsov, S.O.: On the intractability of computing the Duquenne-Guigues base. J. Univ. Comput. Sci. **10**(8), 927–933 (2004)
17. Maier, D.: The Theory of Relational Databases, vol. 11. Computer Science Press, Rockville (1983)
18. Obiedkov, S.A., Duquenne, V.: Attribute-incremental construction of the canonical implication basis. .Ann. Math. Artif. Intell. **49**(1–4), 77–99 (2007)
19. Outrata, J., Vychodil, V.: Fast algorithm for computing fixpoints of Galois connections induced by object-attribute relational data. Inf. Sci. **185**(1), 114–127 (2012)
20. Uno, T., Asai, T., Uchida, Y., Hiroki A.: LCM: an efficient algorithm for enumerating frequent closed item sets. In: FIMI, vol. 90. Citeseer (2003)
21. Uno, T., Asai, T., Uchida, Y., Arimura, H.: An efficient algorithm for enumerating closed patterns in transaction databases. In: Suzuki, E., Arikawa, S. (eds.) DS 2004. LNCS (LNAI), vol. 3245, pp. 16–31. Springer, Heidelberg (2004). https://doi.org/10.1007/978-3-540-30214-8_2
22. Uno, T., Kiyomi, M., Arimura, H.: LCM ver. 2: efficient mining algorithms for frequent/closed/maximal itemsets. In: FIMI, vol. 126 (2004)
23. Uno, T., Kiyomi, M., Arimura, H.: LCM ver. 3: collaboration of array, bitmap and prefix tree for frequent itemset mining. In: Proceedings of the 1st International Workshop on Open Source Data Mining: Frequent Pattern Mining Implementations, pp. 77–86. ACM (2005)
24. Wild, M.: Computations with finite closure systems and implications. In: Du, D.-Z., Li, M. (eds.) COCOON 1995. LNCS, vol. 959, pp. 111–120. Springer, Heidelberg (1995). https://doi.org/10.1007/BFb0030825

# Approximate Computation of Exact Association Rules

Saurabh Bansal[1] , Sriram Kailasam[1] , and Sergei Obiedkov[2] (✉) 

[1] IIT Mandi, Mandi, India
s.obiedkov@hse.ru, sriramk@iitmandi.ac.in
[2] HSE University, Moscow, Russia

**Abstract.** We adapt a polynomial-time approximation algorithm for computing the canonical basis of implications to approximately compute frequent implications, also known as exact association rules. To this end, we define a suitable notion of approximation that takes into account the frequency of attribute subsets and show that our algorithm achieves a desired approximation with high probability. We experimentally evaluate the proposed algorithm on several artificial and real-world data sets.

**Keywords:** Association rule · Implication · PAC learning

## 1 Introduction

Formal concept analysis offers several approaches to computing implication bases of a formal context. Most popular ones are based on NEXT CLOSURE, a general algorithm for computing closed sets of a closure operator [11]. This algorithm is suitable for building the canonical basis of a formal context, since its premises, taken together with closed attribute sets of the context, form a closure system. The attribute-incremental algorithm from [19] is often faster than approaches based on NEXT CLOSURE, but its memory requirements limit its applicability to contexts with a moderate number of closed sets.

Recently, probably approximately correct (PAC) algorithms for computing the implication basis have been considered [8,9]. In this paper, we continue this line of research by proposing a new notion of approximation that focuses on capturing the frequent (in the sense of frequent itemset mining) part of the implication theory behind the context. We adapt a previously proposed PAC-algorithm to compute this frequency-aware approximation and show that it usually produces more accurate approximations than the original PAC-algorithm and thus can be useful even if one is not interested specifically in frequent implications. We also show that the algorithm runs much faster than exact NEXT CLOSURE-based algorithms on dense formal contexts, at least if the size of their canonical

Supported by SPARC, a Government of India Initiative under grant no. SPARC/2018-2019/P682/SL.

A. Braud et al. (Eds.): ICFCA 2021, LNAI 12733, pp. 107–122, 2021.
https://doi.org/10.1007/978-3-030-77867-5_7

basis is significantly smaller than the number of closed sets (which is often the case, as suggested by our experiments in Sect. 5).

The paper is organized as follows. In Sect. 2, we present the necessary definitions from formal concept analysis. Section 3 recalls two notions of Horn approximations and a PAC-algorithm for finding them. In Sect. 4, we introduce frequency-aware approximations and adapt the algorithm from the previous section to compute them. Section 5 describes the results of empirical evaluation of the proposed algorithm.

## 2   Main Definitions

Recall that a *formal context* $\mathbb{K}$ is a triple $(G, M, I)$, where $G$ is a set of objects, $M$ is a set of attributes, and $I \subseteq G \times M$ is an incidence relation specifying which objects have which attributes [13]. For $X \subseteq G$ and $Y \subseteq M$, the following *derivation operators* are defined:

$$X' = \{m \in M \mid \forall g \in X \colon (g, m) \in I\} \qquad Y' = \{g \in G \mid \forall m \in Y \colon (g, m) \in I\}$$

Two closure operators are defined by subsequent application of the two derivation operators; $X''$ and $Y''$ are said to be *closed* in $\mathbb{K}$. Closed subsets of $M$ are called *(concept) intents* of $\mathbb{K}$. The set of all intents of $\mathbb{K}$ is denoted by $\mathrm{Int}\,\mathbb{K}$.

An *implication* is an expression of the form $A \to B$, where $A \subseteq M$ is called the *premise* and $B \subseteq M$ the *conclusion* of the implication. A subset $C \subseteq M$ is a *model* of $A \to B$ if $A \nsubseteq C$ or $B \subseteq C$. The implication $A \to B$ *holds* or *is valid* in $\mathbb{K}$ if, for every $g \in G$, the set $\{g\}'$ is its model. If $\{g\}'$ is not a model of an implication, then $g$ is a *counterexample* to it.

The *support* of an attribute set $A \subseteq M$ is $|A'|$, the number of objects that have all attributes from $A$. The *relative support* of $A$ is $|A'|/|G|$. The *(relative) support* or *frequency* of an implication $A \to B$ is the (relative) support of $A \cup B$. An attribute set or an implication is called *frequent* if its support is above a certain specified threshold. The support parameter is important in association rule mining, where the goal is to find implications with high support that may still have a small number of counterexamples in the context [1]. There, valid implications are known as *exact association rules*.

A set $C \subseteq M$ is a *model* of a set $\mathcal{L}$ of implications over $M$ if it is a model of every implication from $\mathcal{L}$. The set of all models of $\mathcal{L}$ is denoted by $\mathrm{Mod}\,\mathcal{L}$.

An implication set $\mathcal{L}$ over $M$ defines a closure operator that maps $C \subseteq M$ to the smallest subset $\mathcal{L}(C) \in \mathrm{Mod}\,\mathcal{L}$ such that $C \subseteq \mathcal{L}(C)$. If $C$ is a model of $\mathcal{L}$, then $C = \mathcal{L}(C)$ and we say that $C$ *is closed* under $\mathcal{L}$.

An implication $A \to B$ *follows* from an implication set $\mathcal{L}$ if every model of $\mathcal{L}$ is a model of $A \to B$ or, equivalently, $B \subseteq \mathcal{L}(A)$. An implication set $\mathcal{L}$ is *non-redundant* if no implication $A \to B \in \mathcal{L}$ follows from $\mathcal{L} \setminus \{A \to B\}$.

A non-redundant set $\mathcal{L}$ of implications valid in $\mathbb{K}$ is called a *basis* of $\mathbb{K}$ if every implication valid in $\mathbb{K}$ follows from $\mathcal{L}$. A formal context can have several bases of different sizes. The *canonical* or *Duquenne–Guigues basis* of $\mathbb{K}$ is known to contain the smallest number of implications among all bases of $\mathbb{K}$ [14]. It is defined as $\{P \rightarrow P'' \mid P \text{ is a pseudo-intent of } \mathbb{K}\}$, where $P \subseteq M$ is called a *pseudo-intent* if $P \neq P''$ and, for every $Q \subsetneq P$, we have $Q'' \subseteq P$ whenever $Q$ is a pseudo-intent.

## 3   Probably Approximately Correct Computation of Implications

Being able to compute the canonical basis $\mathcal{L}$ of a formal context $\mathbb{K}$ in total polynomial time, i.e., time polynomial in the sizes of $\mathbb{K}$ and $\mathcal{L}$, is a major open problem. Known algorithms that compute $\mathcal{L}$ directly also compute $\text{Int}\,\mathbb{K}$ as a side product [5,11,15,19]. However, $|\text{Int}\,\mathbb{K}|$ can be exponentially larger than $|\mathcal{L}|$, see Example 1 from Sect. 5.3.

This motivates approximation algorithm design for finding the canonical basis. Probably approximately correct computation of the canonical basis of a formal context has been considered in various settings in [8,9,20,21]. The settings differ in whether the context is available directly or via a particular set of queries, as in the query learning framework [2] or in attribute exploration [12].

The notions of approximation that we will use here are more general than in these works. Assuming a probability distribution $\mathcal{D}$ over attribute subsets of $M$, we define the *Horn $\mathcal{D}$-distance* between an implication set $\mathcal{L}$ over $M$ and a context $\mathbb{K} = (G, M, I)$ as the probability of obtaining a subset closed under $\mathcal{L}$ but not in $\mathbb{K}$ or vice versa when choosing it according to $\mathcal{D}$:

$$\text{dist}^{\mathcal{D}}(\mathcal{L}, \mathbb{K}) := \Pr_{\mathcal{D}}(A \in \text{Mod}\,\mathcal{L} \triangle \text{Int}\,\mathbb{K}).$$

Here, $A \triangle B$ is the symmetric difference between the sets $A$ and $B$. An $\epsilon$-*Horn $\mathcal{D}$-approximation* of $\mathbb{K}$, where $0 < \epsilon < 1$, is an implication set $\mathcal{L}$ over $M$ such that $\text{dist}^{\mathcal{D}}(\mathcal{L}, \mathbb{K}) \leq \epsilon$. If $\mathcal{D}$ is the uniform distribution, then $\epsilon$-Horn $\mathcal{D}$-approximation of $\mathbb{K}$ is the $\epsilon$-Horn approximation of $\mathbb{K}$ as defined in the papers referenced above.

These papers also use the notion of an $\epsilon$-strong Horn approximation, which we generalize in a similar way. The *strong Horn $\mathcal{D}$-distance* between $\mathcal{L}$ and $\mathbb{K}$ is the probability of choosing a subset with different closures under $\mathcal{L}$ and in $\mathbb{K}$:

$$\text{dist}^{\mathcal{D}}_{\text{STRONG}}(\mathcal{L}, \mathbb{K}) := \Pr_{\mathcal{D}}(\mathcal{L}(A) \neq A'').$$

$\mathcal{L}$ is an $\epsilon$-*strong Horn $\mathcal{D}$-approximation* of $\mathbb{K}$ if $\text{dist}^{\mathcal{D}}_{\text{STRONG}}(\mathcal{L}, \mathbb{K}) \leq \epsilon$. An $\epsilon$-strong Horn $\mathcal{D}$-approximation is always an $\epsilon$-Horn $\mathcal{D}$-approximation.

An approximation $\mathcal{L}$ is an *upper approximation* of $\mathbb{K}$ if all implications of $\mathcal{L}$ are valid in $\mathbb{K}$, or, equivalently, $\mathcal{L}(A) \subseteq A''$ for all $A \subseteq M$, i.e., if $\text{Int}\,\mathbb{K} \subseteq \text{Mod}\,\mathcal{L}$.

We are interested in algorithms that, given a formal context $\mathbb{K} = (G, M, I)$, a distribution $\mathcal{D}$ over subsets of $M$, and parameters $0 < \epsilon \leq 1$ and $0 < \delta \leq 1$, compute, with probability at least $1 - \delta$, an $\epsilon$- or $\epsilon$-strong Horn $\mathcal{D}$-approximation of $\mathbb{K}$. As discussed in [20], such algorithms exist for the uniform distribution and they work in time polynomial in $|G|$, $|M|$, the size of the canonical basis of $\mathbb{K}$, $1/\epsilon$, and $1/\delta$. Such an algorithm for an upper $\epsilon$-Horn approximation has been presented (in a different setting) already in [16] based on the results for learning Horn formulas with membership and equivalence queries [3]. We present its generalization to an arbitrary distribution as Algorithm 1.

---

**Algorithm 1.** HORNAPPROXIMATION($\mathbb{K}$, $EX_{\mathcal{D}}$, $\epsilon$, $\delta$)

---

**Input:** A formal context $\mathbb{K} = (G, M, I)$, a sampling oracle $EX_{\mathcal{D}}$ that returns a subset of $M$ according to distribution $\mathcal{D}$, $0 < \epsilon \leq 1$, and $0 < \delta \leq 1$.
**Output:** A set of implications $\mathcal{L}$ that, with probability at least $1 - \delta$, is an $\epsilon$-Horn $\mathcal{D}$-approximation of $\mathbb{K}$.

1: $\mathcal{L} := []$
2: $i := 1$
3: **while** ISAPPROXIMATELYEQUIVALENT($\mathcal{L}$, $\mathbb{K}$, $EX_{\mathcal{D}}$, $q_i(\epsilon, \delta)$) returns $X$ **do**
4:      $found := \mathbf{false}$
5:      **for all** $A \to B \in \mathcal{L}$ **do**
6:          $C := A \cap X$
7:          **if** $A \neq C \neq C''$ **then**
8:              $found := \mathbf{true}$
9:              replace $A \to B$ by $C \to C''$ in $\mathcal{L}$
10:             **exit for**
11:     **if not** $found$ **then**
12:         add $X \to X''$ to the end of $\mathcal{L}$
13:     $i := i + 1$
14: **return** $\mathcal{L}$

---

Algorithm 1 receives a sampling oracle $EX_{\mathcal{D}}$ that takes no arguments and, when called, returns a subset of $M$ according to distribution $\mathcal{D}$. Starting with an empty list $\mathcal{L}$ of implications, the algorithm repeatedly calls procedure ISAP-PROXIMATELYEQUIVALENT to check if $\mathcal{L}$ is an $\epsilon$-Horn $\mathcal{D}$-approximation of $\mathbb{K}$. If not, this procedure is expected to return a model $X$ of $\mathcal{L}$ such that $X \neq X''$, which means that there is an implication valid in $\mathbb{K}$ that does not follow from $\mathcal{L}$. This $X$ is called a *negative counterexample* to $\mathcal{L}$, as opposed to a *positive counterexample* $Y$, which is such that $\mathcal{L}(Y) \neq Y = Y''$. The algorithm then either refines one of the implications of $\mathcal{L}$ or simply adds implication $X \to X''$ so as to ensure that $X \subsetneq \mathcal{L}(X) \subseteq X''$. This guarantees that $\mathcal{L}$ always contains only valid implications of $\mathbb{K}$ and thus no positive counterexamples to $\mathcal{L}$ are possible.

If the ISAPPROXIMATELYEQUIVALENT procedure always returns a negative counterexample $X$ when it exists, then Algorithm 1 computes the canonical basis of $\mathbb{K}$. This easily follows from the results presented in [4] regarding the original query-based algorithm from [3]. In this case, the ISAPPROXIMATELYEQUIVA-LENT procedure implements an equivalence oracle in the sense of [2,3]. However, finding such an $X$ given $\mathbb{K}$ and $\mathcal{L}$ is the problem referred to as CMI in [17], where it is shown that it is at least as hard as (the decision version of) the Hypergraph Transversal Problem; no polynomial-time algorithm is known to exist for it.

To achieve an $\epsilon$-Horn approximation with probability at least $1 - \delta$, we use Algorithm 2, which makes a certain number of attempts to generate such an $X$ using $EX_{\mathcal{D}}$ and returns **true** if all attempts fail.

---

**Algorithm 2.** IsApproximatelyEquivalent($\mathcal{L}$, $\mathbb{K}$, $EX_{\mathcal{D}}$, $k$)

---

**Input:** A set $\mathcal{L}$ of implications valid in context $\mathbb{K} = (G, M, I)$, a sampling oracle $EX_{\mathcal{D}}$
  that returns a subset of $M$ according to distribution $\mathcal{D}$, and $k \in \mathbb{N}$.
**Output:** A set $X \subseteq M$ such that $\mathcal{L}(X) = X \neq X''$ if found; **true**, otherwise.
  1: **for** $j := 1$ **to** $k$ **do**
  2:         $X := EX_{\mathcal{D}}()$
  3:         **if** $\mathcal{L}(X) = X \neq X''$ **then**
  4:                 **return** $X$
  5: **return true**

---

How many attempts are needed depends on how far we are in computing a Horn approximation or, more precisely, how many counterexamples we have already produced. In Algorithm 1, we use the function $q_i(\epsilon, \delta)$ to determine the number of calls to $EX_{\mathcal{D}}$ needed to generate the $i$th $X$. It has been known that an algorithm using equivalence queries can be transformed into a PAC algorithm for the same learning problem by replacing the $i$th equivalence query by

$$\left\lceil \frac{1}{\epsilon} \left( \ln \frac{1}{\delta} + i \ln 2 \right) \right\rceil \tag{1}$$

calls to the $EX_{\mathcal{D}}$ oracle and terminating if none of them returns a counterexample [2]. Here, $\epsilon$ is the desired approximation quality and $\delta$ is the upper bound on the probability of failing to achieve an $\epsilon$-approximation.

The quantity (1) grows linearly with $i$; however, a logarithmic dependence on $i$ is sufficient, as shown in [21]. Defining the function $q_i$ as

$$q_i(\epsilon, \delta) = \left\lceil \log_{1-\epsilon} \frac{\delta}{i(i+1)} \right\rceil, \tag{2}$$

we guarantee that Algorithm 1 computes an $\epsilon$-Horn $\mathcal{D}$-approximation of $\mathbb{K}$ with probability at least $1 - \delta$ in time polynomial in $|G|$, $|M|$, $|\mathcal{L}|$, $1/\epsilon$, and $1/\delta$, where $\mathcal{L}$ is the set of implications upon the termination of the algorithm, whose size never exceeds the size of the canonical basis of $\mathbb{K}$. The total number of calls to $EX_{\mathcal{D}}$ is $O(|\mathcal{L}||M|(\log|\mathcal{L}||M| + \log 1/\delta)/\epsilon)$ when using (2) compared with $O(|\mathcal{L}||M|(|\mathcal{L}||M| + \log 1/\delta)/\epsilon)$ when using (1) [21].

To obtain an $\epsilon$-strong Horn $\mathcal{D}$-approximation, we replace the call to IsApproximatelyEquivalent in Algorithm 1 to the call to the IsStronglyApproximatelyEquivalent procedure presented as Algorithm 3 [9].

## 4 Computing Frequency-Aware Approximations

For practical applications, it may be reasonable to assume that attribute subsets are distributed according to their frequency. For a context $\mathbb{K} = (G, M, I)$ with finite $G$ and $A \subseteq M$, this means

$$\Pr(A) = \frac{|A'|}{\sum_{B \subseteq M} |B'|}. \tag{3}$$

---

**Algorithm 3.** IsStronglyApproximatelyEquivalent($\mathcal{L}$, $\mathbb{K}$, $EX_{\mathcal{D}}$, $k$)

---

**Input:** A set $\mathcal{L}$ of implications valid in context $\mathbb{K} = (G, M, I)$, a sampling oracle $EX_{\mathcal{D}}$
 that returns a subset of $M$ according to distribution $\mathcal{D}$, and $k \in \mathbb{N}$.
**Output:** A set $\mathcal{L}(X) \subseteq M$ such that $\mathcal{L}(X) \neq X''$ if found; **true**, otherwise.
 1: **for** $j := 1$ **to** $k$ **do**
 2:      $Y := \mathcal{L}(EX_{\mathcal{D}}())$
 3:      **if** $Y \neq Y''$ **then**
 4:           **return** $Y$
 5: **return true**

---

Plugging this probability into the definition of (strong) Horn distance, we obtain the notion of frequency-aware Horn approximation: an $\epsilon$- ($\epsilon$-strong) Horn $\mathcal{D}$-approximation $\mathcal{L}$ of $\mathbb{K}$ is a *frequency-aware $\epsilon$- ($\epsilon$-strong) Horn approximation* of $\mathbb{K}$ if $\mathcal{D}$ is the probability distribution defined by (3).

The reason for using a frequency-aware approximation is two-fold. On the one hand, such approximation $\mathcal{L}$ ensures that most implications with high support follow from $\mathcal{L}$ and, in the case of the strong approximation, the closures of most frequent subsets under $\mathcal{L}$ coincide with their closures in the context. In the framework of association rule mining [1], such implications and such subsets are usually considered the most important. On the other hand, a frequency-aware approximation ignores attribute subsets that never occur in data. For real-world data sets, these may be the bulk of all the subsets, resulting in a misleadingly low value of $\mathrm{dist}^{\mathcal{D}}(\mathcal{L}, \mathbb{K})$ when $\mathcal{D}$ is the uniform distribution. Using frequency-aware approximation allows one to capture the implications that describe dependencies inside attribute combinations that actually occur in data instead of focusing on implications that describe incompatibilities between attributes (which may also be important, but, in many cases, are a part of background knowledge).

To compute such frequency-aware approximations, we need to simulate a sampling oracle that samples attribute subsets according to (3). This oracle can be simulated by polynomial-time Algorithm 1 "Frequency-based Sampling" from [7], resulting in a total–polynomial time PAC algorithm for computing frequent implications. The algorithm uses the following probability distribution on objects $g \in G$ of context $\mathbb{K} = (G, M, I)$:

$$\Pr(g) = \frac{2^{|\{g\}'|}}{\sum_{h \in G} 2^{|\{h\}'|}}. \tag{4}$$

In other words, the probability of an object $g \in G$ is proportional to the number of subsets of its intent $\{g'\}$.

The algorithm consists of two steps. First, it selects an object $g$ from $G$ according to probability distribution (4), and then it selects a subset of $\{g\}'$ uniformly at random. It is shown in [7] that this algorithm generates an attribute subset according to probability distribution (3).

Therefore, a frequency-aware $\epsilon$- or $\epsilon$-strong Horn approximation can be computed by Algorithm 1 by passing the algorithm just described in place of $EX_{\mathcal{D}}$.

# 5    Experimental Evaluation

In this section, we study the performance of the randomized algorithm on real-life, as well as artificial data sets. We are primarily interested in two characteristics: the runtime and the quality of approximation.

## 5.1    Quality Factor

The randomized algorithms presented here are guaranteed to produce an upper $\epsilon$- or $\epsilon$-strong Horn approximation with the desired probability. In particular, if $\mathcal{L}$ is the implication set obtained from $\mathbb{K}$ when running the algorithm for $\epsilon$-approximation with parameters $\epsilon$ and $\delta$ and a sampling oracle for probability distribution $\mathcal{D}$, then, with probability at least $1 - \delta$, we have $\text{dist}^{\mathcal{D}}(\mathcal{L}, \mathbb{K}) \leq \epsilon$. Since $\mathcal{L}$ contains only implications valid in $\mathbb{K}$ and, consequently, $\text{Int}\,\mathbb{K} \subseteq \text{Mod}\,\mathcal{L}$, this means that

$$\frac{|\text{Mod}\,\mathcal{L}| - |\text{Int}\,\mathbb{K}|}{2^{|M|}} \leq \epsilon$$

when $\mathcal{D}$ is the uniform distribution. In other words, the difference between $|\text{Mod}\,\mathcal{L}|$ and $|\text{Int}\,\mathbb{K}|$ is small when considered on the scale of $2^{|M|}$: $\text{Mod}\,\mathcal{L}$ contains at most $\epsilon 2^{|M|}$ extra subsets in addition to those in $\text{Int}\,\mathbb{K}$.

However, if $\text{Int}\,\mathbb{K}$ is small compared to $2^{|M|}$, this may still allow $\text{Mod}\,\mathcal{L}$ to be several times larger than $\text{Int}\,\mathbb{K}$. To see if this really happens in practice, we introduce the *quality factor* (QF) defined as follows:

$$QF(\mathcal{L}, \mathbb{K}, A) = \frac{|\text{Int}\,\mathbb{K} \cap \mathfrak{P}(A)|}{|\text{Mod}\,\mathcal{L} \cap \mathfrak{P}(A)|},$$

where $A \subseteq M$ and $\mathfrak{P}(A)$ is the power set of $A$. This measures the proportion of subsets of $A$ closed in the context among those closed under the computed implications. When we report the quality factor in the experiments, we set $A$ to be the set of roughly $\alpha|M|$ most frequent attributes of $M$, where $\alpha$ is $1/4$ for real-world data sets and $1/2$ for artificial data sets.

## 5.2    Testbed

The testbed consists of a server Intel Xeon E5-2650 v3 @ 2.30 GHz with 20 cores and 40 threads.

## 5.3    Data Sets

The formal contexts used in the experiments are described in Table 1, where the last five columns correspond to the number of attributes, the number of objects, the size of the canonical basis, the number of intents, and the density, $|I|/|G||M|$, of the context $(G, M, I)$ named in the first column.

The first six data sets are real-world data sets, while the rest are synthetically generated. The real-world data sets have been derived from Census, Shuttle, Mushroom, Connect, and Chess data sets from the UCI machine learning

**Table 1.** Contexts.

| Context | Attributes | Objects | Canonical basis | Intents | Density |
|---|---|---|---|---|---|
| Census | 122 | 48842 | 71787 | 248846 | 0.08 |
| nom10shuttle | 97 | 43500 | 810 | 2931 | 0.10 |
| Mushroom | 119 | 8124 | 2323 | 238710 | 0.19 |
| Connect | 114 | 7222 | 86583 | 50468988 | 0.38 |
| inter10shuttle | 178 | 43500 | 936 | 38199148 | 0.46 |
| Chess | 75 | 3196 | 73162 | 930851337 | 0.49 |
| Example 1 ($n = 5$) | 25 | 3125 | 5 | 28629152 | 0.80 |
| Example 1 ($n = 6$) | 36 | 46656 | 6 | 62523502210 | 0.83 |
| Example 2 ($n = 10$) | 21 | 30 | 1024 | 2038103 | 0.92 |
| Example 2 ($n = 15$) | 31 | 45 | 32768 | 2133134741 | 0.95 |

repository [10]. They have been converted into formal contexts by using nominal scaling for categorical features (one attribute per category) and by scaling numerical features into multiple attributes using equidistant cut points. In the inter10shuttle data set, inter-ordinal rather than nominal scaling is used [15].

Example 1 is a context with $n^n$ objects and $n^2$ attributes $M = M_1 \cup \cdots \cup M_n$ with $|M_i| = n$, $M_i \cap M_j = \varnothing$ for all $1 \leq i < j \leq n$, where the object intents $\{g\}'$ are all possible subsets of $M$ such that $|\{g\}' \cap M_i| = n - 1$ for all $1 \leq i \leq n$ [12]. The canonical basis consists of only $n$ implications of the type $M_i \to M$ for $1 \leq i \leq n$. The number of concept intents is $(2^n - 1)^n + 1$. This context is interesting, because the number of its closed attribute sets is exponential in $|M|$, while the size of the canonical basis is only linear in $|M|$. This is precisely the type of a context that should be hard for NEXT CLOSURE–based algorithms, since they have to compute all closed sets as a side product, and much easier for our randomized algorithm. We ran experiments for $n = 5$ and $n = 6$.

Example 2 is a context with $3n$ objects $g_1, g_2, \ldots, g_{3n}$ and $2n + 1$ attributes $m_0, m_1, \ldots, m_{2n}$, where object $g_i$ has attribute $m_j$ if $i \leq n$ and $j \notin \{0, i, i + n\}$, or if $i > n$ and $j \neq i - n$ [18]. There are exactly $2^n$ pseudo-intents of the form $\{m_{i_1}, m_{i_2}, \ldots, m_{i_n}\}$ where $i_j \in \{j, j + n\}$; thus, the size of the canonical basis is exponential in the context size. We ran experiments for $n = 10$ and $n = 15$.

## 5.4   Experiments

In all the experiments, a parallelized implementation of the randomized algorithm was used. We parallelized the search for a counterexample in Algorithms 2 and 3, as well as the search for an implication $A \to B$ to be refined in the main loop of Algorithm 1. Our implementation and the data sets used for the experiments are available at https://github.com/saurabh18213/Implication-Basis. Unless mentioned otherwise, forty threads were allocated to run the algorithm. The actual number of threads used at different points of the execution of the algorithm was determined using certain heuristics.

In Experiments 1, 3 and 4, we set $\epsilon = 0.1$ for real-world data sets, $\epsilon = 0.01$ for Example 1, $\epsilon = 0.01$ for Example 2 ($n = 10$), and $\epsilon = 0.001$ for Example 2 ($n = 15$). In Experiment 2, we vary the value of $\epsilon$. All the reported results are for $\delta = 0.1$. No significant change in total time, the computed number of implications, and Quality Factor was observed when $\delta$ was varied. For real-world and artificial data sets, all the results are average of three and five measurements, respectively.

**Experiment 1: Comparing Approximations.** In this experiment, we compute $\epsilon$- and $\epsilon$-strong Horn $\mathcal{D}$-approximations for different $\mathcal{D}$, varying the sampling oracle used in Algorithms 2 and 3. We use the *Uniform* oracle that generates subsets of $M$ uniformly at random and the *Frequent* oracle that generates subsets according to the probability distribution specified by (3), as described in Sect. 4. In addition, we test the following combination of the two oracles. We first use the Uniform oracle. If, at some call to Algorithm 3, all $k$ attempts to generate a counterexample with the Uniform oracle fail, instead of terminating the algorithm, we "redo" the $k$ attempts, now with the Frequent oracle. If one of the attempts succeeds, we keep using the Frequent oracle for the remaining part of the computation; otherwise, the algorithm terminates. This approach is denoted by *Both* in the results below.

**Table 2.** Runtime in seconds for different types of approximation.

| Data set | $\epsilon$-strong Horn approximation | | | $\epsilon$-Horn approximation | | |
|---|---|---|---|---|---|---|
| | Uniform | Frequent | Both | Uniform | Frequent | Both |
| Census | 0.18 | 1451.64 | 1184.10 | 0.16 | 5.02 | 0.21 |
| nom10shuttle | 0.15 | 0.73 | 0.71 | 0.14 | 0.43 | 0.44 |
| Mushroom | 0.11 | 1.89 | 1.95 | 0.06 | 0.16 | 0.14 |
| Connect | 0.14 | 307.51 | 307.10 | 0.07 | 0.08 | 0.07 |
| inter10shuttle | 0.59 | 6.77 | 6.47 | 0.58 | 0.60 | 0.60 |
| Chess | 0.07 | 167.96 | 169.77 | 0.04 | 0.04 | 0.03 |
| Example 1 ($n = 5$) | 0.03 | 0.03 | 0.04 | 0.03 | 0.03 | 0.04 |
| Example 1 ($n = 6$) | 0.31 | 0.27 | 0.36 | 0.31 | 0.29 | 0.37 |
| Example 2 ($n = 10$) | 0.27 | 0.17 | 0.27 | 0.21 | 0.19 | 0.26 |
| Example 2 ($n = 15$) | 96.72 | 74.64 | 108.77 | 83.31 | 75.12 | 115.81 |

The time taken, the number of implications computed, and the value of the Quality Factor for each data set are shown in Tables 2, 3, and 4, respectively.

When computing an $\epsilon$-strong Horn approximation, the runtime on the real-world data sets is significantly higher with the Frequent oracle than with the Uniform oracle, but so is the number of implications computed and, usually, the Quality Factor. An exception is Connect, for which QF almost does not

**Table 3.** The number of implications computed for different types of approximation. The last column shows the number of implications in the entire canonical basis.

| Data set | $\epsilon$-strong Horn approximation | | | $\epsilon$-Horn approximation | | | *Total* |
|---|---|---|---|---|---|---|---|
| | Uniform | Frequent | Both | Uniform | Frequent | Both | |
| Census | 48 | 20882 | 19111 | 41 | 1210 | 71 | 71787 |
| nom10shuttle | 76 | 201 | 201 | 76 | 137 | 146 | 810 |
| Mushroom | 95 | 577 | 593 | 7 | 72 | 59 | 2323 |
| Connect | 120 | 10774 | 10730 | 7 | 9 | 9 | 86583 |
| inter10shuttle | 172 | 446 | 430 | 171 | 171 | 171 | 936 |
| Chess | 64 | 6514 | 6542 | 48 | 48 | 48 | 73162 |
| Example 1 ($n = 5$) | 5 | 0 | 5 | 5 | 0 | 5 | 5 |
| Example 1 ($n = 6$) | 6 | 0 | 6 | 6 | 0 | 6 | 6 |
| Example 2 ($n = 10$) | 357 | 269 | 340 | 321 | 262 | 347 | 1024 |
| Example 2 ($n = 15$) | 7993 | 6813 | 8375 | 7612 | 6970 | 8424 | 32768 |

**Table 4.** Quality Factor (QF) for different types of approximation.

| Data set | $\epsilon$-strong Horn approximation | | | $\epsilon$-Horn approximation | | |
|---|---|---|---|---|---|---|
| | Uniform | Frequent | Both | Uniform | Frequent | Both |
| Census | 0.0003 | 0.0184 | 0.0180 | 0.0003 | 0.0014 | 0.0004 |
| nom10shuttle | 0.0004 | 0.0695 | 0.0613 | 0.0004 | 0.0157 | 0.0208 |
| Mushroom | 0.0004 | 0.1454 | 0.1482 | 0.0001 | 0.0032 | 0.0014 |
| Connect | 0.9979 | 0.9979 | 0.9979 | 0.0001 | 0.0016 | 0.0016 |
| inter10shuttle | 0.4900 | 0.5533 | 0.5429 | 0.4900 | 0.4900 | 0.4900 |
| Chess | 0.6927 | 1.0000 | 0.9830 | 0.6927 | 0.6927 | 0.6927 |
| Example 1 ($n = 5$) | 1.0000 | 0.9692 | 1.0000 | 1.0000 | 0.9692 | 1.0000 |
| Example 1 ($n = 6$) | 1.0000 | 0.9844 | 1.0000 | 1.0000 | 0.9844 | 1.0000 |
| Example 2 ($n = 10$) | 1.0000 | 1.0000 | 1.0000 | 1.0000 | 1.0000 | 1.0000 |
| Example 2 ($n = 15$) | 1.0000 | 1.0000 | 1.0000 | 1.0000 | 1.0000 | 1.0000 |

change despite a sharp increase in the number of implications. With $\epsilon$-Horn approximation, the situation is generally similar.

The runtime, the number of implications, and QF are lower for Example 1 when using the Frequent oracle than when using the Uniform oracle. This is because all non-trivial valid implications (i.e., implications $A \to B$ with $B \nsubseteq A$) have zero support and are ignored by frequency-aware approximations. The randomized algorithm computes no implications when using the Frequency oracle.

In Example 2, all non-trivial valid implications have non-zero support. In particular, all $2^n$ implications from the canonical basis have a rather high support $n$ (while the total number of objects in the context is $3n$), and their premises are also rather large ($n$ out of $2n + 1$ attributes). Because of this, all three metrics are almost the same for the Uniform and Frequent oracles in Example 2: chances

**Table 5.** Time in seconds for different values of $\epsilon$.

| Data set | 0.3 | 0.2 | 0.1 | 0.05 | 0.01 |
|---|---|---|---|---|---|
| Census | 0.19 | 37.63 | 1184.10 | 2345.26 | 2336.88 |
| nom10shuttle | 0.44 | 0.47 | 0.71 | 0.82 | 1.43 |
| Mushroom | 0.82 | 1.27 | 1.95 | 2.75 | 5.03 |
| Connect | 308.69 | 307.54 | 307.10 | 306.97 | 307.44 |
| inter10shuttle | 4.41 | 5.34 | 6.47 | 7.91 | 12.72 |
| Chess | 169.23 | 169.50 | 169.77 | 168.04 | 168.99 |
| Example 1 ($n = 5$) | 0.02 | 0.02 | 0.03 | 0.03 | 0.04 |
| Example 1 ($n = 6$) | 0.23 | 0.23 | 0.29 | 0.30 | 0.36 |
| Example 2 ($n = 10$) | 0.002 | 0.002 | 0.002 | 0.01 | 0.27 |
| Example 2 ($n = 15$) | 0.002 | 0.002 | 0.002 | 0.002 | 0.63 |

to generate a negative counterexample to a current implication set $\mathcal{L}$ are similar for the two oracles.

As expected, for real-world data sets, most of the metrics are higher in the case of $\epsilon$-strong approximation than in the case of $\epsilon$-approximation. For the artificial data sets, there is not much difference due to the fact that the closure of all non-closed sets there is $M$ and, therefore, Algorithm 1 adds to $\mathcal{L}$ only implications of the form $X \to M$ whether it computes $\epsilon$- or $\epsilon$-strong approximation.

When using both oracles, as described above, we usually obtain results similar to what we get with the Frequent oracle alone. An important exception is Example 1, where all implications have zero support. In general, using both oracles lets us capture such zero-support implications in addition to frequent implications.

**Experiment 2: The Quality of Approximation.** In this experiment, we vary the value of the $\epsilon$ parameter. The results in Tables 5–7 are for $\epsilon$-strong Horn approximation with counterexamples generated following the approach labeled as "Both" in the description of Experiment 1. As expected, for most data sets, the run time, the number of implications computed, and the value of the quality factor increase as the value of $\epsilon$ is decreased. The exceptions are Chess and Connect, where a very good approximation (QF $\approx 1$) is computed even at $\epsilon = 0.3$. The decrease in $\epsilon$ has no substantial effect on any of the metrics, even though the number of implications computed is several times smaller than the size of the canonical basis. It seems that the implication set $\mathcal{L}$ computed by the randomized algorithm forms the essential part of the implication theory behind the context, while valid implications that do not follow from $\mathcal{L}$ must have limited applicability due to large premises with low support.

It should also be said that the Quality Factor as we compute it is not very relevant to Examples 1 and 2. Recall that, for artificial data sets, we select the $|M|/2$ most frequent attributes of $M$ and then check how many subsets of these

**Table 6.** The number of implications computed for different values of $\epsilon$. The last column shows the number of implications in the entire canonical basis.

| Data set | 0.3 | 0.2 | 0.1 | 0.05 | 0.01 | Total |
|---|---|---|---|---|---|---|
| Census | 49 | 2865 | 19111 | 26257 | 26253 | 71787 |
| nom10shuttle | 136 | 149 | 201 | 231 | 303 | 810 |
| Mushroom | 349 | 440 | 593 | 749 | 1036 | 2323 |
| Connect | 10790 | 10746 | 10730 | 10735 | 10759 | 86583 |
| inter10shuttle | 356 | 383 | 430 | 479 | 582 | 936 |
| Chess | 6563 | 6572 | 6542 | 6537 | 6578 | 73162 |
| Example 1 ($n = 5$) | 3 | 4 | 5 | 5 | 5 | 5 |
| Example 1 ($n = 6$) | 1 | 2 | 6 | 6 | 6 | 6 |
| Example 2 ($n = 10$) | 1 | 2 | 4 | 28 | 340 | 1024 |
| Example 2 ($n = 15$) | 0 | 0 | 0 | 1 | 422 | 32768 |

**Table 7.** Quality Factor (QF) for different values of $\epsilon$.

| Data set | 0.3 | 0.2 | 0.1 | 0.05 | 0.01 |
|---|---|---|---|---|---|
| Census | 0.0004 | 0.0034 | 0.0180 | 0.0208 | 0.0208 |
| nom10shuttle | 0.0090 | 0.0140 | 0.0613 | 0.1017 | 0.1753 |
| Mushroom | 0.0382 | 0.0692 | 0.1482 | 0.2726 | 0.4504 |
| Connect | 0.9979 | 0.9979 | 0.9979 | 0.9979 | 0.9979 |
| inter10shuttle | 0.4956 | 0.5202 | 0.5429 | 0.6451 | 0.8910 |
| Chess | 0.9981 | 1.0000 | 0.9830 | 0.9963 | 1.0000 |
| Example 1 ($n = 5$) | 0.9692 | 0.9815 | 1.0000 | 1.0000 | 1.0000 |
| Example 1 ($n = 6$) | 0.9844 | 0.9875 | 0.9969 | 1.0000 | 1.0000 |
| Example 2 ($n = 10$) | 1.0000 | 1.0000 | 1.0000 | 1.0000 | 1.0000 |
| Example 2 ($n = 15$) | 1.0000 | 1.0000 | 1.0000 | 1.0000 | 1.0000 |

attributes are closed under the computed implications but not in the context. However, for Example 2, any selection of $|M|/2$ attributes contains at most one subset that is not closed in the context; so the value of the Quality Factor is bound to be high no matter what implications we compute. This explains why we have QF = 1 even for cases when no implications have been computed. The situation is similar for Example 1, though less dramatic. There, the number of implications computed is a better indicator of the approximation quality than QF. Tables 5 and 6 show that, for Example 1, we compute the basis exactly at $\epsilon \leq 0.05$ in a fraction of a second.

**Experiment 3: The Frequency of Implications.** In this experiment, we compute the support of implications in an $\epsilon$-strong Horn approximation obtained with counterexamples generated using the *Frequent* oracle. The results are shown in Table 8. We show relative supports as percentages of $|G|$. In the second column, the average support of the implications is shown. For columns 3–6, if the value of $P_x$ is $y$, then at least $x\%$ of the implications in the approximation have support greater than or equal to $y$. Example 1 is not shown because it has no implications with non-zero support.

**Table 8.** The relative support of the implications in the approximation.

| Data set | Average | $P_{10}$ | $P_{50}$ | $P_{90}$ | $P_{99}$ | $|G|$ |
|---|---|---|---|---|---|---|
| Census | 0.0335 | 0.0409 | 0.0082 | 0.0020 | 0.0020 | 48842 |
| nom10shuttle | 5.4038 | 12.1456 | 1.2345 | 0.0023 | 0.0000 | 43500 |
| Mushroom | 11.7045 | 25.6360 | 7.9764 | 1.1489 | 0.5252 | 8124 |
| Connect | 0.9276 | 1.3385 | 0.2446 | 0.0415 | 0.0277 | 7222 |
| inter10shuttle | 35.7838 | 99.9831 | 11.7080 | 0.0092 | 0.0069 | 43500 |
| Chess | 6.7690 | 15.7697 | 4.0989 | 0.9387 | 0.5423 | 3196 |
| Example 2 ($n = 10$) | 32.2586 | 33.3333 | 33.3333 | 30.0000 | 26.6667 | 30 |
| Example 2 ($n = 15$) | 32.6944 | 33.3333 | 33.3333 | 31.1111 | 28.8889 | 45 |

**Experiment 4: Runtime.** In this experiment, we compare the performance of the randomized algorithm for computing an $\epsilon$-strong approximation with that of two algorithms computing the canonical basis exactly: the optimized version of NEXT CLOSURE from [5] (referred to as "6 + 1" there) and LinCbO from [15], which combines this optimized version with the LinClosure algorithm for computing the closures under implications [6] and introduces further optimizations. The randomized algorithm is run with the values of $\epsilon$ and $\delta$ specified in the beginning of Sect. 5.4. Counterexamples were generated following the approach labeled as "Both" in the description of Experiment 1.

We show the results in Table 9. To make comparison fair, we give the runtime of both the parallel version of the randomized algorithm with forty threads and the version with one thread only. The number of threads does not affect the value of the quality factor, which is therefore shown only once. The exact algorithms have taken excessive time on Example 1 ($n = 6$); so we had to terminate them before the basis was computed.

Between the two exact algorithms, LinCbO is consistently faster. The randomized algorithm, both with forty threads and one thread, runs faster than the exact algorithms on all contexts except the two sparse contexts (Census and nom10shuttle) and the medium-density context (Mushroom). For Mushroom context, the randomized algorithm runs faster than the exact algorithms, when forty threads are used. On Mushroom context, the value of the quality factor of the implication set obtained by the randomized algorithm is rather low, although

Table 9. Runtime in seconds (Experiment 4)

| Data set | 1 thread | 40 threads | QF | NEXT CLOSURE | LINCBO |
|---|---|---|---|---|---|
| Census | 29608 | 1184.10 | 0.0180 | 522 | 177 |
| nom10shuttle | 3.34 | 0.71 | 0.0613 | 1.25 | 0.44 |
| Mushroom | 25.92 | 1.95 | 0.1482 | 49 | 10.8 |
| Connect | 6239.75 | 307.10 | 0.9979 | 23310 | 19420 |
| inter10shuttle | 42.52 | 6.47 | 0.5429 | 19223 | 16698 |
| Chess | 1955.12 | 169.77 | 0.9830 | 325076 | 234309 |
| Example 1 ($n = 5$) | 0.05 | 0.04 | 1.0000 | 384 | 65 |
| Example 1 ($n = 6$) | 0.55 | 0.36 | 1.0000 | – | – |
| Example 2 ($n = 10$) | 0.22 | 0.27 | 1.0000 | 5.94 | 2.8 |
| Example 2 ($n = 15$) | 84.97 | 108.77 | 1.0000 | 203477 | 29710 |

higher than for the two sparse contexts. For the dense contexts, the value of the quality factor is close to 1 except for inter10shuttle, where it is around 0.54.

This behavior is consistent with our expectations. In dense contexts, we usually have a large number of concept intents, which have to be enumerated as a side product by the exact algorithms. This slows them down considerably. The randomized algorithm does not have this weakness. In general, the randomized algorithm is preferable when the size of the basis is small with respect to the number of concept intents. If however the size of the canonical basis is comparable with the number of intents (as in the case of Census and nom10shuttle), the randomized algorithm tends to perform much worse, both in terms of runtime and quality. Context density can be a good (even if not always reliable) indicator for the applicability of the randomized algorithm.

Note also that the forty-thread version of the algorithm is up to twenty times faster than the single-thread version on hard instances (such as Census and Connect).

## 6  Conclusion

Finding the canonical basis of a formal context is a computationally hard problem, which makes it reasonable to search for relatively efficient approximate solutions. To this end, an approach within the framework of probably approximately correct learning has been recently proposed [8,20] based on older works in machine learning and knowledge compilation [3,16]. The main contribution of this paper is two-fold. On the one hand, we extend the previously proposed approach by introducing frequency (or support) into approximation so as to shift the focus to approximating frequent implications. On the other hand, we present the first experimental evaluation of this approach in terms of its efficiency compared to the exact computation of the canonical basis.

Loosely speaking, a frequency-aware Horn approximation of a formal context $\mathbb{K}$, as considered in this paper, is a subset $\mathcal{L}$ of implications valid in $\mathbb{K}$ from which most valid frequent implications of $\mathbb{K}$ follow. Somewhat more precisely, such $\mathcal{L}$ is biased towards ensuring, for $A \subseteq M$ with large support $|A'|$, that $A = \mathcal{L}(A)$ if and only if $A = A''$, or, in the case of strong approximation, that $\mathcal{L}(A) = A''$. In many application settings, frequent implications, also known as exact association rules, are regarded as the most important. We present a total–polynomial time algorithm to compute such an approximation with high probability. For certain practical purposes, it may be even more valuable than a full basis of frequent implications (whichever way it is defined), whose size can be exponential in the size of the input context.

A frequency-aware approximation can be relevant even if we are not interested specifically in frequent implications. If most attribute subsets are of zero support, which often happens in real-world data sets, then $\mathcal{L}$ would be regarded as a good Horn approximation provided that $\mathcal{L}(A) = M$ for all (or most) subsets $A$ that never occur in data—no matter where $\mathcal{L}$ maps those relatively few subsets that do occur. Taking frequency into account solves this problem by making it hard to ignore such subsets. This results in a much more meaningful approximation.

Our experiments show that, if the size of the canonical basis is small compared to the number of concept intents, a high-quality approximation can be computed in significantly less time than it takes Next Closure–based algorithms to compute the basis exactly. The randomized algorithm that we propose for this purpose is very easy to parallelize, which can further decrease the total runtime. It remains to be seen how well, in terms of efficiency, the algorithm performs against algorithms that are not related to Next Closure.

**Acknowledgments.** We thank Aimene Belfodil for letting us know of the paper [7].

# References

1. Agrawal, R., Imielinski, T., Swami, A.N.: Mining association rules between sets of items in large databases. In: Buneman, P., Jajodia, S. (eds.) Proceedings of the 1993 ACM SIGMOD International Conference on Management of Data, Washington, DC, pp. 207–216. ACM Press (1993). https://doi.org/10.1145/170035.170072
2. Angluin, D.: Queries and concept learning. Mach. Learn. **2**(4), 319–342 (1988)
3. Angluin, D., Frazier, M., Pitt, L.: Learning conjunctions of Horn clauses. Mach. Learn. **9**, 147–164 (1992)
4. Arias, M., Balcázar, J.L.: Construction and learnability of canonical Horn formulas. Mach. Learn. **85**(3), 273–297 (2011). https://doi.org/10.1007/s10994-011-5248-5
5. Bazhanov, K., Obiedkov, S.: Optimizations in computing the Duquenne-Guigues basis of implications. Annal. Mathe. Artif. Intell. **70**(1), 5–24 (2014)
6. Beeri, C., Bernstein, P.: Computational problems related to the design of normal form relational schemas. ACM TODS **4**(1), 30–59 (1979)

7. Boley, M., Lucchese, C., Paurat, D., Gärtner, T.: Direct local pattern sampling by efficient two-step random procedures. In: Proceedings of the 17th ACM SIGKDD International Conference on Knowledge Discovery and Data Mining. p. 582–590. KDD'11, Association for Computing Machinery, New York, NY, USA (2011). https://doi.org/10.1145/2020408.2020500, https://doi.org/10.1145/2020408.2020500

8. Borchmann, D., Hanika, T., Obiedkov, S.: On the usability of probably approximately correct implication bases. In: Bertet, K., Borchmann, D., Cellier, P., Ferré, S. (eds.) ICFCA 2017. LNCS (LNAI), vol. 10308, pp. 72–88. Springer, Cham (2017). https://doi.org/10.1007/978-3-319-59271-8_5

9. Borchmann, D., Hanika, T., Obiedkov, S.: Probably approximately correct learning of Horn envelopes from queries. Discrete Appl. Math. **273**, 30–42 (2020). https://doi.org/10.1016/j.dam.2019.02.036

10. Dua, D., Graff, C.: UCI Machine Learning Repository (2017)

11. Ganter, B.: Two basic algorithms in concept analysis. In: Kwuida, L., Sertkaya, B. (eds.) ICFCA 2010. LNCS (LNAI), vol. 5986, pp. 312–340. Springer, Heidelberg (2010). https://doi.org/10.1007/978-3-642-11928-6_22

12. Ganter, B., Obiedkov, S.: More expressive variants of exploration. In: Obiedkov, S. (ed.) Conceptual Exploration, pp. 107–315. Springer, Heidelberg (2016). https://doi.org/10.1007/978-3-662-49291-8

13. Ganter, B., Wille, R.: Formal Concept Analysis: Mathematical Foundations. Springer, Berlin/Heidelberg (1999). https://doi.org/10.1007/978-3-642-59830-2

14. Guigues, J.L., Duquenne, V.: Famille minimale d'implications informatives résultant d'un tableau de données binaires. Mathématiques et Sciences Humaines **24**(95), 5–18 (1986)

15. Janostik, R., Konecny, J., Krajča, P.: LinCbO: fast algorithm for computation of the Duquenne-Guigues basis (2021). https://arxiv.org/abs/2011.04928

16. Kautz, H.A., Kearns, M.J., Selman, B.: Horn approximations of empirical data. Artif. Intell. **74**(1), 129–145 (1995)

17. Khardon, R.: Translating between Horn representations and their characteristic models. J. Artif. Intell. Res. (JAIR) **3**, 349–372 (1995)

18. Kuznetsov, S.: On the intractability of computing the Duquenne-Guigues base. J. Univ. Comput. Sci. **10**(8), 927–933 (2004)

19. Obiedkov, S., Duquenne, V.: Attribute-incremental construction of the canonical implication basis. Ann. Math. Artif. Intell. **49**(1), 77–99 (2007)

20. Obiedkov, S.: Learning implications from data and from queries. In: Cristea, D., Le Ber, F., Sertkaya, B. (eds.) ICFCA 2019. LNCS (LNAI), vol. 11511, pp. 32–44. Springer, Cham (2019). https://doi.org/10.1007/978-3-030-21462-3_3

21. Yarullin, R., Obiedkov, S.: From equivalence queries to PAC learning: the case of implication theories. Int. J. Approx. Reason. **127**, 1–16 (2020)

# Methods and Applications

# An Incremental Recomputation of From-Below Boolean Matrix Factorization

Martin Trnecka$^{(\boxtimes)}$ and Marketa Trneckova

Department of Computer Science, Palacký University Olomouc,
Olomouc, Czech Republic

**Abstract.** The Boolean matrix factorization (BMF) is a well-established and widely used tool for preprocessing and analyzing Boolean (binary, yes-no) data. In many situations, the set of factors is already computed, but some changes in the data occur after the computation, e.g., new entries to the input data are added. Recompute the factors from scratch after each small change in the data is inefficient. In the paper, we propose an incremental algorithm for (from-below) BMF which adjusts the already computed factorization according to the changes in the data. Moreover, we provide a comparison of the incremental and non-incremental algorithm on real-world data.

**Keywords:** Boolean matrix factorization · Incremental algorithm

## 1 Introduction

Boolean matrix factorization (BMF) is generally considered as a fundamental method, which, on the one hand, is closely connected to several basic problems such as factor enumeration, data dimensionality, theoretical understanding of relational data, and the like, and which, on the other hand, has many direct real-world applications, e.g., computational biology [17], recommendation systems [12], logic circuit synthesis [11], classification [1], formal concept analysis [9], tiling databases [10], role-based access control [7], and computer network analysis [13].

The aim of the BMF methods is to find factors (new variables) hidden in the data that can be used to provide a new fundamental view on the input data in a concise and presumably comprehensible way. In general, BMF is a set covering problem, i.e., the BMF methods cover the input data, more precisely the entries of the input data matrix, by factors which are represented by rectangular areas (or rectangles for short) in the data. Especially interesting is a class of BMF methods, which is also a subject of study of this paper, called from-below BMF, where the factors are represented via rectangles full of ones (factors cover only nonzero entries of the input data matrix) instead of general rectangles (factors cover nonzero and zeros entries of the input data matrix). A strong connection between from-below BMF and Formal concept analysis [9] was established in [4] and later extended in [2]. The main advantage of from-below methods is that

© Springer Nature Switzerland AG 2021
A. Braud et al. (Eds.): ICFCA 2021, LNAI 12733, pp. 125–137, 2021.
https://doi.org/10.1007/978-3-030-77867-5_8

these are able to achieve an exact factorization of the input data matrix, i.e., the factors covers exactly all input entries.

Very common situation is that the set of factors is already established and the data are changing over time. In such case a recomputation of all factors from scratch is very inefficient, especially when only a small change occurs.

There are many (general) BMF methods, e.g., [2–5,7,10,14,16,19,20]. Surprisingly, none of them is incremental.

In the paper, we formalize an incremental version of the BMF problem and we introduce an incremental algorithm for it. The algorithm adjusts the already computed factors according to the changes in the data. Moreover, we show later, in many cases the incremental algorithm can provide even better results than its non-incremental counterpart and significant speed up.

The rest of the paper is organized as follows. In the next section, we provide a brief overview of the notation used through the paper and a formalization of the incremental BMF problem. In Sect. 3 we describe our incremental algorithm. In Sect. 4 we experimentally evaluate the proposed algorithm and we provide its comparison with the state-of-the-art BMF algorithm GRECOND [4]. Section 5 summarizes the paper and outlines potential research directions.

## 2    Problem Description

Through the paper, we use a matrix terminology. Matrices are denoted by upper-case bold letters ($\mathbf{A}$). $\mathbf{A}_{ij}$ denotes the entry corresponding to the row $i$ and the column $j$ of $\mathbf{A}$. The $i$th row and $j$th column vector of $\mathbf{A}$ is denoted by $\mathbf{A}_{i\_}$ and $\mathbf{A}_{\_j}$, respectively. The set of all $m \times n$ Boolean (binary) matrices is denoted by $\{0,1\}^{m \times n}$.

If $\mathbf{A} \in \{0,1\}^{m \times n}$ and $\mathbf{B} \in \{0,1\}^{m \times n}$, we have the following element-wise matrix operations. The *Boolean sum* $\mathbf{A} \vee \mathbf{B}$ which is the matrix sum where $1 + 1 = 1$. The *Boolean element-wise product* $\mathbf{A} \wedge \mathbf{B}$ which is the element-wise matrix product. The *Boolean subtraction* $\mathbf{A} \ominus \mathbf{B}$ which is the matrix subtraction, where $0 - 1 = 0$.

If $\langle i, j \rangle$ is a pair of nonnegative integers, $\mathbf{A} \in \{0,1\}^{m \times n}$ and $i \leq m$ and $j \leq n$, we write $\langle i, j \rangle \in \mathbf{A}$ if $\mathbf{A}_{ij} = 1$ and $\langle i, j \rangle \notin \mathbf{A}$ otherwise. Operation $\mathbf{A} + \langle i, j \rangle$ sets the entry $\mathbf{A}_{ij}$ to 1. Similarly, the operation $\mathbf{A} - \langle i, j \rangle$ sets the entry $\mathbf{A}_{ij}$ to 0.

We use the same notation for Boolean vectors that are denoted by lower-case bold letters ($\mathbf{b}$). If $i$ is a nonnegative integer, $i \leq m$, $\mathbf{b} \in \{0,1\}^{1 \times m}$ is a Boolean vector, we write $i \in \mathbf{b}$ if the entry $\mathbf{b}_i = 1$ and $i \notin \mathbf{b}$ otherwise. Operation $\mathbf{b} + i$ sets the entry $\mathbf{b}_i$ to 1, and operation $\mathbf{b} - i$ sets the entry $\mathbf{b}_i$ to 0.

Two Boolean vectors $\mathbf{b} \in \{0,1\}^{1 \times m}$ and $\mathbf{c} \in \{0,1\}^{1 \times m}$ are equal if $\forall i \in \{1, \ldots, m\}, \mathbf{b}_i = \mathbf{c}_i$. Similarly, we define $\leq$ relation.

For a set $\mathcal{O}$ of $m$ objects and a set $\mathcal{A}$ of $n$ attributes, we identify the input data matrix as a matrix $\mathbf{A} \in \{0,1\}^{m \times n}$, i.e. rows and columns of $\mathbf{A}$ represent the objects and attributes respectively. The entry $\mathbf{A}_{ij} = 1$ if the object $i$ has the attribute $j$, otherwise $\mathbf{A}_{ij} = 0$, i.e. $\mathbf{A}$ captures a Boolean relation between $\mathcal{O}$ and $\mathcal{A}$.

For the input data matrix $\mathbf{A}$ one may construct function $\uparrow\colon 2^{\mathcal{O}} \to 2^{\mathcal{A}}$ and $\downarrow\colon 2^{\mathcal{A}} \to 2^{\mathcal{O}}$ assigning to sets $O \subseteq \mathcal{O}$ and $A \subseteq \mathcal{A}$ the sets: $O^{\uparrow} = \{j \in A \mid \forall i \in O, \mathbf{A}_{ij} = 1\}$, and $A^{\downarrow} = \{i \in \mathcal{O} \mid \forall j \in A, \mathbf{A}_{ij} = 1\}$. We represent these sets via their characteristic vectors, i.e., $O^{\uparrow} \in \{0,1\}^{1 \times n}$, where $O^{\uparrow}_j = 1$ if all objects in $O$ share attribute $j$, and $A^{\downarrow} \in \{0,1\}^{m \times 1}$, where $A^{\downarrow}_i = 1$ if object $i$ has all attributes in $A$.

## 2.1  Non-incremental From-Below BMF

The *basic from-below BMF problem* [4] is defined as follows.

*Problem 1.* For a given matrix $\mathbf{A} \in \{0,1\}^{m \times n}$ find matrices $\mathbf{C} \in \{0,1\}^{m \times k}$ (object-factor matrix) and $\mathbf{B} \in \{0,1\}^{k \times n}$ (factor-attribute matrix), with the number of factors $k$ as small as possible, for which $\mathbf{A} = \mathbf{C} \otimes \mathbf{B}$.

$\otimes$ denotes *Boolean matrix multiplication*, i.e. $(\mathbf{C} \otimes \mathbf{B})_{ij} = \max_{l=1}^{k} \min(\mathbf{C}_{il}, \mathbf{B}_{lj})$. An example of from-below BMF follows.

$$\mathbf{A} = \begin{pmatrix} 1\,1\,1\,0\,0 \\ 0\,0\,0\,1\,1 \\ 0\,0\,1\,1\,1 \\ 1\,1\,1\,0\,0 \\ 1\,1\,0\,0\,0 \end{pmatrix} = \begin{pmatrix} 1\,0\,1\,1 \\ 0\,1\,0\,0 \\ 0\,1\,0\,1 \\ 1\,0\,1\,1 \\ 0\,0\,1\,0 \end{pmatrix} \otimes \begin{pmatrix} 1\,1\,1\,0\,0 \\ 0\,0\,0\,1\,1 \\ 1\,1\,0\,0\,0 \\ 0\,0\,1\,0\,0 \end{pmatrix} = \mathbf{C} \otimes \mathbf{B}.$$

Note, the $\mathbf{A}$ can be expressed as the Boolean sum of particular factors, i.e. $\mathbf{A} = \bigvee_{i=1}^{k}(\mathbf{C}_{-i} \otimes \mathbf{B}_{i-})$, where $k$ is the number of factors.

$$\mathbf{A} = \begin{pmatrix} 1\,1\,1\,0\,0 \\ 0\,0\,0\,1\,1 \\ 0\,0\,1\,1\,1 \\ 1\,1\,1\,0\,0 \\ 1\,1\,0\,0\,0 \end{pmatrix} = \begin{pmatrix} \mathbf{1\,1\,1}\,0\,0 \\ 0\,0\,0\,0\,0 \\ 0\,0\,0\,0\,0 \\ \mathbf{1\,1\,1}\,0\,0 \\ 0\,0\,0\,0\,0 \end{pmatrix} \oplus \begin{pmatrix} 0\,0\,0\,0\,0 \\ 0\,0\,0\,\mathbf{1\,1} \\ 0\,0\,0\,\mathbf{1\,1} \\ 0\,0\,0\,0\,0 \\ 0\,0\,0\,0\,0 \end{pmatrix} \oplus \begin{pmatrix} \mathbf{1\,1}\,0\,0\,0 \\ 0\,0\,0\,0\,0 \\ 0\,0\,0\,0\,0 \\ \mathbf{1\,1}\,0\,0\,0 \\ \mathbf{1\,1}\,0\,0\,0 \end{pmatrix} \oplus \begin{pmatrix} 0\,0\,\mathbf{1}\,0\,0 \\ 0\,0\,0\,0\,0 \\ 0\,0\,\mathbf{1}\,0\,0 \\ 0\,0\,\mathbf{1}\,0\,0 \\ 0\,0\,0\,0\,0 \end{pmatrix}.$$

We say that the entry $\mathbf{A}_{ij}$ is assigned to the factor $l$ if $i \in \mathbf{C}_{-l}$ and $j \in \mathbf{B}_{l-}$.

## 2.2  Incremental From-Below BMF

We defined the incremental variant of the basic from-below BMF problem in the following way.

*Problem 2.* For given $\mathbf{A}$ and matrices $\mathbf{C}$ and $\mathbf{B}$, for which $\mathbf{C} \otimes \mathbf{B} = \mathbf{A}$, and $\langle i, j \rangle$, where $i \in \{1, \ldots, m\}$ and $j \in \{1, \ldots, n\}$ find matrices $\mathbf{C}'$ and $\mathbf{B}'$ for which $\mathbf{C}' \otimes \mathbf{B}' = \mathbf{A}'$, with the number of factors as small as possible, where $\mathbf{A}'$ is constructed from $\mathbf{A}$ as follows:

$$\mathbf{A}'_{pq} = \begin{cases} \mathbf{A}_{pq}, & \text{if } \langle p, q \rangle \neq \langle i, j \rangle \\ 0, & \text{if } \langle p, q \rangle = \langle i, j \rangle \text{ and } \langle p, q \rangle \in \mathbf{A} \\ 1, & \text{if } \langle p, q \rangle = \langle i, j \rangle \text{ and } \langle p, q \rangle \notin \mathbf{A} \end{cases}$$

The indexes $i$ and $j$ are restricted to the dimension of $\mathbf{A}$ only for the sake of simplicity. Problem 2 can be extended to the case, where $i$, $j$ or both are beyond the dimension of $\mathbf{A}$. For more details see the end of Sect. 3.2.

The BMF problem is computationally hard problem (NP-hard) [18] and the incremental version is no easier. Therefore, unless P=NP, an algorithm with polynomial time complexity that solves the BMF problem, or its incremental variant does not exist.

## 3    Incremental Algorithm

In what follows, we present an incremental heuristic algorithm that for a given input data matrix $\mathbf{A}$, object-factor and factor-attribute matrices $\mathbf{C}$ and $\mathbf{B}$ and $\langle i, j \rangle$, computes matrices $\mathbf{A}'$, $\mathbf{C}'$ and $\mathbf{B}'$. $\mathbf{A}'$ is $\mathbf{A}$ where the entry $\mathbf{A}_{ij}$ is changed (zero is set to one and vice versa), $\mathbf{C}'$ and $\mathbf{B}'$ are modification of $\mathbf{C}$ and $\mathbf{B}$ for which $\mathbf{C}' \otimes \mathbf{B}' = \mathbf{A}'$. The algorithm follows the search strategy of the state-of-the-art algorithm GRECOND [4] and it is divided into two parts.

### 3.1    Adding of New Entries

The first part (Algorithm 1) handles the case where the entry $\mathbf{A}_{ij} = 0$ is set to 1.

The algorithm iterates over the factors in matrices $\mathbf{C}'$ and $\mathbf{B}'$ (loop on lines 4–25). There are several cases that the algorithm analyses.

The first case (lines 5–12), handles the situation where the new entry $\langle i, j \rangle$ can be assigned to an already existing factor $l$. Namely, if all objects in factor $l$ have attribute $j$ (line 5), attribute $j$ is assigned to factor $l$, i.e. $l$-th row of matrix $\mathbf{B}'$ is modified ($\mathbf{B}'_{lj} = 1$). In such situation, we must check all remaining factors, if there is no other factor $p$, with the same set of attributes (lines 7–11). If such factor exists (line 8), there is only one and can be removed (all entries of the factor $p$ are now assigned to the factor $l$). To demonstrate how the first case works let us consider matrices $\mathbf{A}$, $\mathbf{C}$ and $\mathbf{B}$ depicted bellow:

$$\begin{pmatrix} 1\,1\,1 \\ 1\,1\,0 \\ 1\,0\,0 \end{pmatrix} = \begin{pmatrix} 1\,1\,1 \\ 1\,1\,0 \\ 1\,0\,0 \end{pmatrix} \otimes \begin{pmatrix} 1\,0\,0 \\ 1\,1\,0 \\ 1\,1\,1 \end{pmatrix}.$$

Changing the entry $\langle 2, 3 \rangle$ of $\mathbf{A}$, i.e. $\mathbf{A}' \leftarrow \mathbf{A} + \langle 2, 3 \rangle$, the second factor in matrices $\mathbf{C}'$ and $\mathbf{B}'$ meet the condition on line 5 so the entry $\mathbf{B}'_{23}$ is set to 1. The third factor is removed (loop on lines 7–11), since it has the same set of attributes. According to the first case, the new matrices $\mathbf{A}'$, $\mathbf{C}'$ and $\mathbf{B}'$ look as follows:

$$\begin{pmatrix} 1\,1\,1 \\ 1\,1\,1 \\ 1\,0\,0 \end{pmatrix} = \begin{pmatrix} 1\,1 \\ 1\,1 \\ 1\,0 \end{pmatrix} \otimes \begin{pmatrix} 1\,0\,0 \\ 1\,1\,1 \end{pmatrix}.$$

**Algorithm 1.** New entry adding

**Input:** Matrices $\mathbf{A} \in \{0,1\}^{m \times n}$, $\mathbf{C} \in \{0,1\}^{m \times k}$, $\mathbf{B} \in \{0,1\}^{k \times n}$, for which $\mathbf{C} \otimes \mathbf{B} = \mathbf{A}$,
  and $\langle i,j \rangle \notin \mathbf{A}, i \le m, j \le n$.
**Output:** Matrices $\mathbf{A}'$, $\mathbf{C}'$ and $\mathbf{B}'$ for which $\mathbf{C}' \otimes \mathbf{B}' = \mathbf{A}'$.

```
 1: A' ← A + ⟨i, j⟩, B' ← B, C' ← C
 2: c ← {0}^{m×1} + i
 3: d ← {0}^{1×n} + j
 4: for l = 1 to k do
 5:     if C'_l = d↓ then                                     # case 1
 6:         B'_l_ ← B'_l_ ∨ d
 7:         for p = 1, p ≠ l to k do
 8:             if B'_p_ = c↑ then
 9:                 remove factor p from C' and B'
10:             end if
11:         end for
12:     end if
13:     if B'_i_ = c↑ then                                    # case 2
14:         C'_l ← C'_l ∨ c
15:         for p = 1, p ≠ l to k do
16:             if C'_p = d↓ then
17:                 remove factor p from C' and B'
18:             end if
19:         end for
20:     end if
21:     if C'_l ≤ d↓ and B'_l_ ≤ c↑ then                     # case 3
22:         C'_l ← C'_l ∨ c
23:         B'_l_ ← B'_l_ ∨ d
24:     end if
25: end for
26: if no case occurs then
27:     add c as new column to C'
28:     add d as new row to B'
29: else
30:     remove redundant factors from C' and B'
31: end if
```

The second case (lines 13–20), handles a symmetric situation to the first case. Namely, if object $i$ has all attributes assigned to factor $l$ (line 13), then object $i$ is assigned to factor $l$. In this case, we must check all remaining factors, if there is no other factor $p$, to which all objects that share attribute $j$ are assigned (lines 15–19). Again, if such factor exists (line 16), there is only one and can be removed.

The third case (lines 21–24), handles the situation where both object $i$ and attribute $j$ extend factor $l$ (line 21). Namely, all objects in factor $l$ share attribute $j$ and all attributes in factor $l$ are attributes of object $i$.

We demonstrate how the third case works on the following matrices $\mathbf{A}$, $\mathbf{C}$ and $\mathbf{B}$:

$$\begin{pmatrix} 0\,1\,0 \\ 1\,0\,1 \\ 0\,1\,0 \end{pmatrix} = \begin{pmatrix} 1\,0 \\ 0\,1 \\ 1\,0 \end{pmatrix} \otimes \begin{pmatrix} 0\,1\,0 \\ 1\,0\,1 \end{pmatrix}.$$

Let $\mathbf{A}' \leftarrow \mathbf{A} + \langle 2, 2 \rangle$. Both the first and the second factor in matrices $\mathbf{C}'$ and $\mathbf{B}'$ meet the condition on line 21, i.e., entries $\mathbf{C}'_{21}, \mathbf{C}'_{22}, \mathbf{B}'_{12}$ and $\mathbf{B}'_{22}$ are set to 1. New matrices follows:

$$\begin{pmatrix} 0\,1\,0 \\ 1\,1\,1 \\ 0\,1\,0 \end{pmatrix} = \begin{pmatrix} 1\,0 \\ 1\,1 \\ 1\,0 \end{pmatrix} \otimes \begin{pmatrix} 0\,1\,0 \\ 1\,1\,1 \end{pmatrix}.$$

Unless one of the above cases occurs, a new factor including $\mathbf{A}_{ij}$ is added (lines 26–29).

During an execution of case 1, 2 and 3 a factor that is no longer necessary may occur. More precisely, all entries assigned to the factor are also assigned to some other existing factors. If one of the cases occurs the remove redundant factors procedure (Algorithm 3) is executed. The procedure iterates over all factors and tries to remove factors whose all entries are assigned to some other factor. For each factor $l$ an internal copy (stored in $\mathbf{D}$) of $\mathbf{A}$ is initialized (line 2). Loop over all remaining factors (lines 3–11) removes all entries that are assigned to these factors from $\mathbf{D}$. If all entries assigned to factor $l$ are not in $\mathbf{D}$, then factor $l$ can be removed (lines 7–10).

For example, in the following matrices, the second factor can be removed because all entries assigned to the factor are also assigned to some other factors, namely to the first and the third factor.

$$\begin{pmatrix} 1\,1\,0 \\ 1\,1\,1 \\ 0\,1\,1 \end{pmatrix} = \begin{pmatrix} 1\,0\,0 \\ 1\,1\,1 \\ 0\,0\,1 \end{pmatrix} \otimes \begin{pmatrix} 1\,1\,0 \\ 1\,1\,1 \\ 0\,1\,1 \end{pmatrix}.$$

## 3.2  Removing of Existing Entries

The second part of the algorithm (Algorithm 2) handles the case where the entry $\mathbf{A}_{ij} = 1$ is set to 0. This part of the algorithm is a little easier than the first part and follows the below-described logic.

The algorithm iterates over the factors in matrices $\mathbf{C}'$ and $\mathbf{B}'$ (loop on lines 4–21). There are several cases that the algorithm analyses.

The first case (lines 5–7), removes a factor including only object $i$ and attribute $j$.

The second case (lines 8–10), removes an object from the factor $l$ if a particular factor contains attribute $j$ only. Similarly, the third case (lines 11–13) removes the attribute $j$ from factor $l$ if only object $i$ is assigned to factor $l$.

The fourth case (lines 14–20), handles the situation, where the factor to which the removed entry is assigned splits into two parts. Namely, if the entry $\langle i, j \rangle$ is assigned to factor $l$ (line 14), then the algorithm creates vector $\mathbf{b}$ including

---

**Algorithm 2.** Existing entry removing

---

**Input:** Matrices $\mathbf{A} \in \{0,1\}^{m \times n}$, $\mathbf{C} \in \{0,1\}^{m \times k}$, $\mathbf{B} \in \{0,1\}^{k \times n}$, for which $\mathbf{C} \otimes \mathbf{B} = \mathbf{A}$, and $\langle i, j \rangle \in \mathbf{A}$.
**Output:** Matrices $\mathbf{C}'$ and $\mathbf{B}'$ for which $\mathbf{C}' \otimes \mathbf{B}' = \mathbf{A}'$.

1: $\mathbf{A}' \leftarrow \mathbf{A} - \langle i, j \rangle, \mathbf{B}' \leftarrow \mathbf{B}, \mathbf{C}' \leftarrow \mathbf{C}$
2: $\mathbf{c} \leftarrow \{0\}^{m \times 1} + i$
3: $\mathbf{d} \leftarrow \{0\}^{1 \times n} + j$
4: **for** $l = 1$ **to** $k$ **do**
5:    **if** $\mathbf{C}'_{\_l} = \mathbf{c}$ **and** $\mathbf{B}'_{l\_} = \mathbf{d}$ **then**             # case 1
6:       remove factor $l$ from $\mathbf{C}'$ and $\mathbf{B}'$
7:    **end if**
8:    **if** $\mathbf{B}'_{l\_} = \mathbf{d}$ **then**                      # case 2
9:       $\mathbf{C}' \leftarrow \mathbf{C}' - \langle i, l \rangle$
10:   **end if**
11:   **if** $\mathbf{C}'_{\_l} = \mathbf{c}$ **then**                   # case 3
12:      $\mathbf{B}' \leftarrow \mathbf{B}' - \langle l, j \rangle$
13:   **end if**
14:   **if** $\langle i, l \rangle \in \mathbf{C}'$ **and** $\langle l, j \rangle \in \mathbf{B}'$ **then**    # case 4
15:      $\mathbf{b} \leftarrow \mathbf{B}'_{l\_} - j$
16:      $\mathbf{a} \leftarrow \mathbf{c}^{\uparrow l}$
17:      $\mathbf{C}' \leftarrow \mathbf{C}' - \langle i, l \rangle$
18:      add $\mathbf{a}$ as new column to $\mathbf{C}'$
19:      add $\mathbf{b}$ as new row to $\mathbf{B}'$
20:   **end if**
21: **end for**
22: remove redundant factors from $\mathbf{C}'$ and $\mathbf{B}'$

---

the same attributes assigned to factor $l$ with an exception of attribute $j$ (line 15). Then the algorithm computes all objects that share the same attributes as object $i$ (vector $\mathbf{a}$). Vector $\mathbf{a}$ and $\mathbf{b}$ form a new factor which is added to matrices $\mathbf{C}'$ and $\mathbf{B}'$ (lines 18–19). From the original factor $l$ object $i$ is removed (line 17). Let us consider matrices $\mathbf{A}$, $\mathbf{C}$ and $\mathbf{B}$ depicted bellow:

$$\begin{pmatrix} 1 & 1 & 1 \\ 1 & 1 & 1 \\ 1 & 0 & 0 \end{pmatrix} = \begin{pmatrix} 1 & 1 \\ 1 & 1 \\ 1 & 0 \end{pmatrix} \otimes \begin{pmatrix} 1 & 0 & 0 \\ 1 & 1 & 1 \end{pmatrix}.$$

Let $\mathbf{A}' \leftarrow \mathbf{A} - \langle 2, 3 \rangle$. The second factor in matrices $\mathbf{C}'$ and $\mathbf{B}'$ meet the condition on line 14. The vector $\mathbf{b}$ is set to $(1\ 1\ 0)$ and added as a new row to $\mathbf{B}'$, and the vector $\mathbf{a}$ is set to transposed vector $(1\ 1\ 0)$ and added as a new column to $\mathbf{C}'$. The entry $\mathbf{C}'_{22}$ is set to 0. According to the fourth case of Algorithm 2, the new matrices $\mathbf{A}'$, $\mathbf{C}'$ and $\mathbf{B}'$ look as follows.

$$\begin{pmatrix} 1 & 1 & 1 \\ 1 & 1 & 0 \\ 1 & 0 & 0 \end{pmatrix} = \begin{pmatrix} 1 & 1 & 1 \\ 1 & 0 & 1 \\ 1 & 0 & 0 \end{pmatrix} \otimes \begin{pmatrix} 1 & 0 & 0 \\ 1 & 1 & 1 \\ 1 & 1 & 0 \end{pmatrix}.$$

In each call of Algorithm 2 the remove redundant factors procedure is executed (line 22).

---

**Algorithm 3.** Remove redundant factors

---

**Input:** Matrices $\mathbf{A} \in \{0,1\}^{m \times n}$, $\mathbf{C} \in \{0,1\}^{m \times k}$, $\mathbf{B} \in \{0,1\}^{k \times n}$, for which $\mathbf{C} \otimes \mathbf{B} = \mathbf{A}$.
**Output:** Matrices $\mathbf{C}'$ and $\mathbf{B}'$ for which $\mathbf{C}' \otimes \mathbf{B}' = \mathbf{A}$.

1: **for** $l = k$ **to** 1 **do**
2:     $\mathbf{D} \leftarrow \mathbf{A}, \mathbf{C}' \leftarrow \mathbf{C}, \mathbf{B}' \leftarrow \mathbf{B}$
3:     **for** $p = k, p \neq l$ **to** 1 **do**
4:         **for all** $\langle i,j \rangle, i \in \mathbf{C}'_{\_p}, j \in \mathbf{B}'_{p\_}$ **do**
5:             $\mathbf{D} \leftarrow \mathbf{D} - \langle i,j \rangle$
6:         **end for**
7:         **if** $\forall \langle i,j \rangle$, with $i \in \mathbf{C}'_{\_l}$ and $j \in \mathbf{B}'_{l\_}$, $\langle i,j \rangle \notin \mathbf{D}$ **then**
8:             remove factor $l$ from $\mathbf{C}'$ and $\mathbf{B}'$
9:             **break**
10:        **end if**
11:    **end for**
12: **end for**

---

As we referred in Sect. 2.2 Problem 2 can be easily extended to the case where $i > m$, $j > n$ or both. In this case the dimensions of $\mathbf{A}'$, $\mathbf{C}'$ and $\mathbf{B}'$ must be enlarged. All newly added entries of the matrices are set to 0. After this step, the algorithm can be applied as it is.

### 3.3   Time Complexity

We now discuss an upper bound of the worst case time complexity of the above-described incremental algorithm. Time complexity of both Algorithm 1 and Algorithm 2 is mostly affected by removing redundant factors which is $\mathcal{O}(k^2 mn)$. Each of the cases described in Sect. 3 can be done in $\mathcal{O}(mn)$.

Note, $k$ is usually much smaller than both $m$ and $n$, $m$ is smaller than $n$. Therefore, the observed time complexity is significantly better. Moreover, the time complexity of our algorithm is much better than the time complexity of an usual BMF algorithm. More details will be given in Sect. 4.4.

## 4   Experimental Evaluation

In the following section, we provide an experimental comparison of the incremental and the non-incremental algorithm. To make the comparison fair, we choose as the baseline algorithm the GRECOND whose search strategy is implemented in our incremental algorithm.

## 4.1 Datasets

We used various real-world datasets, namely, breast [6], dblp [15], iris [6], paleo [8], page-blocks [6], post [6], zoo [6]. The basic characteristics of the datasets are displayed in Table 1. Namely, the number of objects, the number of attributes, the number of factors computed via the non-incremental algorithm and the number of non-zero entries.

**Table 1.** Characteristic of real-world datasets.

| Dataset | No. of objects | No. of attributes | No. of factors | No. of entries |
|---|---|---|---|---|
| breast | 699 | 20 | 19 | 6974 |
| dblp | 6980 | 19 | 21 | 17173 |
| iris | 150 | 19 | 20 | 750 |
| paleo | 501 | 139 | 151 | 3537 |
| page-blocks | 5473 | 46 | 48 | 60203 |
| post | 90 | 20 | 29 | 720 |
| zoo | 101 | 28 | 30 | 862 |

## 4.2 Adding of New Entries

To simulate an addition of new entries, we randomly selected a prescribed number of non-zero entries, namely, 1%, 2%, 3%, 4%, and 5% and set them to 0. After this, we computed the initial set of factors via the non-incremental algorithm. Then we gradually set (in random order) the selected entries back to the original value and computed the new set of factors (after each change) via the incremental algorithm. This is repeated until the original data are achieved. The described procedure enables a direct comparison with the non-incremental algorithm.

Table 2 summarizes the average number of factors over 10 iterations as well as the standard deviation (the numbers after ± symbol). We observe that the incremental algorithm produces comparable results to these produced by the non-incremental algorithm (see the column "factors" in Table 1) especially if the number of changed entries is small. Moreover, in some cases (dblp, iris, paleo, post) the incremental algorithm slightly outperforms the non-incremental one.

## 4.3 Removing of Existing Entries

Table 3 shows the results from an experiment, where a prescribed number of non-zero entries, namely, 1%, 2%, 3%, 4%, and 5% is gradually (in random order) set to 0. In this case, a direct comparison of incremental and non-incremental algorithm is not possible. For this reason, we report in Table 3 also the results

**Table 2.** Adding of new entries. The average number of factors over 10 iterations and standard deviation are presented. Bold marks the cases where the number of factor is smaller than the number of factor obtained via the non-incremental algorithm column "factors".

| Dataset | No. of factors | Number of added entries | | | | |
|---|---|---|---|---|---|---|
| | | 1% | 2% | 3% | 4% | 5% |
| breast | 19 | 23.70 ± 1.05 | 24.60 ± 1.26 | 25.40 ± 1.42 | 27.10 ± 0.99 | 27.30 ± 1.41 |
| dblp | 21 | **19.40 ± 0.51** | **19.20 ± 0.42** | **19.10 ± 0.31** | **19.20 ± 0.42** | **19.30 ± 0.48** |
| iris | 20 | **18.20 ± 0.63** | **18.70 ± 1.70** | **18.20 ± 1.31** | **19.10 ± 1.44** | **19.40 ± 1.64** |
| paleo | 151 | **139.20 ± 0.42** | **139.10 ± 0.31** | **139.10 ± 0.31** | **139.00 ± 0.00** | **139.10 ± 0.31** |
| page-blocks | 48 | 50.70 ± 1.56 | 52.30 ± 2.16 | 55.20 ± 1.93 | 57.80 ± 2.74 | 56.30± 1.63 |
| post | 29 | **26.70 ± 0.94** | **28.00 ± 1.24** | **27.10 ± 2.46** | **27.40 ± 1.83** | **28.00 ± 1.97** |
| zoo | 30 | **29.50 ± 1.26** | 30.90 ± 0.99 | 30.60 ± 2.06 | 31.80 ± 1.39 | 31.10 ± 1.85 |

obtained via the non-incremental algorithm. The initial set of factors for the incremental algorithm was computed from the original data.

Table 3 shows a difference in the number of computed factors. One may clearly see that the incremental algorithm outperforms the non-incremental algorithm in almost all cases.

Overall, the incremental algorithm is very competitive with the non-incremental one, especially in the cases where small changes in the data are made. The main benefit of any incremental algorithm is the running time. We discuss this in the following section.

## 4.4   Running Times

Run-time of our incremental algorithm is very low, despite the slow implementation in MATLAB. In the evaluation, we used an implementation of the non-incremental algorithm in C language with a high level of optimization. We compared the running times of the non-incremental and the incremental algorithm on an ordinary PC with Intel Core i7-3520M 2.9 GHz and 8 GB RAM.

Table 4 shows the time required for the computation of factors in the case where 1 % of randomly chosen entries in the input data are flipped, i.e., ones to zeros and vice versa. We measured the sum of times required for achieving the final factorization. Namely, we flip one entry in the data and compute the

**Table 3.** Removing of existing entries. The average number of factors over 10 iterations and standard deviation are presented. Bold marks the better result. The incremental algorithm produces (sometimes significantly) smaller number of factors than the non-incremental algorithm.

| Dataset | Algorithm | Number of removed entries | | | | |
|---|---|---|---|---|---|---|
| | | 1% | 2% | 3% | 4% | 5% |
| breast | Inc. | **17.30± 0.67** | **17.30 ± 0.99** | **17.40 ± 0.84** | **17.50 ± 0.84** | **18.00 ± 0.81** |
| | Non-inc. | 19.30 ± 1.33 | 18.80 ± 1.22 | 18.10 ± 1.37 | 18.60 ± 1.83 | 19.40 ± 0.96 |
| dblp | Inc. | **18.00 ± 0.00** | **18.00 ± 0.00** | **18.00 ± 0.00** | **18.00 ± 0.00** | **18.00 ± 0.00** |
| | Non-inc. | 20.00 ± 0.00 | 20.00 ± 0.00 | 20.00 ± 0.00 | 20.00 ± 0.00 | 20.00 ± 0.00 |
| iris | Inc. | **18.00 ± 0.81** | **18.30 ± 0.67** | **18.50 ± 1.08** | **18.50 ± 0.84** | **18.20 ± 0.78** |
| | Non-inc. | 18.80 ± 0.63 | 19.00 ± 0.47 | 19.20 ± 0.63 | 19.20 ± 0.42 | 18.80 ± 0.63 |
| paleo | Inc. | **140.00 ± 0.00** | **138.90 ± 0.31** | **136.60 ± 0.84** | **135.80 ± 0.46** | **134.90 ± 0.31** |
| | Non-inc. | 148.70 ± 0.67 | 147.40 ± 0.96 | 144.30 ± 1.33 | 142.60 ± 1.57 | 142.60 ± 1.34 |
| page-blocks | Inc. | **46.00 ± 0.47** | **46.50 ± 0.52** | **45.50 ± 0.70** | **45.00 ± 0.94** | **45.50 ± 0.71** |
| | Non-inc. | 49.30 ± 0.94 | 49.70 ± 1.33 | 47.70 ± 1.56 | 47.60 ± 1.34 | 48.70 ± 0.95 |
| post | Inc. | **24.60 ± 1.26** | **24.40 ± 1.17** | **25.00 ± 1.41** | **24.10 ± 0.31** | **22.90 ± 0.56** |
| | Non-inc. | 26.20 ± 0.91 | 26.50 ± 1.35 | 26.40 ± 1.07 | 25.60 ± 1.34 | 25.70 ± 2.49 |
| zoo | Inc. | **29.10 ± 0.56** | **29.10 ± 0.73** | **29.20 ± 0.42** | 28.40 ± 0.699 | 28.40 ± 0.84 |
| | Non-inc. | 29.40 ± 0.96 | 29.20 ± 1.39 | 29.20 ± 0.78 | **28.10 ± 1.59** | **27.70 ± 0.82** |

factorization (from scratch) via the non-incremental algorithm and via our incremental algorithm (from the factorization established in the previous step).

The incremental algorithm is significantly faster than the non-incremental one, although the implementations are considerably unbalanced to the detriment of the incremental algorithm.

**Table 4.** Comparison of running times of the incremental and non-incremental algorithm in seconds for 1 % of randomly flipped entries. The average running times over 10 iterations and standard deviation are presented. The incremental algorithm is significantly faster than the non-incremental one.

| Dataset | Non-inc. | Inc. |
|---|---|---|
| breast | 6.7268 ± 1.244 | **0.0063 ± 0.0008** |
| dblp | 17.926 ± 3.862 | **0.1918 ± 0.0095** |
| iris | 0.6759 ± 0.1254 | **0.0017 ± 0.0002** |
| paleo | 3.9879 ± 0.2483 | **0.0190 ± 0.0063** |
| page-block | 118.3173 ± 13.5330 | **0.4768 ± 0.1254** |
| post | 0.6513 ± 0.0124 | **0.0012 ± 0.0001** |
| zoo | 0.7424 ± 0.0372 | **0.0006 ± 0.0000** |

## 5 Conclusions

In the paper, we formulated the incremental from-below BMF problem reflecting a realistic scenario of BMF using and we presented an algorithm for them. Our experimental evaluation shows that the incremental algorithm is not only significantly faster, but in some cases delivers even better results, i.e., a smaller number of factors, than its non-incremental counterpart.

Future research shall include the following issues. Efficient implementation (in C language) of the incremental algorithm, its parallelization and application on the big data. Additionally, in many cases, the incremental algorithm is affected by the quality of the initial set of factors. It seems to be promising to create a special method for calculating the set, which will take into account the future incremental computation.

**Acknowledgment.** The paper was supported by the grant JG 2020 of Palacký University Olomouc, No. JG_2020_003. Support by Grant No. IGA_PrF_2020_019 and No. IGA_PrF_2021_022 of IGA of Palacký University are also acknowledged. The authors would like to thank Jan Outrata for providing an efficient implementation of the non-incremental algorithm.

## References

1. Belohlavek, R., Grissa, D., Guillaume, S., Mephu Nguifo, E., Outrata, J.: Boolean factors as a means of clustering of interestingness measures of association rules. Ann. Math. Artif. Intell. **70**(1), 151–184 (2013). https://doi.org/10.1007/s10472-013-9370-x
2. Belohlavek, R., Trnecka, M.: From-below approximations in Boolean matrix factorization: geometry and new algorithm. J. Comput. Syst. Sci. **81**(8), 1678–1697 (2015). https://doi.org/10.1016/j.jcss.2015.06.002
3. Belohlavek, R., Trnecka, M.: A new algorithm for Boolean matrix factorization which admits overcovering. Discrete Appl. Math. **249**, 36–52 (2018). https://doi.org/10.1016/j.dam.2017.12.044

4. Belohlavek, R., Vychodil, V.: Discovery of optimal factors in binary data via a novel method of matrix decomposition. J. Comput. Syst. Sci. **76**(1), 3–20 (2010). https://doi.org/10.1016/j.jcss.2009.05.002
5. Claudio, L., Salvatore, O., Raffaele, P.: A unifying framework for mining approximate top-k binary patterns. IEEE Trans. Knowl. Data Eng. **26**(12), 2900–2913 (2014). https://doi.org/10.1109/TKDE.2013.181
6. Dua, D., Graff, C.: UCI machine learning repository (2017). http://archive.ics.uci.edu/ml
7. Ene, A., Horne, W.G., Milosavljevic, N., Rao, P., Schreiber, R., Tarjan, R.E.: Fast exact and heuristic methods for role minimization problems. In: Ray, I., Li, N. (eds.) Proceedings of the 13th ACM Symposium on Access Control Models and Technologies, SACMAT 2008, Estes Park, CO, USA, 11–13 June 2008, pp. 1–10. ACM (2008). https://doi.org/10.1145/1377836.1377838
8. Fortelius, M., et al.: Neogene of the old world database of fossil mammals (now) (2003). http://www.helsinki.fi/science/now
9. Ganter, B., Wille, R.: Formal Concept Analysis. Springer, Heidelberg (1999). https://doi.org/10.1007/978-3-642-59830-2
10. Geerts, F., Goethals, B., Mielikäinen, T.: Tiling databases. In: Suzuki, E., Arikawa, S. (eds.) DS 2004. LNCS (LNAI), vol. 3245, pp. 278–289. Springer, Heidelberg (2004). https://doi.org/10.1007/978-3-540-30214-8_22
11. Hashemi, S., Tann, H., Reda, S.: Approximate logic synthesis using Boolean matrix factorization. In: Reda, S., Shafique, M. (eds.) Approximate Circuits, pp. 141–154. Springer, Cham (2019). https://doi.org/10.1007/978-3-319-99322-5_7
12. Ignatov, D.I., Nenova, E., Konstantinova, N., Konstantinov, A.V.: Boolean matrix factorisation for collaborative filtering: an FCA-based approach. In: Agre, G., Hitzler, P., Krisnadhi, A.A., Kuznetsov, S.O. (eds.) AIMSA 2014. LNCS (LNAI), vol. 8722, pp. 47–58. Springer, Cham (2014). https://doi.org/10.1007/978-3-319-10554-3_5
13. Kocayusufoglu, F., Hoang, M.X., Singh, A.K.: Summarizing network processes with network-constrained Boolean matrix factorization. In: IEEE International Conference on Data Mining, ICDM 2018, Singapore, 17–20 November 2018, pp. 237–246. IEEE Computer Society (2018). https://doi.org/10.1109/ICDM.2018.00039
14. Lucchese, C., Orlando, S., Perego, R.: Mining top-k patterns from binary datasets in presence of noise. In: Proceedings of the SIAM International Conference on Data Mining, SDM 2010, Columbus, Ohio, USA, 29 April–1 May 2010, pp. 165–176. SIAM (2010). https://doi.org/10.1137/1.9781611972801.15
15. Miettinen, P.: Matrix decomposition methods for data mining: computational complexity and algorithms (2009)
16. Miettinen, P., Mielikäinen, T., Gionis, A., Das, G., Mannila, H.: The discrete basis problem. IEEE Trans. Knowl. Data Eng. **20**(10), 1348–1362 (2008). https://doi.org/10.1109/TKDE.2008.53
17. Nau, D.S., Markowsky, G., Woodbury, M.A., Amos, D.B.: A mathematical analysis of human leukocyte antigen serology. Math. Biosci. **40**(3–4), 243–270 (1978)
18. Stockmeyer, L.J.: The Set Basis Problem is NP-complete. IBM Thomas J. Watson Research Division, Research reports (1975)
19. Trnecka, M., Trneckova, M.: Data reduction for Boolean matrix factorization algorithms based on formal concept analysis. Knowl. Based Syst. **158**, 75–80 (2018). https://doi.org/10.1016/j.knosys.2018.05.035
20. Xiang, Y., Jin, R., Fuhry, D., Dragan, F.F.: Summarizing transactional databases with overlapped hyperrectangles. Data Min. Knowl. Discov. **23**(2), 215–251 (2011). https://doi.org/10.1007/s10618-010-0203-9

# Clustering and Identification of Core Implications

Domingo López-Rodríguez$^{(\boxtimes)}$ , Pablo Cordero, Manuel Enciso,
and Ángel Mora

Universidad de Málaga, Málaga, Spain
{dominlopez,pcordero,enciso,amora}@uma.es

**Abstract.** FCA exhaustively uses the notion of cluster by grouping attributes and objects and providing a solid algebraic structure to them through the concept lattice. Our proposal explores how we can cluster implications. This work opens a research line to study the knowledge inside the clusters computed from the Duquenne-Guigues basis. Some alternative measures to induce the clusters are analysed, taking into account the information that directly appears in the *appearance* and the *semantics* of the implications. This work also allows us to show the fcaR package, which has the main methods of FCA and the Simplification Logic. The paper ends with a motivation of the potential applications of performing clustering on the implications.

## 1 Introduction

Formal Concept Analysis (FCA) has established itself at the theoretical level and is increasingly used in real-life problems [6,7,33]. Our community explores how to solve real problems in data science, machine learning, social network analysis, etc. Solving problems from these areas and developing new tools could be a way to open a window to researchers outside FCA.

Since the early eighties, when R. Wille and B. Ganter [16] developed Formal Concept Analysis, the community has been growing. The interest in the use of this well-founded tool has increased considerably. The continuous development of the theoretical foundations and generalisations of the classical framework [3,4,13,24,28,30] and the enthusiasm of how to put in practice this progress [6–8,18,33] have formed a solid community formally linked. However, as U. Priss mentioned in [32], "FCA is widely unknown among information scientists in the USA even though this technology has a significant potential for applications". The community recognises that it is necessary an additional effort and perhaps new tools to make FCA more appealing. Books about machine learning, big data, and data science, in general, have not included anything about FCA, notwithstanding its powerful knowledge and its considerable potential in applications.

Supported by Grants TIN2017-89023-P, UMA2018-FEDERJA-001 and PGC2018-095869-B-I00 of the Junta de Andalucia, and European Social Fund.

A. Braud et al. (Eds.): ICFCA 2021, LNAI 12733, pp. 138–154, 2021.
https://doi.org/10.1007/978-3-030-77867-5_9

Developing new tools and libraries could be a valuable resource to open our community. In this work, we present a library in the R language, named fcaR, which implements of the most popular methods and algorithms in FCA. Here we show how to extract interesting knowledge by computing implication clusters from the Duquenne-Guigues basis to illustrate the benefits of such a tool. fcaR vertebrates this proposal by introducing some code throughout the paper.

FCA is firmly based on the (bi)clustering of attributes and objects. The Concept Lattice provides a formal structure of these clusters. Nevertheless, the application of clustering to the set of implications has not been explored as far as we know. In the following section, we briefly survey the relationship between Clustering and FCA.

The first step in this paper is to propose several dissimilarity measures between implications to cluster them in different ways. In some of these measures, to compute the distance matrix between implications, we will use Simplification Logic and its attribute closure operator, included in the package fcaR. From this distance matrix, clusters of implications arise representing new knowledge. A $K$-medoid algorithm, specifically the PAM algorithm [23], is used to compute each cluster's central implications and further generate the clusters of implications.

Since our starting point is the Duquenne-Guigues basis, we analyse the different dissimilarity measures taking into account the pseudointents, the right-hand side of the implications, and the closed sets computed from the pseudointents. We end the paper with an experiment result and drawing up some potential applications from the implication clustering.

The rest of the paper is organised as follows: in Sect. 2, we analyse how clustering is used in FCA in the literature. The central notions of FCA and the fcaR package are briefly outlined in Sect. 3. Section 4 shows the new research line proposed in this work, along with its formulations and possible developments, also defining the idea of implication dissimilarity in terms of distance functions and how implication clustering is related to cluster pseudointents and their closures. This proposal's promising result is shown in Sect. 5, developing an experiment centred on a dataset well-known in the machine learning community. Finally, Sect. 6 presents some conclusions and future works.

## 2  Previous Works on FCA and Clustering

FCA carries out clustering of objects by itself. However, it is well-known that the size of the concept lattice is possibly exponential with respect to the size of the formal context, even for a small context. Diatta in [14] ensured that pattern concepts [17] coincide with clusters associated with dissimilarity measures.

Beyond that, techniques based on clustering have been explored to group the closest concepts. For instance, in [29], Melo *et al.* presented a tool to apply visual analytics to cluster concepts using a $K$-means algorithm [27] to identify clusters. Bocharov *et al.* [5] group the objects by the $K$-means algorithm and propose modifying the Close-by-One algorithm for consensus clustering to reduce the concept lattice. In [38], the authors compute attribute clusters using similarity and dissimilarity functions to reduce the concept lattice.

Stumme *et al.* [37] proposed Iceberg lattices as a clustering method of the original lattice to reduce computing bases of association rules and their visualisation. Kumar in [26] uses clustering to reduce the formal context and, therefore, the number of association rules extracted from it.

Other authors have used FCA in a variety of approaches to detect objects with similar properties. In [22], triclustering, based on FCA, was developed to detect groups of objects with similar properties under similar conditions and used it to develop recommender systems. Cigarrán *et al.* have some interesting works [7,9] applying FCA in Social Network Analysis to detect topics in Twitter. These authors proved that FCA could solve real problems with better results than classical techniques, generating clusters of topics less subject to cluster granularity changes.

Other works deal with the idea of clustering association rules (in transactional databases) to reduce the number of rules extracted [2,19,36]. These works do not define rule dissimilarity as a function of each rule's terms (items). Instead, they define the dissimilarity between rules in terms of the sets of transactions supporting each rule. Thus, the knowledge present in the rule clustering is explicitly related to the database and cannot be abstracted from it.

## 3   Background and the `fcaR` package

Over the years, U. Priss has collected a list of the main FCA-based tools on its website https://www.upriss.org.uk/fca/fca.html. We emphasise that the most used for FCA are ConExp, ToscanaJ, Galicia, FcaStone, and some libraries developed in C, Python, etc. In this work, we take the opportunity to present the `fcaR` package[1] as a valuable tool to solve real problems and bring FCA closer to other communities.

In the following, we briefly summarise the main concepts in Formal Concept Analysis (FCA) we need for this work, showing with a running example how the `fcaR` package is used. For more detailed reading about FCA, see [18].

**Definition 1 (Formal Context).** *A formal context is a triplet* $\mathbb{K} := \langle G, M, I \rangle$ *where $G$ and $M$ are non-empty finite sets and $I \subseteq G \times M$ is a binary relation between $G$ and $M$.*

The elements in $G$ and $M$ are named objects and attributes, respectively. In addition, $(g, m) \in I$ is read as the object $g$ has the attribute $m$.

*Example 1.* We consider this example appearing in [18] where $G$ is the set of planets and $M$ the set of some properties of these planets (Table 1).

In the R language, we will use the following to introduce this matrix with the name `planets` in the `fc_planets` formal context object. The sets $G$ and

---

[1] As far as we know, no package using the R language has been developed and published in CRAN repository for FCA, even when the R language together with Python are considered the main languages in data science, machine learning, big data, etc. To this date, `fcaR` has more than 8,000 downloads.

**Table 1.** Properties of the planets of the solar system.

|         | Small | Medium | Large | Near | Far | Moon | No_moon |
|---------|-------|--------|-------|------|-----|------|---------|
| Mercury | ×     |        |       | ×    |     |      | ×       |
| Venus   | ×     |        |       | ×    |     |      | ×       |
| Earth   | ×     |        |       | ×    |     | ×    |         |
| Mars    | ×     |        |       | ×    |     | ×    |         |
| Jupiter |       |        | ×     |      | ×   | ×    |         |
| Saturn  |       |        | ×     |      | ×   | ×    |         |
| Uranus  |       | ×      |       |      | ×   | ×    |         |
| Neptune |       | ×      |       |      | ×   | ×    |         |
| Pluto   | ×     |        |       |      | ×   | ×    |         |

$M$, and also subsequently computed concepts and implications, are stored inside this formal context object[2].

```
> library(fcaR)
> fc_planets <- FormalContext$new(planets)
> fc_planets$attributes
[1] "small"   "medium"  "large"   "near"   "far"    "moon"   "no_moon"
> fc_planets$objects
[1] "Mercury" "Venus" "Earth" "Mars" "Jupiter" "Saturn" "Uranus"
"Neptune" "Pluto"
```

Each formal context $\mathbb{K}$ defines two derivation operators, which form a Galois connection between $\langle 2^G, \subseteq \rangle$ and $\langle 2^M, \subseteq \rangle$. They are the following:

$$(-)' : 2^G \to 2^M \text{ where } A' = \{m \in M \mid (g, m) \in I \text{ for all } g \in A\}.$$
$$(-)' : 2^M \to 2^G \text{ where } B' = \{g \in G \mid (g, m) \in I \text{ for all } m \in B\}.$$

*Example 2.* In fcaR we use sparse matrices and sparse sets to provide efficiency to the algorithms, then to use an object variable or an attribute variable, first we create a new sparse variable (SparseSet$new method), and then we assign the value 1 (variable$assign method). To compute intent and extent in R language, with the planets example, we will do the following:

```
> # The planets  are stored in a vector
> myPlanets <- c("Earth","Mars")
> # A new sparse object variable is created
> mySparsePlanets <- SparseSet$new(attributes = fc_planets$objects)
> # Assigning to myPlanets the value 1 in the variable sparse
> mySparsePlanets$assign(myPlanets,values = 1)
> # The content of the sparse variable is
> mySparsePlanets
```

---

[2] In this work, we do not use all the methods in the fcaR package to manage the formal context, the concept lattice, the concepts, the implications, etc. See https:// neuroimaginador.github.io/fcaR/ for more details.

```
{Earth, Mars}
> fc_planets$intent(mySparsePlanets) # Computing the intent
{small, near, moon}

# In a similar way for attributes
> myAttributes <- c("medium", "far","moon")
> mySparseAttributes <- SparseSet$new(attributes = fc_planets$attributes)
> mySparseAttributes$assign(myAttributes,values = 1)
> mySparseAttributes
{medium, far, moon}
> fc_planets$extent(mySparseAttributes)
{Uranus, Neptune}
```

The main aim of this area is to extract knowledge from the context allowing to reason. One of the ways to represent knowledge is utilising the concept lattice. Another equivalent alternative knowledge representation, more suitable to define reasoning methods, is given in terms of attribute implications.

**Definition 2 (Attribute Implication).** *Given a formal context* $\mathbb{K}$, *an attribute implication is an expression* $A \rightarrow B$ *where* $A, B \subseteq M$ *and we say that* $A \rightarrow B$ *holds in* $\mathbb{K}$ *whenever* $B' \subseteq A'$.

That is, $A \rightarrow B$ holds in $\mathbb{K}$ if every object that has all the attributes in $A$ also has all the attributes in $B$. The closeness of these expressions with propositional logic formulas leads to a logical style way to manage them. Although the most used syntactic inference system is the so-called Armstrong's Axioms, we will use the Simplification Logic, $\mathbb{SL}$, introduced in [10]. This logic allows the design of automated reasoning methods [10–12,31] and it is guided by the idea of simplifying the set of implications by efficiently removing redundant attributes. In [31], the results and proofs about $\mathbb{SL}$ are presented.

*Example 3.* We use the fcaR package to extract the set of implications from the formal context in Example 1, by using the Next_Closure algorithm [16], using the command fc_planets$find_implications(). The set of implications is

$$
\begin{aligned}
\Gamma = \{ \ &\{\text{no\_moon}\} &&\Rightarrow \{\text{small, near}\} \\
&\{\text{far}\} &&\Rightarrow \{\text{moon}\} \\
&\{\text{near}\} &&\Rightarrow \{\text{small}\} \\
&\{\text{large}\} &&\Rightarrow \{\text{far, moon}\} \\
&\{\text{medium}\} &&\Rightarrow \{\text{far, moon}\} \\
&\{\text{medium, large, far, moon}\} &&\Rightarrow \{\text{small, near, no\_moon}\} \\
&\{\text{small, near, moon, no\_moon}\} &&\Rightarrow \{\text{medium, large, far}\} \\
&\{\text{small, near, far, moon}\} &&\Rightarrow \{\text{medium, large, no\_moon}\} \\
&\{\text{small, large, far, moon}\} &&\Rightarrow \{\text{medium, near, no\_moon}\} \\
&\{\text{small, medium, far, moon}\} &&\Rightarrow \{\text{large, near, no\_moon}\}\}
\end{aligned}
$$

An interesting argument of the `find_implications()` function, when the number of implications is large, is `parallelize` to take advantage of the cores in the machine. The functions `size, cardinality` can be applied to the `imps` variable to check the number of implications and the size of the attributes on them. The package eases the manipulation of implications using the typical operations of subsetting in R language (`imp[2:3]`, for instance).

To conclude this section, we introduce the outstanding notion of closure of a set of attributes with respect to a set of implications, which is strongly related to the syntactic treatment of implications. Note that the algorithms developed in `fcaR` package to manipulate implications and to compute closures are based on Simplification Logic [31]. For a set of implications, `apply_rules` and `closure` functions can be respectively applied to remove redundancy and to compute the closures of attributes[3]. We make clear that the results in this paper are independent of the closure algorithm used.

**Definition 3.** *Given $\Gamma \subseteq \mathcal{L}_M$ and $X \subseteq M$, the (syntactic) closure of $X$ with respect to $\Gamma$ is the largest subset of $M$, denoted $X_\Gamma^+$, such that $\Gamma \vdash X \to X_\Gamma^+$.*

The mapping $(-)_\Gamma^+ : 2^M \to 2^M$ is a closure operator on $\langle 2^M, \subseteq \rangle$. This notion is the key to designing automatic reasoning methods due to the following equivalence:

$$\Gamma \vdash A \to B \quad \text{iff} \quad \{\varnothing \to A\} \cup \Gamma \vdash \varnothing \to B \quad \text{iff} \quad B \subseteq A_\Gamma^+$$

From now on, we omit the subindex (i.e. we write $X^+$) when no confusion arises.

*Example 4.* We will use the following to compute the closure of the attribute named `small` in Example 1 using our `fcaR` package:

```
> S <- SparseSet$new(attributes = fc_planets$attributes)
> S$assign("small"=1)
> imps$closure(S)
{small, far, moon}
```

## 4    Proposed Research Line

In this section, we propose a new research line accompanied by preliminary results. In this line, we aim to study the potential use and applications of performing (unsupervised) clustering on the Duquenne-Guigues basis of implications. We present this idea using a running example, and we have used the `fcaR` package to help automate the computations and perform experiments.

Given a formal context $\mathbb{K} = (G, M, I)$, and a set of valid implications $\Gamma$, we can interpret $\Gamma$ as a partition (disjoint by definition), i.e., $\Gamma = \Gamma_1 \cup \Gamma_2 \cup \ldots \cup \Gamma_K$, where each set $\Gamma_i$ is called a *cluster of implications*, and it is defined such that

$$\phi(\Gamma_1, \ldots, \Gamma_K) = \sum_{i=1}^{K} \delta(\Gamma_i)$$

---

[3] See https://neuroimaginador.github.io/fcaR/articles/implications.html.

is minimum, where $\delta(\Gamma_i)$ represents an internal dissimilarity measure in $\Gamma_i$. Thus, our motivation is to group similar implications in the same cluster, building homogeneous groups of implications.

In a similar way to classical clustering techniques, $\delta(\Gamma_i)$ can be defined in terms of the distances between implications in the same cluster $\Gamma_i$. Therefore, we propose defining a distance function between implications that can adequately capture and differentiate the essential aspects of their appearance and semantics. Thus, given two implications $P \to Q$ and $R \to T$ from the Duquenne-Guigues basis, we propose to quantify their different appearance by measuring the dissimilarity between $P$ and $R$ and/or between $Q$ and $T$. Their possibly different semantic information can be quantified by comparing their syntactic attribute closures $P^+$ and $R^+$. Our intuition is that the pseudo-intents and the closed sets play an essential role in the clusters, but we want to explore the possibilities.

In order to measure the (dis)similarity between two sets of attributes, we can consider several options. Let us suppose $A, B \subset M$. The following measures are based on well-known distances:

- Hamming (or Manhattan) distance [20]: $d_M(A, B) = |A \triangle B|$ (where $\triangle$ denotes the symmetric set difference operator) measures the amount of attributes that are present in only one of $A$ and $B$.
- Jaccard index [21]: $d_J(A, B) = 1 - \frac{|A \cap B|}{|A \cup B|}$ measures the proportion of common attributes in $A$ and $B$.
- Cosine distance: $d_{\cos}(A, B) = 1 - \frac{|A \cap B|}{\sqrt{|A| \cdot |B|}}$.

Thus, the dissimilarity $\mathrm{dis}(P \to Q, R \to T)$ between two implications $P \to Q$ and $R \to T$, following the previous comment, can be defined in terms of $d(P, R)$, $d(Q, T)$ and $d(P^+, R^+)$, where $d$ is any of $d_M$, $d_J$ or $d_{\cos}$. The use of one or another of these terms is subject to the interpretation and could partially depend on the problem to solve.

Initially, we aim at studying these different possibilities:

$$\mathrm{dis}_1(P \to Q, R \to T) := d(P, R)$$
$$\mathrm{dis}_2(P \to Q, R \to T) := d(P^+, R^+)$$
$$\mathrm{dis}_3(P \to Q, R \to T) := d(P, R) + d(Q, T)$$
$$\mathrm{dis}_4(P \to Q, R \to T) := d(P, R) + d(P^+, R^+)$$
$$\mathrm{dis}_5(P \to Q, R \to T) := d(P, R) + d(Q, T) + d(P^+, R^+)$$

Remember that $d(P, R)$ is a term that quantifies the difference between the pseudointents forming the left-hand sides of the corresponding implications and that $d(P^+, R^+)$ measures the difference in the closed sets that are produced by using the implications. Pseudointents and closed sets represent two levels in the biclustering of the formal context. Therefore it is reasonable to think about clustering the implications by using those distinctive components.

Once all the pairwise distances are computed, we can use a clustering algorithm to generate the implication clusters. For each cluster, we determine an implication, named *central implication*, $P \to Q$, providing the measure of internal dissimilarity in the cluster as follows:

$$\delta(\Gamma_i) := \frac{1}{|\Gamma_i|} \sum_{R \to T \in \Gamma_i} \mathrm{dis}(P \to Q, R \to T)$$

which can be interpreted as a measure of within $\Gamma_i$ dispersion.

The clustering algorithm should provide a proper partition of $\Gamma$ such that $\phi(\Gamma_1, \ldots, \Gamma_K)$ is minimum. Another characteristic of clustering is that each implication $R \to T$ is assigned to a cluster $\Gamma_i$ if, by definition, its dissimilarity to the central implication of $\Gamma_i$ (which we will call $P_i \to Q_i$) is lower than its dissimilarity to the central implications of the other clusters, that is, if

$$\mathrm{dis}(P_i \to Q_i, R \to T) \leq \mathrm{dis}(P_j \to Q_j, R \to T) \quad \forall j \neq i$$

Given the definition of dissimilarity above, the proposed clustering aims at building coherent groups of implications that have similar pseudointents or produce similar closed sets.

There are many possible choices of clustering algorithms. In this paper, we propose the use of the PAM (partitioning around medoids) algorithm [23] to compute the clusters and their central implications, which, in this context are called the *medoids* of the clusters, since it is more robust to the presence of noise and isolated components in the data than the $K$-means algorithm [27], widely used in machine learning.

*Example 5.* Following our running example, we will find clusters in the implications of Example 3. And, for instance, we consider the dissimilarity function

$$\mathrm{dis}(P \to Q, R \to T) := |P \triangle R| + |P^+ \triangle R^+| \tag{1}$$

The next R code computes the dissimilarity matrix, that is, the matrix $D = (D_{i,j})$ where the entry $D_{i,j} := \mathrm{diss}(P_i \to Q_i, P_j \to Q_j)$ is the dissimilarity between the $i$th and the $j$th implications.

```
> diss <- implication_distance(imps)
> D <- as.matrix(diss)
> D
    1  2  3 4 5  6  7 8  9 10
1   0  7  3 8 8  9  7 9  9  9
2   7  0  6 3 3  8 10 8  8  8
3   3  6  0 7 7 10  8 8 10 10
4   8  3  7 0 4  7  9 9  7  9
5   8  3  7 4 0  7  9 9  9  7
6   9  8 10 7 7  0  6 4  2  2
7   7 10  8 9 9  6  0 2  4  4
8   9  8  8 9 9  4  2 0  2  2
9   9  8 10 7 9  2  4 2  0  2
10  9  8 10 9 7  2  4 2  2  0
```

Then, we use the PAM algorithm of the `cluster` R package to compute the clusters using $K = 2$ (two clusters) and their central implications.

```
> cluster <- cluster::pam(diss, k = 2)
> # The following are the central implications
> imps[cluster$id.med]
Implication set with 2 implications.
Rule 1: {far} -> {moon}
Rule 2: {small, near, far, moon} -> {medium, large, no_moon}
```

Therefore, we already have the implications in each cluster:

```
> imps[cluster$clustering == 1]
Implication set with 5 implications.
Rule 1: {no_moon} -> {small, near}
Rule 2: {far} -> {moon}
Rule 3: {near} -> {small}
Rule 4: {large} -> {far, moon}
Rule 5: {medium} -> {far, moon}
> imps[cluster$clustering == 2]
Implication set with 5 implications.
Rule 1: {medium, large, far, moon} -> {small, near, no_moon}
Rule 2: {small, near, moon, no_moon} -> {medium, large, far}
Rule 3: {small, near, far, moon} -> {medium, large, no_moon}
Rule 4: {small, large, far, moon} -> {medium, near, no_moon}
Rule 5: {small, medium, far, moon} -> {large, near, no_moon}
```

Note that the second cluster is formed by implications that present all the attributes. This cluster can be disregarded as uninformative since its implications present combinations of attributes that are not found in any object of the formal context. In terms of association rules, they would be considered as implications with zero-support and not interesting for our proposal. Thus, in what follows, we will consider only implications that do not present all attributes.

```
> # Take the first 5 implications
> imps <- imps[1:5]
> diss <- implication_dist(imps)
> D <- as.matrix(diss)
> rownames(D) <- seq(imps$cardinality())
> D
   1  2  3  4  5
1  0 10  4 12 12
2 10  0  8  4  4
3  4  8  0 10 10
4 12  4 10  0  4
5 12  4 10  4  0
```

Furthermore, for these implications, the computation of the clusters produces the following clusters:

```
> cluster <- cluster::pam(diss, k = 2)
> # The central implications
> imps[cluster$id.med]
```

```
Implication set with 2 implications.
Rule 1: {near} -> {small}
Rule 2: {far} -> {moon}
> # The cluster 1 is:
> imps[cluster$clustering == 1]
Implication set with 2 implications.
Rule 1: {no_moon} -> {small, near}
Rule 2: {near} -> {small}
> # The cluster 2 is:
> imps[cluster$clustering == 2]
Implication set with 3 implications.
Rule 1: {far} -> {moon}
Rule 2: {large} -> {far, moon}
Rule 3: {medium} -> {far, moon}
```

The computed clustering can be viewed as the result of minimising the clustering algorithm's objective function. It will be minimum the mean dissimilarity of each implication in its cluster with respect to its central implication.

As a conclusion, we can observe that the clustering renders a natural result describing the clusters as a set of implications with specific knowledge about:

1. Planets near the Sun which are therefore small.
2. Distant planets that, therefore, have satellites.

Now, we approach to experiment what happens when we change the dissimilarity function to:

$$\mathrm{dis}(P \to Q, R \to T) := |P \triangle R| \tag{2}$$

that is, when we consider only the *difference* in the pseudointents. Similarly, we compute the new dissimilarity matrix and obtain:

```
> D_LHS
  1 2 3 4 5
1 0 2 2 2 2
2 2 0 2 2 2
3 2 2 0 2 2
4 2 2 2 0 2
5 2 2 2 2 0
```

We can observe that any two implications are at distance 2. The consequence is that the clusters will be uninformative. Any possible partition into two clusters $\Gamma_1$ and $\Gamma_2$, using this dissimilarity matrix, has the same mean dissimilarity; therefore, clusters can be considered as generated by randomness. We can check the central implications:

```
> clusterLHS <- cluster::pam(dissLHS, k = 2)
> # The central implications
> imps[clusterLHS$id.med]
Implication set with 2 implications.
Rule 1: {large} -> {far, moon}
Rule 2: {medium} -> {far, moon}
```

Note the overlap in the closed sets defined by these central implications. In this case, the implications in the first cluster and the unique implication in the second cluster (that is also its central implication) does not provide any further insight, making it clear that, in this case, the clusters are random guesses.

```
# First cluster - Implication set with 4 implications.
Rule 1: {no_moon} -> {small, near}
Rule 2: {far} -> {moon}
Rule 3: {near} -> {small}
Rule 4: {large} -> {far, moon}
# Second cluster - Implication set with 1 implications.
Rule 1: {medium} -> {far, moon}
```

It seems clear that to consider the dissimilarity measure proposed in Eq. (1), representing the difference in the knowledge provided by pseudo-intents and closed sets, is more appropriate than the proposed in Eq. (2), representing only the differences in the pseudointents.

To conclude this section, we explain the line of research we have in mind. The clustering relationship on implications to object and attribute clustering or to concept clustering seems to be interesting. We devise potential future applications in reducing the computational cost of computing closures in specific scenarios or the possible application to FCA's factorisation techniques. Also, it will be of interest to study the different properties of implication clustering when performed on different types of bases (direct-optimal [34] and ordered-direct bases [1] and sets of implications without attribute redundancies, for instance). Last, it will be of interest to extend the study to determine the properties of clustering of association rules with this new proposal, in contrast to what has already been studied [2, 19, 36].

## 5   Experimental Results

This section presents results to illustrate how the obtained clustering of implications is consistent with the formal context's observed data.

We apply our proposal to the data from the so-called MONK's problems [15], a well-known set of 3 datasets used in machine learning competitions. Each of the 3 datasets consists of 6 categorical attributes, a1 to a6, taking integer values, and a binary class attribute. For this work, all categorical variables have been binarized, making an aggregate of 19 binary attributes, including the two class attributes, class = 0 and class = 1.

For each of these three problems, we have computed the Duquenne-Guigues basis, consisting of 524, 723 and 489 implications, respectively. After removing the implications that incorporate all the attributes, as commented before, the final sets of implications consisted of 505, 704 and 471 implications for problems MONKS-1, MONKS-2 and MONKS-3, respectively.

Then, we apply a dissimilarity function (one of $dis_1, \ldots, dis_5$, or any other combination) to obtain a dissimilarity matrix. To determine the optimal number

**Fig. 1.** Bi-dimensional representation of the implication space. Each dot represents an implication. In order to obtain this plot, the multidimensional scaling technique has been used to map implications to $\mathbb{R}^2$ points, preserving their mutual dissimilarities.

of clusters present in the implication set, we use the *silhouette index* [35]. In all these problems, the optimal number of clusters determined by this index was 2.

Thus, to continue with our proposal, for each problem we have applied the method to partition the implication set into two clusters. For instance, if we use the dissimilarity function $dis_4$, which incorporated the distance between the pseudointents and between the closures, with the Hamming distance, on the MONKS-1 problem, we obtain the following central implications:

$$\{a5 = 1\} \Rightarrow \{\text{class} = 1\}$$
$$\{\text{class} = 0,\, a2 = 1,\, a5 = 2\} \Rightarrow \{a6 = 2\}$$

The clustering results can be visually inspected by applying an algorithm of multidimensional scaling [25], whose results can be plotted to obtain a graphical bi-dimensional representation of the *implications space*. This plot can also be used to inspect the potential number of clusters present in the implications. The results of the clustering can be checked in Fig. 1.

In that Fig. 1, we can check that 2 clusters seem to be a good proposal. The clustering algorithm has almost correctly identified the two implication groups, confirming the estimated value using the *silhouette index*.

We have explored the consistency of the clustering performed using the different measures of dissimilarity. proposed in Sect. 4 ($dis_1, \ldots, dis_5$), based on different distance functions (Hamming, Jaccard and Cosine indexes).

First, we study if the implications leading to the same closed set are grouped in the same cluster. Thus, we introduce the notion of *closure purity*. Let us consider the set of *equivalence classes* in the Duquenne-Guigues basis $\Gamma$, as

$$[P \to Q] = \{R \to T \in \Gamma : P^+ = R^+\}$$

Two implications belong to the same equivalence class if the closures of their respective pseudointents are the same. Then, we can define *closure purity* as the proportion of those equivalence classes whose implications are all assigned to the same cluster. The ideal situation is that this index equals 1, meaning that whole equivalence classes form clusters.

**Table 2.** *Closure purity* for different dissimilarity measures and distance functions.

| Problem | Dissimilarity | Hamming | Jaccard | Cosine |
|---------|---------------|---------|---------|--------|
| MONKS-1 | $dis_1$ | 0.953 | 1.000 | 0.983 |
|         | $dis_2$ | 1.000 | 1.000 | 1.000 |
|         | $dis_3$ | 0.962 | 0.953 | 0.953 |
|         | $dis_4$ | 1.000 | 1.000 | 0.971 |
|         | $dis_5$ | 1.000 | 0.988 | 0.962 |
| MONKS-2 | $dis_1$ | 0.928 | 0.966 | 0.942 |
|         | $dis_2$ | 1.000 | 1.000 | 1.000 |
|         | $dis_3$ | 0.986 | 0.966 | 0.974 |
|         | $dis_4$ | 0.994 | 1.000 | 0.998 |
|         | $dis_5$ | 0.996 | 0.954 | 0.974 |
| MONKS-3 | $dis_1$ | 0.923 | 1.000 | 0.972 |
|         | $dis_2$ | 1.000 | 1.000 | 1.000 |
|         | $dis_3$ | 0.935 | 0.985 | 0.978 |
|         | $dis_4$ | 1.000 | 0.997 | 0.994 |
|         | $dis_5$ | 1.000 | 0.994 | 0.966 |

The results of this comparison are presented in Table 2. It is evident that $dis_2$ achieves closure purity equal to 1 since it is defined as the dissimilarity between the closures given by two implications. Thus, any two implications in the same equivalence class have dissimilarity 0 and therefore are assigned to the same cluster. Interestingly, other dissimilarity measures, such as $dis_4$, taking into account also the difference in the pseudointents, in many occasions achieve closure purity equal to 1, meaning that they can also separate the equivalence classes coherently.

Also, we study if there are common attributes inside the implications in a given cluster. Table 3 shows the attributes that appear in at least 80% of the implications in each cluster. Note that there is always a cluster with no common attributes using both Jaccard and Cosine indexes, indicating greater heterogeneity in that cluster's implications. With the Hamming distance, we obtain that the dissimilarity measures $dis_2$ (considering only $P^+$ and $R^+$) and $dis_4$ (considering, besides, the difference between pseudointents) always find common attributes in each cluster. Remarkably, the common attributes found are the class attributes mentioned earlier. Hence, the clustering procedure has been able to locate key attributes in a completely unsupervised manner, provided the knowledge present in the implication set. It is evident that if we reduce the threshold to be less than 80%, we will find a greater number of common attributes. We have used a threshold of 80% to retain just representative attributes in each cluster.

**Table 3.** Sets of common attributes in each of the clusters found for different distance functions and dissimilarity measures. A ∅ symbol indicates that no common attributes are found in the implications of the given cluster.

| Problem | Diss. | Hamming | | Jaccard | | Cosine | |
|---|---|---|---|---|---|---|---|
| | | Cluster 1 | Cluster 2 | Cluster 1 | Cluster 2 | Cluster 1 | Cluster 2 |
| MONKS-1 | $dis_1$ | {class = 1} | ∅ | {class = 1, a5 = 1} | ∅ | {class = 1, a5 = 1} | ∅ |
| | $dis_2$ | {class = 1} | {class = 0} | {class = 1, a5 = 1} | ∅ | {class = 1, a5 = 1} | ∅ |
| | $dis_3$ | {class = 1} | ∅ | {class = 1} | ∅ | {class = 1} | ∅ |
| | $dis_4$ | {class = 1} | {class = 0} | {class = 1, a5 = 1} | ∅ | {class = 1} | ∅ |
| | $dis_5$ | {class = 1} | {class = 0} | {class = 1} | ∅ | {class = 1} | ∅ |
| MONKS-2 | $dis_1$ | ∅ | ∅ | {a5 = 1} | ∅ | {a4 = 1} | ∅ |
| | $dis_2$ | {class = 0} | {class = 1} | {class = 0, a6 = 1} | ∅ | {class = 0, a5 = 1, a6 = 1} | ∅ |
| | $dis_3$ | {class = 0} | ∅ | {class = 0} | ∅ | {class = 0} | ∅ |
| | $dis_4$ | {class = 0} | {class = 1} | {class = 0, a4 = 1, a6 = 1} | ∅ | {class = 0} | ∅ |
| | $dis_5$ | {class = 0} | ∅ | {class = 0} | ∅ | {class = 0} | ∅ |
| MONKS-3 | $dis_1$ | ∅ | {class = 1} | {class = 0, a5 = 4} | ∅ | {class = 0, a5 = 4} | ∅ |
| | $dis_2$ | {class = 0} | {class = 1} | {class = 0, a5 = 4} | ∅ | {class = 0, a5 = 4} | ∅ |
| | $dis_3$ | {class = 0} | ∅ | {class = 0} | ∅ | {class = 0} | ∅ |
| | $dis_4$ | {class = 0} | {class = 1} | {class = 0} | ∅ | {class = 0} | ∅ |
| | $dis_5$ | {class = 0} | {class = 1} | {class = 0} | ∅ | {class = 0} | ∅ |

This leads us to think that implication clustering could be used with promising results in classification tasks in datasets, and/or, for instance, in recommender systems for medical diagnosis in which some attributes play the role of identifying diseases from symptoms (the rest of the attributes). To finish this section, it seems clear that very compelling results are emerging from the hidden knowledge in the clusters of implications.

## 6   Conclusions

We have presented the fcaR package developed in the R language throughout this work. The package has two objectives. The first one is to provide a tool to the FCA community and make FCA works visible to other areas as machine learning, data science, etc., where the use of the R language is widely extended. Thus, in this line, to promote the so-called reproducible research and the sharing of knowledge, the scripts to replicate the results in this work, as well as the results themselves, are hosted in https://github.com/Malaga-FCA-group/FCA-ImplicationClustering.

From the theoretical point of view, the paper proposes a method to cluster implications, hence extracting interesting knowledge about the central implications, which reveal groups of objects with a special meaning and shared characteristics. This work opens the windows to new interesting research in current areas of interest as Social Network Analysis. The identification of topics could be addressed by our clustering implication method based on logic.

Natural clusters (consistent with the data) seem to emerge from the implication clusters, and this could have potential applications to reduce the concept lattice, the bases of implications, etc. Key attributes arise from the clusters, with potential applications revealing attributes and object clusters and their leaders. It could also be of interest to study the relationship between the concept lattice obtained directly from a formal context and obtained after clustering objects. The study of *closure purity* can reveal interesting properties about closed sets and their features.

# References

1. Adaricheva, K., Nation, J., Rand, R.: Ordered direct implicational basis of a finite closure system. Discrete Appl. Math. **161**(6), 707–723 (2013)
2. An, A., Khan, S., Huang, X.: Hierarchical grouping of association rules and its application to a real-world domain. Int. J. Syst. Sci. **37**(13), 867–878 (2006)
3. Belohlavek, R., De Baets, B., Outrata, J., Vychodil, V.: Computing the lattice of all fixpoints of a fuzzy closure operator. IEEE Trans. Fuzzy Syst. **18**(3), 546–557 (2010)
4. Bělohlávek, R., Vychodil, V.: Attribute implications in a fuzzy setting. In: Missaoui, R., Schmidt, J. (eds.) ICFCA 2006. LNCS (LNAI), vol. 3874, pp. 45–60. Springer, Heidelberg (2006). https://doi.org/10.1007/11671404_3
5. Bocharov, A., Gnatyshak, D., Ignatov, D.I., Mirkin, B.G., Shestakov, A.: A lattice-based consensus clustering algorithm. In: International Conference on Concept Lattices and Their Applications, vol. CLA2016, pp. 45–56 (2016)
6. Carbonnel, J., Bertet, K., Huchard, M., Nebut, C.: FCA for software product line representation: mixing configuration and feature relationships in a unique canonical representation. Discrete Appl. Math. **273**, 43–64 (2020)
7. Castellanos, A., Cigarrán, J., García-Serrano, A.: Formal concept analysis for topic detection: a clustering quality experimental analysis. Inf. Syst. **66**, 24–42 (2017)
8. Chemmalar Selvi, G., Lakshmi Priya, G.G., Joseph, R.B.: A FCA-based concept clustering recommender system. In: Vinh, P.C., Rakib, A. (eds.) ICCASA/ICTCC -2019. LNICST, vol. 298, pp. 178–187. Springer, Cham (2019). https://doi.org/10.1007/978-3-030-34365-1_14
9. Cigarrán, J., Castellanos, Á., García-Serrano, A.: A step forward for Topic Detection in Twitter: an FCA-based approach. Expert Syst. Appl. **57**, 21–36 (2016)
10. Cordero, P., Enciso, M., Mora, A., de Guzmán, I.P.: SL$_{FD}$ logic: elimination of data redundancy in knowledge representation. In: Garijo, F.J., Riquelme, J.C., Toro, M. (eds.) IBERAMIA 2002. LNCS (LNAI), vol. 2527, pp. 141–150. Springer, Heidelberg (2002). https://doi.org/10.1007/3-540-36131-6_15
11. Cordero, P., Enciso, M., Bonilla, A.M., Ojeda-Aciego, M.: Bases via minimal generators. In: Proceedings of the International Workshop "What can FCA do for Artificial Intelligence?" (FCA4AI at IJCAI 2013), Beijing, China, 5 August 2013, pp. 33–36 (2013)
12. Cordero, P., Enciso, M., Mora, Á., Ojeda-Aciego, M.: Computing minimal generators from implications: a logic-guided approach. In: Proceedings of Concept Lattices and Applications, CLA 2012. pp. 187–198 (2012)
13. Demko, C., Bertet, K., Faucher, C., Viaud, J.F., Kuznetsov, S.O.: NextPriority Concept: a new and generic algorithm computing concepts from complex and heterogeneous data. Theor. Comput. Sci. **845**, 1–20 (2020)
14. Diatta, J.: A relation between the theory of formal concepts and multiway clustering. Pattern Recogn. Lett. **25**(10), 1183–1189 (2004)

15. Dua, D., Graff, C.: UCI machine learning repository (2017). http://archive.ics.uci.edu/ml
16. Ganter, B., Wille, R.: Formal Concept Analysis. Springer, Heidelberg (1999). https://doi.org/10.1007/978-3-642-59830-2
17. Delugach, H.S., Stumme, G. (eds.): ICCS-ConceptStruct 2001. LNCS (LNAI), vol. 2120. Springer, Heidelberg (2001). https://doi.org/10.1007/3-540-44583-8
18. Ganter, B., Rudolph, S., Stumme, G.: Explaining data with formal concept analysis. In: Krötzsch, M., Stepanova, D. (eds.) Reasoning Web. Explainable Artificial Intelligence. LNCS, vol. 11810, pp. 153–195. Springer, Cham (2019). https://doi.org/10.1007/978-3-030-31423-1_5
19. Hahsler, M.: Grouping association rules using lift. In: Proceedings of 11th INFORMS Workshop on Data Mining and Decision Analytics (DMDA 2016) (2016)
20. Hamming, R.W.: Error detecting and error correcting codes. Bell Syst. Tech. J. **29**(2), 147–160 (1950)
21. Jaccard, P.: The distribution of the flora in the alpine zone. New Phytol. **11**(2), 37–50 (1912)
22. Kashnitsky, Y., Ignatov, D.I.: Can FCA-based recommender system suggest a proper classifier? In: Proceedings of the International Workshop "What can FCA do for Artificial Intelligence?" (FCA4AI at IJCAI 2014), p. 17 (2014)
23. Kaufman, L., Rousseeuw, P.J.: Partitioning around medoids (program PAM). In: Finding Groups in Data: An Introduction to Cluster Analysis, vol. 344, pp. 68–125 (1990)
24. Konecny, J.: Attribute implications in L-concept analysis with positive and negative attributes: validity and properties of models. Int. J. Approx. Reason. **120**, 203–215 (2020)
25. Kruskal, J.B.: Multidimensional Scaling, no. 11. Sage, Los Angeles (1978)
26. Kumar, C.A.: Fuzzy clustering-based formal concept analysis for association rules mining. Appl. Artif. Intell. **26**(3), 274–301 (2012)
27. MacQueen, J., et al.: Some methods for classification and analysis of multivariate observations. In: Proceedings of the Fifth Berkeley Symposium on Mathematical Statistics and Probability, Oakland, CA, USA, vol. 1, pp. 281–297 (1967)
28. Medina, J., Ojeda-Aciego, M., Ruiz-Calviño, J.: Formal concept analysis via multi-adjoint concept lattices. Fuzzy Sets Syst. **160**(2), 130–144 (2009)
29. Melo, C., Mikheev, A., Le Grand, B., Aufaure, M.A.: Cubix: a visual analytics tool for conceptual and semantic data. In: Proceedings - 12th IEEE International Conference on Data Mining Workshops, ICDMW 2012, pp. 894–897 (2012)
30. Missaoui, R., Ruas, P.H.B., Kwuida, L., Song, M.A.J.: Pattern discovery in triadic contexts. In: Alam, M., Braun, T., Yun, B. (eds.) ICCS 2020. LNCS (LNAI), vol. 12277, pp. 117–131. Springer, Cham (2020). https://doi.org/10.1007/978-3-030-57855-8_9
31. Mora, Á., Cordero, P., Enciso, M., Fortes, I., Aguilera, G.: Closure via functional dependence simplification. Int. J. Comput. Math. **89**(4), 510–526 (2012)
32. Priss, U.: Formal concept analysis in information science. Ann. Rev. Inf. Sci. Technol. **40**(1), 521–543 (2006)
33. Ravi, K., Ravi, V., Prasad, P.S.R.K.: Fuzzy formal concept analysis based opinion mining for CRM in financial services. Appl. Soft Comput. J. **60**, 786–807 (2017)
34. Rodríguez-Lorenzo, E., Bertet, K., Cordero, P., Enciso, M., Mora, A., Ojeda-Aciego, M.: From implicational systems to direct-optimal bases: a logic-based approach. Appl. Math. Inf. Sci. **2**, 305–317 (2015)

35. Rousseeuw, P.J.: Silhouettes: a graphical aid to the interpretation and validation of cluster analysis. J. Comput. Appl. Math. **20**, 53–65 (1987)
36. Strehl, A., Gupta, G.K., Ghosh, J.: Distance based clustering of association rules. In: Proceedings ANNIE, vol. 9, pp. 759–764 (1999)
37. Stumme, G., Maedche, A.: FCA-MERGE: bottom-up merging of ontologies. In: IJCAI International Joint Conference on Artificial Intelligence, pp. 225–230 (2001)
38. Sumangali, K., Aswani Kumar, Ch.: Concept lattice simplification in formal concept analysis using attribute clustering. J. Ambient Intell. Humaniz. Comput. **10**(6), 2327–2343 (2018). https://doi.org/10.1007/s12652-018-0831-2

# Extracting Relations in Texts with Concepts of Neighbours

Hugo Ayats[(✉)], Peggy Cellier, and Sébastien Ferré

Univ Rennes, INSA, CNRS, IRISA Campus de Beaulieu, 35042 Rennes, France
{hugo.ayats,peggy.cellier,sebastien.ferre}@irisa.fr

**Abstract.** During the last decade, the need for reliable and massive Knowledge Graphs (KG) increased. KGs can be created in several ways: manually with forms or automatically with Information Extraction (IE), a natural language processing task for extracting knowledge from text. Relation Extraction is the part of IE that focuses on identifying relations between named entities in texts, which amounts to find new edges in a KG. Most recent approaches rely on deep learning, achieving state-of-the-art performances. However, those performances are still too low to fully automatize the construction of reliable KGs, and human interaction remains necessary. This is made difficult by the statistical nature of deep learning methods that makes their predictions hardly interpretable. In this paper, we present a new symbolic and interpretable approach for Relation Extraction in texts. It is based on a modeling of the lexical and syntactic structure of text as a knowledge graph, and it exploits *Concepts of Neighbours*, a method based on Graph-FCA for computing similarities in knowledge graphs. An evaluation has been performed on a subset of TACRED (a relation extraction benchmark), showing promising results.

## 1 Introduction

During the last decade, the need for reliable and massive knowledge bases, represented as Knowledge Graphs (KG), increased. KGs allow to structure, organize and share knowledge. A challenge is to build KGs that are at the same time reliable and large. There exist several ways to create those KGs: manually with forms (e.g., wikidata[1]), providing reliability (assuming the producer is reliable), or automatically by using Information Extraction (IE) techniques [8], allowing to easily build very large KG by using the large amount of existing textual data (e.g., scientific papers, books, official websites). IE is a natural language processing task for extracting knowledge from text. Relation Extraction is a sub-task of IE that focuses on identifying relations between named entities in texts, which amounts to find new edges between KG entities. It is a classification task, where given two named entities in a sentence, the goal is to predict the relation type between the two.

---

[1] https://www.wikidata.org/.

© Springer Nature Switzerland AG 2021
A. Braud et al. (Eds.): ICFCA 2021, LNAI 12733, pp. 155–171, 2021.
https://doi.org/10.1007/978-3-030-77867-5_10

Among existing approaches for the relation extraction task, the deep learning methods are the most efficient. First, those method were based on convolutional neural networks [15]. Other deep learning approaches, such as [17], work on pruned dependency trees. Today, methods based on language models such as BERT [18] provide the best results. However, those deep learning methods have two main limitations. First, they are not sufficiently reliable for a full automation. Second, they suffer, as most of numerical approaches, of a lack of explanations for predictions, which hinders human interaction.

Formal Concept Analysis (FCA) has already been shown effective for classification tasks [9], and has the advantage to provide symbolic representations of predictions that can be used as explanations. However, it faces a problem of tractability when all concepts need to be computed. An interesting approach is to adopt lazy learning, where this computation is delayed until there is an instance to be classified, and only the concepts that are relevant to that instance are computed. This lazy approach was applied early to the incremental building of a logical context [4], then later advocated in [10] with Pattern Structures [6], and applied, for instance, to relation extraction in biomedical texts [11]. More recently, this has been formalized as *Concepts of Neighbours* [2] based on Graph-FCA [1]. The particularity of the Concepts of Neighbours method, such as this contribution, is the use of an efficient anytime algorithm based on the partitioning of the set of entities into concepts. This has for consequence that there is no need for a sampling strategy: all entities appear in the extensions of the returned concepts. The use of Concepts of Neighbours has been applied to the completion of knowledge graphs [3], and has shown competitive results compared to deep learning approaches, and state-of-the-art results compared to rule-based approaches such as AnyBURL [13] or AMIE+ [5]. Rule-based approaches are also interpretable but they are not lazy because they compute a set of rules before seeing any test instance. Therefore, they have to strongly restrict the considered graph patterns for combinatorial reasons, unlike in Concepts of Neighbours where all connected graph patterns with constants are supported. AnyBURL only mines path rules and AMIE+ focuses on small connected and closed rules.

In this paper, we introduce a symbolic and lazy approach for relation extraction based on Concepts of Neighbours. A first contribution is the representation of sentences by graphs, covering both lexical and syntactic information. Other contributions, compared to previous application of Concepts of Neighbours, are that the instances to be classified are couples of graph nodes, instead of single nodes, and that node labels are organized into a taxonomy (e.g., word hypernyms). We validate our approach with experiments on TACRED [20], a dataset for evaluating relation extraction methods.

In the sequel, Sect. 2 gives preliminaries about Graph-FCA and Concepts of Neighbours. Then, Sect. 3 introduces the modeling of sentences as graphs. Section 4 details how Concepts of Neighbours can be used for relation extraction. Finally, Sect. 5 presents the experiments conducted on dataset TACRED.

$O = \{Charles, Diana, William, Harry, Kate, George, Charlotte, Louis, male, female\}$
$A = \{parent, spouse, female, male\}$
$I = \{parent(\{William, Harry\}, \{Charles, Diana\}),$
$\qquad parent(\{George, Charlotte, Louis\}, \{William, Kate\}),$
$\qquad spouse(Charles, Diana), spouse(William, Kate),$
$\qquad male(\{Charles, William, Harry, George, Louis\}),$
$\qquad female(\{Diana, Kate, Charlotte\})\}$

**Fig. 1.** Example Graph-FCA context $K = (O, A, I)$ describing part of the British royal family. Notation $p(\{a, b\}, \{c, d\})$ stands for $p(a, c), p(a, d), p(b, c), p(b, d)$.

## 2 Preliminaries

In this section, we recall the main definitions and results of *Concepts of Neighbours* [3]. We start by defining *graph contexts* and *graph concepts*, introduced in Graph-FCA [1], a generalization of Formal Concept Analysis (FCA) [7] to graphs. A *graph context* $K = (O, A, I)$ is a labeled and directed multi-hypergraph, where objects are nodes, attributes are edge labels, and incidence elements are edges (noted like atoms in predicate logic $a(o_1, \ldots, o_k)$). As a running example, Fig. 1 defines a small context describing (part of) the British royal family.

**Definition 1.** *A* graph concept *is defined as a pair $C = (R, Q)$, where $R$ is a set of $k$-tuples of objects and $Q$ is a conjunctive query such that $R = res(Q)$ is the set of results of $Q$, and $Q = msq(R)$ is the most specific query that verifies $R = res(Q)$. $R$ is called the* extension *$ext(C)$, and $Q$ is called the* intension *$int(C)$.*

The most specific query $Q = msq(R)$ is the conjunctive query representing what the tuples of objects in $R$ have all in common. For the sake of simplicity, we restrict the following examples to 1-tuples, aka. *singletons*. In the running example, the singletons $(William)$ and $(Charlotte)$ have in common the following query, $Q_{WC}$, that says that both have married parents:

$$Q_{WC} = msq(\{(William), (Charlotte)\})$$
$$= (x) \leftarrow parent(x, y), \; female(y), \; parent(x, z), \; male(z), \; spouse(y, z).$$

We have $R_{WC} = res(Q_{WC}) = \{(William), (Harry), (George), (Charlotte), (Louis)\}$ so that $C_{WC} = (R_{WC}, Q_{WC})$ is a graph concept.

A concept $C_1 = (R_1, Q_1)$ is more specific than a concept $C_2 = (R_2, Q_2)$, in notation $C_1 \leq C_2$, if $R_1 \subseteq R_2$. For example, a concept more specific than $C_{WC}$ is the concept of *the children of Kate and William*, whose extension is $\{(George), (Charlotte), (Louis)\}$, and whose intension is:

$$(x) \leftarrow parent(x, y), (y = Kate), parent(x, z), (z = William).$$

The total number of graph concepts in a knowledge graph is finite but in the worst case, it is exponential in the number of objects, and in arity $k$. It is therefore not feasible in general to compute the set of all concepts. Instead of considering concepts generated by subsets of tuples, we consider concepts generated by pairs of tuples, and use them as a symbolic form of distance between objects.

**Definition 2.** *Let $t_1, t_2 \in O^k$ be two k-tuples of objects. The* conceptual distance $\delta(t_1, t_2)$ *between $t_1$ and $t_2$ is the most specific graph concept whose extension contains both tuples, i.e. $\delta(t_1, t_2) = (R, Q)$ with $Q = msq(\{t_1, t_2\})$, and $R = res(Q)$.*

For example, the above concept $C_{WC}$ is the conceptual distance between $(William)$ and $(Charlotte)$. The "distance values" have therefore a symbolic representation through the concept intension $Q$ that represents what the two tuples have in common. The concept extension $R$ contains in addition to the two tuples all tuples $t_3$ that match the common query $(t_3 \in res(Q))$. Such a tuple $t_3$ can be seen as "between" $t_1$ and $t_2$: in formulas, for all $t_3 \in ext(\delta(t_1, t_2))$, $\delta(t_1, t_3) \leq \delta(t_1, t_2)$ and $\delta(t_3, t_2) \leq \delta(t_1, t_2)$. Note that order $\leq$ on conceptual distances is a partial ordering, unlike classical distance measures.

A numerical distance $dist(t_1, t_2) = |ext(\delta(t_1, t_2))|$ can be derived from the size of the concept extension, because the closer $t_1$ and $t_2$ are, the more specific their conceptual distance is, and the smaller the extension is. For example, the numerical distance is 5 between $(William)$ and $(Charlotte)$ (see $C_{WC}$), and 3 between $(George)$ and $(Charlotte)$.

The number of conceptual distances $\delta(t_1, t_2)$ is no more exponential but quadratic in the number of objects $|O|$. In the context of lazy learning, tuple $t_1$ is fixed, and the number of concepts become linear. For $k$-tuples, that number is bounded by $|O|^k$. Those concepts are called *Concepts of Neighbours*.

**Definition 3.** *Let $t \in O^k$ be a k-tuple of objects. The* Concepts of Neighbours *of $t$ are all the conceptual distances between $t$ and every tuple $t' \in O^k$.*

$$C\text{-}N(t, K) = \{\delta(t, t') \mid t' \in O^k\}$$

Figure 2 shows the 6 Concepts of Neighbours of the singleton $(Charlotte)$, and their partial ordering as a Venn diagram. For instance, the concept containing $(Charlotte)$ only is included in the concept that also contains $(Louis)$ and $(George)$, which is included in the concept that also contains $(William)$ and $(Harry)$. This implies that George is semantically closer to Charlotte than William is. Although $(Louis)$ and $(Diana)$ are both nearest neighbours of $(Charlotte)$, and at the same extensional distance 3, they are so for different reasons as they belong to different Concepts of Neighbours with different intensions. Louis has the same parents as Charlotte, while Diana has the same gender.

The *proper extension* of a concept of neighbours $\delta_l$ is the part of its extension that does not appear in sub-concepts. The proper extensions define a partition over the set of objects $O$, where two objects are in the same proper extension if and only if they are at the same conceptual distance. For instance, Diana and Kate are in the same proper extension.

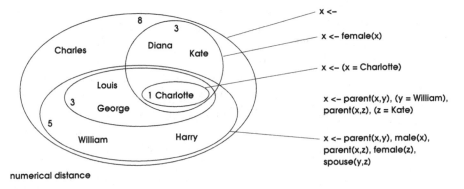

**Fig. 2.** Venn diagram of the extensions of the 6 Concepts of Neighbours of (*Charlotte*), labelled by their intension (right) and numerical distance (inside).

## 3    Modeling Sentences as Graphs

The input data of relation extraction is a set of annotated sentences. Each sentence is annotated by two entities, the subject and object of the relation, and by the type of those entities. The input sentences are split in two parts: training sentences for which the relation is known, and test sentences for which the relation is to be predicted. In this section we explain and discuss the modeling of annotated sentences as graphs because this is the required input of Concepts of Neighbours, and more precisely as RDF graphs because this is the expected input of the existing implementation of Concepts of Neighbours.

### 3.1    NLP Treatments

Before building the RDF graph, several NLP treatments are applied on each sentence[2]. First, the sentence is split into tokens. Second, *part-of-speech* (POS) tags[3] and lemmas are computed for each token. Third, the syntactic structure of the sentence is extracted as a dependency tree[4]. Finally, named entities are identified, and a named entity type is associated to each of them.

Table 1 shows the result of the processing of the sentence *"The University of Rennes is French"*. For example, it shows that the 4th token is *Rennes*, has for lemma *Rennes* and for POS tag *NNP* (a proper noun). It is part of a named entity of type *ORGANIZATION*. It has for parent in the dependency tree the 2nd token and it is linked to this token via relation *nmod* (linking an nominal modifier to its parent noun). We then apply a few post-treatments in order to simplify and improve the sentence representation.

---

[2] We use the *CoreNLP* tool [12] but other tools could be used.
[3] We use the 58 POS tags of English Penn Treebank [16].
[4] We use the dependency grammar proposed by Treebank Universal Dependencies.

**Table 1.** Example of a processed sentence.

| ID | Token | Lemma | POS | NER | Head | Deprel |
|---|---|---|---|---|---|---|
| 1 | The | The | DT | – | 2 | det |
| 2 | University | University | NNP | ORGANIZATION | 6 | nsubj |
| 3 | of | of | IN | ORGANIZATION | 4 | case |
| 4 | Rennes | Rennes | NNP | ORGANIZATION | 2 | nmod |
| 5 | is | be | VBZ | – | 6 | cop |
| 6 | French | french | JJ | NATIONALITY | – | ROOT |
| 7 | – | – | – | – | 6 | punct |

*Removing Punctuation.* Punctuation tokens (e.g., token 7 in Table 1) and their links are removed from the dependency tree. This is easy as they only occur as leaves. Note that the parser takes into account punctuation when extracting the dependency tree.

*Compound Named Entities.* A named entity can overlap several contiguous tokens, for instance *University of Rennes* overlaps tokens 2–4 in the example. However, a named entity is a semantic unit: it holds its own meaning, which can be very different from the meaning of its individual tokens. Therefore, manipulating a named entity as a succession of tokens can cause an important loss of semantics. Except when there is a parse error, the tree structure of an entity is a subtree of the dependency tree. We call it a *factor* by analogy with the definition of a string factor. The proposed solution is to collapse the subtree into its root. Then the sentence retains a valid syntactic and semantic structure (no dangling link for instance). For example, in the sentence presented in Table 1, the named entity *"University of Rennes"* is collapsed into token 2, and tokens 3 and 4 disappear. The expression *"University of Rennes"* can indeed be seen as a proper noun (POS tag *NNP*). In case of a parse error, the named entity is collapsed to the last token as a fallback.

### 3.2 Sentences as an RDF Graph

In order to model a set of processed sentences as one RDF graph, each token is represented by an RDF node (e.g., id:1_2 for the 2nd token of the 1st sentence), and each dependency link is represented by an RDF edge. The lemmas, POS tags, and named entity types of a token are represented by RDF types on the corresponding node (see discussion below). Figure 3 gives the RDF representation for the example in Table 1. The 2nd token is modeled by node *id:1_2*, which has as types lemma *University of Rennes*, POS tag *NNP*, and named entity type *organization*, and is linked to node *id:1_6* by relation *nsubj*.

*Representation of Lemmas and POS Tags as RDF Types.* As specified above, we use relation *rdf:type* instead of defining specific relations for linking a node

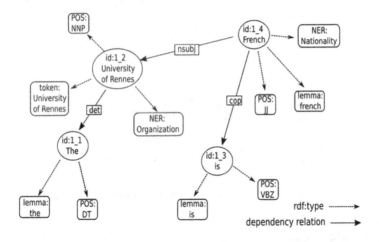

**Fig. 3.** Example of a sentence modeled as an RDF graph.

to its lemma or POS tag. This choice has been made for three reasons. First, by having lemmas and POS tags represented by RDF types rather than RDF nodes, we avoid to have two sentences get connected as soon as they share a lemma or a POS tag. Second, for the computation of Concepts of Neighbours, it prevents from having intensions including dummy patterns such as *"has an unspecified POS tag"*. Third, as presented in Sect. 3.3, it allows us to create a type hierarchy for lemmas and POS tags.

*Lemmatisation of Named Entities.* Note that, if in the general case the lemma of a token is a good representation, it does not stand in the case of named entities: e.g., *unite state of America* vs *United States of America*. Therefore, for named entities, we use the original words instead of the lemmas in the RDF graph.

*Optimization of the Modeling.* Although the algorithm computing Concepts of Neighbours is anytime, the size of the RDF graph has an impact on the number of computed concepts, and hence on the quality of predictions. We can prune the RDF graph according to the position of the subject and object, in a way that reduces its size without loosing too much information. Indeed, Zhang et al. [19] states that not all dependencies are of same interest for extracting relations. Only those close to the path between the subject and the object carry useful information. However, it can be easily seen that reducing the dependency tree to a path would remove essential information for relation extraction, e.g. in the case of a negation attached to a verb that is on the path. Our solution is to prune the dependency tree to keep only the path between the subject and the object, plus the tokens up to maximal distance $K$ from this path. Several values of $K$ were tested, and the value $K = 1$ appears to be a good trade-off between size reduction and performance.

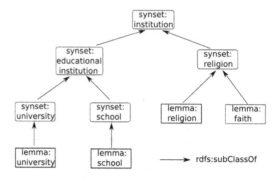

**Fig. 4.** Fragment of a type hierarchy obtained from WordNet

### 3.3 Type Hierarchies

An RDF graph can be enriched with inferred types and edges by declaring domain knowledge. The most common form of domain knowledge is hierarchies of types, based on the RDFS property `rdfs:subClassOf`. The inference rule is that: if node X has type A and A is a subclass of B, then X also has type B. We use several type hierarchies to increase the generalization power of Concepts of Neighbours. First, the set of POS-tags [16] is fine-grained enough to create a hierarchy on a few POS-tags: for example, the gerund of a verb (POS-tag *VBG*) is a subclass of verb (POS-tag *VB*). In total, we have 11 `rdfs:subClassOf` declarations[5].

Second, in order to add semantic knowledge to the modeling, the lexical database WordNet [14] is used for creating a lemma hierarchy for nouns and verbs. Each synset of a lemma is considered as a superclass of the lemma, and each hypernym of a given synset is considered as a superclass of this synset. Figure 4 shows a fragment of this lemma hierarchy. For instance, synset *educational institution* allows to generalize over lemmas *university* and *school*. The lemma hierarchy thus increases the chance to find similarities between sentences using words that have different lemmas but close meanings.

## 4    Relation Extraction with Concepts of Neighbours

Once the modeling is done, for each test example we want to compute which examples of the training corpus are the more similar, and do a prediction from their annotation. In order to do so, an RDF graph regrouping the modeling of all the sentences of the dataset is made, the Concepts of Neighbours method is used to group the examples by similarities, and then a decision method is used to do a prediction from those Concepts of Neighbours. In the following, we first present how the Concepts of Neighbours have been adapted to this specific task.

---

[5] Full hierarchy at https://gitlab.inria.fr/hayats/jena-conceptsofneighbours/-/blob/master/src/conceptualKNN/utils/postag.ttl.

Then, we describe two decision methods for making prediction from Concepts of Neighbours.

## 4.1  Concepts of Neighbours for Relation Extraction in Texts

In order to use Concepts of Neighbours on the modeling presented in Sect. 3, a few aspects needs to be addressed. First, we need to clearly identify the RDF nodes representing the subject and the object in each sentence. Then, we need to compute Concepts of Neighbours on a $(subject, object)$ couple, compared to a single node in previous applications. This strongly increases the number of potential neighbours. Finally, we show how to reduce this number of in the specific case of relation extraction.

**Identification of the Subjects and Objects.** There is a need to unambiguously identify the nodes of the RDF graph forming the $(subject, object)$ couples. The issue is that subjects or objects can overlap several tokens. The collapse of the named entities presented in Sect. 3.1 solves this problem in the vast majority of cases as subjects and objects are most of the time either named entities or one-token expressions. However, ambiguous cases still exist in a small proportion of the sentences. This problem can appear in three cases: first, when the subject or object is expressed as a nominal group (e..g, *"the man"*, *"the university"*); second, when the subject or object includes a named entity but is longer than it (e.g., *"the President of the United States of America"* is a subject, whereas only *"United States of America"* is tagged as a named entity); third, when the subject or object is a named entity that has not been recognized as a named entity.

The solution proposed to solve these cases is similar to the solution presented in Sect. 3.1 to collapse named entities: as a subject or an object is necessarily a group of contiguous tokens with a particular meaning, it must form a tree factor in the dependency tree. Therefore, it can be considered that the root of this factor carries the semantic and syntactic information, and then can be pointed out as the subject or the object. Like for named entities, if the tokens do not form a tree factor because of a wrong annotation or a parse error, the last token is used instead of the root. It can be pointed out that, unlike for named entities, the choice has been made not to merge the tokens constituting the subject or the object as their syntactic structure (if any) is generally informative, like in *"the President of X"*.

**Concepts of Neighbours for Couples of Nodes.** The Concepts of Neighbours of the identified couples $(subject, object)$ are then computed. The method was originally designed to generate Concepts of Neighbours for tuples of arbitrary size. However, until now, it was only applied for unary concepts. The switch from unary concepts to binary ones (and by extension n-ary) has two main consequences discussed in the following.

*Intension.* First of all, unlike unary concepts, binary concepts can have an intension that is not connected. For example, in the sentence presented in Fig. 3, both the connected intension (relating the object to the subject via dependency *nsubj*):

$$(s, o) \leftarrow nsubj(o, s), UnivRennes(s), french(o), cop(o, x), is(x)$$

and the disjoint one (no path between subject and object):

$$(s, o) \leftarrow UnivRennes(s), french(o), cop(o, x), is(x)$$

can appear during the computation of the Concepts of Neighbours. In both cases, the pruning strategy presented in Sect. 3.2 ensures that the intension focuses on and around the path between the subject and the object.

*Reduction of the Set of Couples.* Another issue is the large number of potential neighbours: as detailed in [3], to compute the Concepts of Neighbours for a tuple of $k$ objects in a graph involving $n$ objects, the algorithm has to generate and partition $n^k$ tuples. Therefore, for a large graph, the computation of concepts of arity greater than 1 is rapidly intractable. In the present case, if we consider a dataset composed of ten thousands of sentences, there are tens of billions of potential neighbours. However, the use of the Concepts of Neighbours method is in this context for extracting relations by comparing a *(subject, object)* couple from a test sentence to annotated couples from the training dataset, and it appears that the number of annotated couples is far smaller: only one per example in the training dataset. Therefore, in the following, we use this set of couples for the computation of Concepts of Neighbours, as it permits to simultaneously reduce drastically the computation cost and remove noise from the computed concepts while keeping all the knowledge of interest.

In addition, as evoked in Sect. 3, in a relation extraction dataset each subject and object has a type, and these types can be used to reduce the set of potential neighbours further. For example, if an example has for subject a person and for object a location, it can be seen that the relation expressed by this example could be *place_ of_ birth* or *place_ of_ living*, but can not be *age* or *parent*. Therefore, for a given couple *(subject_type, object_type)*, a set of compatible relations can be deduced from the training dataset. If there is only one compatible relation, this relation can be predicted without computing Concepts of Neighbours for this example, and if there are several possible relations, the set of *(subject, object)* couples from the training dataset that are annotated with compatible relation types can be used as the set of possible neighbours in the algorithm.

## 4.2   Scoring Methods

The computation of Concepts of Neighbours of a *(subject, object)* pair from the test dataset returns a set of concepts, each concept is associated to a set of neighbour couples, and to an extensional distance. In addition, the specialization presented in the previous section ensures that each neighbour couple is annotated with a compatible relation type. From this result, in order to be able to predict a relation, we need to associate a score to each relation type. In the following we present two scoring methods: one based on a weighted votation, and another based on the confidence measure.

*Exponential-Weighted Vote (EV).* Each neighbour $(s, o)$ of each concept "votes" for its annotated relation type $r(s, o)$. However, if not weighted this method will only reflect the proportion of each relation among the training examples annotated with a compatible relation. To avoid this problem, the extensional distance $dist(c)$ of concept $c$ can be used to weight each vote. The extensional distance of a concept of neighbours measures the degree of similarity between the couple from which the concept has been computed and the neighbours that the concept contains: the lower the distance, the higher the similarity. Therefore, each vote is weighted by a decreasing function of the extensional distance. We use the following formula to score a relation type $r$ based on a set of Concepts of Neighbours $C$.

$$score(r, C) := \sum_{c \in C} \sum_{(s,o) \in proper(c)} w(c) \, \mathbf{1}_{r(s,o)=r} \quad \text{where } w(c) = e^{-dist(c)}$$

We have chosen the inverse exponential function to define each weight $w(c)$ because of its rapid decrease, which privileges nearest neighbours. This way, the relation of one very similar example is preferred to the relation of a large number of vaguely similar examples.

*Maximum Confidence (MC).* The second method is similar to the method used by AnyBURL [13], and has been successfully used with Concepts of Neighbours for link prediction [3]. The idea is to consider the intension $int(c) = (s, o) \leftarrow P_c$ of each concept $c$, to use pattern $P_c$ as the body of a rule, and for each relation type $r$ to compute the confidence of the rule $\mathbf{R}_{c,r} : P_c \rightarrow r(s, o)$, defined as usual as:

$$conf(\mathbf{R}_{c,r}) = \frac{|\{(s, o) \mid r(s, o)\} \cap ext(c)|}{|ext(c)|}$$

For each relation type $r$, the score is the list of the confidences of all rules $\mathbf{R}_{c,r}$ predicting that relation, in descending order.

$$score(r, C) := (conf(\mathbf{R}_{c,r}))_{c \in C} \text{in descending order}$$

Such scores are ranked according to inverse lexicographic ordering. That is, the predicted relation type is the relation type with the higher maximal confidence. If several relation types have the same maximal confidence, the relation type with the higher second maximal confidence is predicted, and so on.

## 5    Experiments

In this section we present the experiments conducted on TACRED [20], a standard relation extraction benchmark, to evaluate the proposed method.

## 5.1 Dataset and Baseline

TACRED is a dataset made of 106,264 annotated examples, split into a training corpus (68,124 examples), a development corpus (22,631 examples) and a test dataset (15,509 examples). Each example is a sentence with two identified entity mentions (a subject and an object), typed among 23 possible types (the types used by the Stanford NER system [12]), and annotated with a relation type among 41 effective classes and a *no_relation* class denoting an absence of relation between the two mentions. In order to reflect what can be found in real-world texts, 79.5% of the examples are in the *no_relation* class.

Several remarks can be made about this dataset. First, as the classification with Concepts of Neighbours is a lazy learning method, there is no validation step, then the development dataset can be merged with the training dataset to form a bigger training dataset. Second, the negative examples (those in the *no_relation* class) have no reasons to look like each other but can look like examples in other classes, because they express random situations. Therefore, as our method looks for similarities between a test example and training examples, those negative examples cannot easily be handled in our method. That is why negative examples are removed from the dataset in the experiments, and we focus on discriminating between the 41 relation types rather than discriminating between the presence and absence of a relation. This shrinks the training corpus to 18,446 examples and the test corpus to 3,325 examples.

The evaluation of an approach on TACRED is usually made using the micro-averaged F1 score. However, as the negative examples have been removed from the dataset, this score is equivalent to accuracy. Therefore, accuracy is the measure we use in this evaluation.

The fact that experiments are made on a subset of TACRED causes that direct comparison with existing approaches is no longer possible. In order to evaluate whether our modeling and use of concepts of neighbours are beneficial, we introduce a simple baseline based on named entity types. It predicts the most frequent relation type among the training sentences that have the same subject type and the same object type as the test sentence.

## 5.2 Experimental Settings

As the algorithm to compute the Concepts of Neighbours is anytime, a timeout has to be chosen. In order to see the influence of the computation time on the classification task, eight experiments have been run with respective timeouts: 10, 20, 30, 60, 120, 300, 600 and 1200 s. For each timeout we compare four configurations combining unpruned/pruned modelings (Sect. 3.2) and the two scoring methods (Sect. 4.2).

We have implemented our approach in Java[6], and we use library *ConceptualKNN*[7] for the computation of Concepts of Neighbours, which is based on

---

[6] Code available at https://gitlab.inria.fr/hayats/conceptualknn-relex.

[7] https://gitlab.inria.fr/hayats/jena-conceptsofneighbours.

**Table 2.** Accuracy of our approach depending on timeout, pruning, and scoring method, compared to the baseline.

| Approach | Timeout (seconds) | | | | | | | |
|---|---|---|---|---|---|---|---|---|
| | 10 | 20 | 30 | 60 | 120 | 300 | 600 | 1200 |
| Baseline | 80.4 | | | | | | | |
| Unpruned, EV | 78.6 | 79.1 | 79.2 | 79.4 | 79.2 | 79.2 | 79.6 | 79.6 |
| Unpruned, MC | 78.0 | 78.2 | 78.3 | 78.9 | 79.9 | 79.8 | 80.2 | 80.4 |
| Pruned, EV | 79.5 | 79.7 | 79.6 | 80.1 | 80.4 | 80.3 | 80.4 | 80.4 |
| Pruned, MC | 79.1 | 80.1 | 80.3 | **81.3** | **82.1** | **82.5** | **82.6** | **82.5** |

Apache Jena[8], a Java library for semantic web applications. Experiments have been run on Grid5000[9] to exploit parallel computation.

## 5.3 Quantitative Results

Accuracy of the baseline and of the four versions of the proposed method are presented in Table 2. For a timeout of at least 60s, the proposed approach with the Maximum Confidence scoring and the pruned modeling, has a better accuracy than the baseline, surpassing it by 1.7 point with a timeout of 120s and by 2.2 points with a timeout of over 600s. It can be observed that the pruning of the dependency trees in the modeling is necessary in order to beat the baseline. We assume that, without pruning, the search space for concepts of neighbours gets much larger, and so their computation cannot focus on the useful parts of the dependency trees in the allocated timeout. The EV scoring method shows negative results compared to Maximum Confidence and the baseline. Two conclusions can be made. First, the exponential decrease of vote weights seems to neglect too much distant concepts as Maximum Confidence does not penalize them. Second, it seems better to select a few high-confidence concepts than trying to aggregate predictions from all concepts.

Considering the last line of Table 2 (pruned and MC method), we observe a saturation phenomenon. Indeed, there is an important gain when timeout gets from 10s to 120s, but a far smaller gain from 120s to1200 s. Most of the concepts of neighbours are computed in less than 120 s, and over 120 s, only a few concepts are added. The same phenomenon is seen in Fig. 5. The first chart clearly shows that most of the concepts are obtained in less than 120s. In addition, the second chart shows that, in the case of a model with pruned dependency trees, over 99% of the test sentences have a fully computed set of Concepts of Neighbours – and therefore the predictions are made on the Concepts of Neighbours themselves and not an approximation – for a timeout over 600s, while for a model with full dependency trees, only about 70% of the examples have fully computed set of Concepts of Neighbours.

---

[8] https://jena.apache.org/.
[9] https://www.grid5000.fr/w/Grid5000:Home.

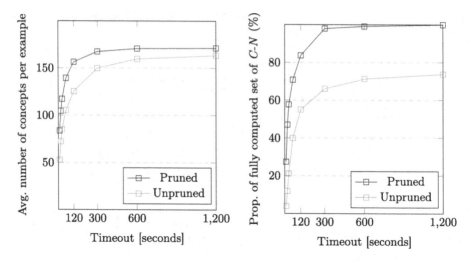

**Fig. 5.** Average number of concepts per example and proportion of fully computed sets of Concepts of Neighbours

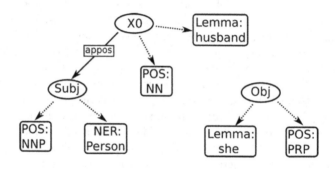

**Fig. 6.** Example of intension of a concept

### 5.4  Qualitative Results

An important asset of this relation extraction method is its interpretability: from a prediction, it can be retrieved which Concepts of Neighbours were used to do this prediction, and for each concept what graph pattern is expressed by the intension and which sentences of the training dataset match this concept.

For illustration, we examine the following example: *"**Her**    husband, **Brad Hagemo**, is an optometrist and Scientologist"*. First, as the subject and the object (in bold) designate persons, there are six possible relations: *per:spouse*, *per:siblings*, *per:parents*, *per:children*, *per:other_family* and *per:alternate_names*. After computation of the Concepts of Neighbours, the relation *per:spouse* is predicted with Maximum Confidence. This prediction is based on the fact that three Concepts of Neighbours predict this relations with a confidence of 1, while at most two concepts predict any other relation with such a confidence. The graph pattern used as intension for these concepts can be made explicit. For example, the intension of one of these concepts is shown in Fig. 6. It expresses that the subject is a person designated as a husband, and the object is a personal pronoun of lemma *"she"*. This intension can effectively predict with quite a high confidence a marital relation. The training examples matching this intension are the sentences *"Kissel had [...] accused **her** husband, Merrill Lynch investment banker **Robert Kissel** of [...] domestic violence."* and *"Jane Callahan Gude, 84, [...] a tireless campaigner for **her** husband, former U.S. Rep. **Gilbert Gude** , died March 24 [...]."*

This example shows that, with this method, an interpretation can be extracted from a prediction. In addition, this shows that a prediction can be made with good confidence from concepts with a disjoint intension. In practice, we observed that concept intensions are rarely connected. A reason is that they may form patterns that are too specific to match any sentence of the training corpus.

# 6 Conclusion

We have presented a lazy-learning method for relation extraction, based on the modeling of linguistic data as a graph and on the computation of Concepts of Neighbours on this graph. This approach has been evaluated and validated against a baseline on the subset of positive examples of the TACRED benchmark. In addition, this comes with the advantage that this approach is interpretable as, for each prediction, detailed information about how this prediction has been made is given in the form of graph patterns over the linguistic structure.

Several aspects of this contribution lead to tracks for future work. First, this method could be coupled with another method – potentially Deep Learning – in order to be able to distinguish the positive examples from the negative ones. The Concepts of Neighbours method can also be improved in order to provide more flexible and expressive patterns, useful in the case of natural language processing. Finally, our approach could be adapted to other NLP tasks that require search for linguistic similarity.

# References

1. Ferré, S., Cellier, P.: Graph-FCA: an extension of formal concept analysis to knowledge graphs. Discrete Appl. Math. **273**, 81–102 (2019)
2. Ferré, S.: Answers partitioning and lazy joins for efficient query relaxation and application to similarity search. In: Gangemi, A., et al. (eds.) ESWC 2018. LNCS, vol. 10843, pp. 209–224. Springer, Cham (2018). https://doi.org/10.1007/978-3-319-93417-4_14
3. Ferré, S.: Application of concepts of neighbours to knowledge graph completion. Data Sci. (2020). https://content.iospress.com/articles/data-science/ds200030, to appear

4. Ferré, S., Ridoux, O.: The use of associative concepts in the incremental building of a logical context. In: Priss, U., Corbett, D., Angelova, G. (eds.) ICCS-ConceptStruct 2002. LNCS (LNAI), vol. 2393, pp. 299–313. Springer, Heidelberg (2002). https://doi.org/10.1007/3-540-45483-7_23

5. Galárraga, L., Teflioudi, C., Hose, K., Suchanek, F.M.: Fast rule mining in ontological knowledge bases with AMIE+. VLDB J. **24**(6), 707–730 (2015)

6. Ganter, B., Kuznetsov, S.O.: Pattern structures and their projections. In: Delugach, H.S., Stumme, G. (eds.) ICCS-ConceptStruct 2001. LNCS (LNAI), vol. 2120, pp. 129–142. Springer, Heidelberg (2001). https://doi.org/10.1007/3-540-44583-8_10

7. Ganter, B., Wille, R.: Formal Concept Analysis: Mathematical Foundations. Springer, Heidelberg (1999). https://doi.org/10.1007/978-3-642-59830-2

8. Grishman, R.: Twenty-five years of information extraction. Nat. Lang. Eng. **25**, 677–692 (2019)

9. Kuznetsov, S.O.: Machine learning and formal concept analysis. In: Eklund, P. (ed.) ICFCA 2004. LNCS (LNAI), vol. 2961, pp. 287–312. Springer, Heidelberg (2004). https://doi.org/10.1007/978-3-540-24651-0_25

10. Kuznetsov, S.O.: Fitting pattern structures to knowledge discovery in big data. In: Cellier, P., Distel, F., Ganter, B. (eds.) ICFCA 2013. LNCS (LNAI), vol. 7880, pp. 254–266. Springer, Heidelberg (2013). https://doi.org/10.1007/978-3-642-38317-5_17

11. Leeuwenberg, A., Buzmakov, A., Toussaint, Y., Napoli, A.: Exploring pattern structures of syntactic trees for relation extraction. In: Baixeries, J., Sacarea, C., Ojeda-Aciego, M. (eds.) ICFCA 2015. LNCS (LNAI), vol. 9113, pp. 153–168. Springer, Cham (2015). https://doi.org/10.1007/978-3-319-19545-2_10

12. Manning, C.D., Surdeanu, M., Bauer, J., Finkel, J., Bethard, S.J., McClosky, D.: The Stanford CoreNLP natural language processing toolkit. In: Proceedings of the 52nd Annual Meeting of the Association for Computational Linguistics: System Demonstrations, pp. 55–60 (2014)

13. Meilicke, C., Chekol, M.W., Ruffinelli, D., Stuckenschmidt, H.: Anytime bottom-up rule learning for knowledge graph completion. In: Proceedings of the Twenty-Eighth International Joint Conference on Artificial Intelligence, pp. 3137–3143 (2019)

14. Miller, G.A.: WordNet: An Electronic Lexical Database. MIT Press, Cambridge (1998)

15. Nguyen, T.H., Grishman, R.: Relation extraction: perspective from convolutional neural networks. In: Proceedings of the 1st Workshop on Vector Space Modeling for Natural Language Processing, pp. 39–48 (2015)

16. Toutanova, K., Klein, D., Manning, C.D., Singer, Y.: Feature-rich part-of-speech tagging with a cyclic dependency network. In: Proceedings of the 2003 Conference of the North American Chapter of the Association for Computational Linguistics on Human Language Technology - NAACL 2003, vol. 1, pp. 173–180 (2003)

17. Wu, F., Zhang, T.: Simplifying graph convolutional networks. In: Proceedings of the 36th International Conference on Machine Learning, p. 11 (2019)

18. Yamada, I., Asai, A., Shindo, H., Takeda, H., Matsumoto, Y.: LUKE: deep contextualized entity representations with entity-aware self-attention. In: Proceedings of the Conference on Empirical Methods in Natural Language Processing (EMNLP), pp. 6442–6454. ACL (2020)

19. Zhang, Y., Qi, P., Manning, C.D.: Graph convolution over pruned dependency trees improves relation extraction. In: Proceedings of the 2018 Conference on Empirical Methods in Natural Language Processing, pp. 2205–2215 (2018)
20. Zhang, Y., Zhong, V., Chen, D., Angeli, G., Manning, C.D.: Position-aware attention and supervised data improve slot filling. In: Proceedings of the 2017 Conference on Empirical Methods in Natural Language Processing, pp. 35–45 (2017)

# Exploration and Visualisation

# Triadic Exploration and Exploration with Multiple Experts

Maximilian Felde[1,2]([✉]) [iD] and Gerd Stumme[1,2] [iD]

[1] Knowledge and Data Engineering Group, University of Kassel, Kassel, Germany
[2] Interdisciplinary Research Center for Information System Design,
University of Kassel, Kassel, Germany
{felde,stumme}@cs.uni-kassel.de

**Abstract.** Formal Concept Analysis (FCA) provides a method called
*attribute exploration* which helps a domain expert discover structural
dependencies in knowledge domains that can be represented by a formal context (a cross table of objects and attributes). Triadic Concept
Analysis is an extension of FCA that incorporates the notion of conditions. Many extensions and variants of attribute exploration have been
studied but only few attempts at incorporating multiple experts have
been made. In this paper we present *triadic exploration* based on Triadic Concept Analysis to explore *conditional attribute implications* in
a triadic domain. We then adapt this approach to formulate attribute
exploration with multiple experts that have different views on a domain.

**Keywords:** Formal concept analysis · Triadic concept analysis ·
Attribute exploration

## 1 Introduction

Attribute exploration [3] is a well established knowledge acquisition method from
the field of Formal Concept Analysis (FCA) [8]. Attribute exploration works on
domains that can be represented as binary tabular data of objects and attributes
(also called features or properties). It helps a domain expert to uncover the
dependency structure of attributes of the domain. For non-binary tabular data
the method of *conceptual scaling*, cf. [7], can be used to transform non-binary
attributes into binary ones.

Attribute exploration is based on the idea that we extend domain information through a domain expert. To this end, attribute exploration uses a question-answer scheme to extract dependency information about attributes. The questions are in the form of *implications*, for example, *do attributes A and B imply
attribute C?* (also written as $AB \rightarrow C$?). The expert's task is to confirm or
refute the validity of such implications in the domain. If the expert refutes the
validity of an implication she has to offer a counterexample, for example, in case
of the question $AB \rightarrow C$? an object of the domain that has the attributes $A$ and
$B$ but lacks attribute $C$.

© Springer Nature Switzerland AG 2021
A. Braud et al. (Eds.): ICFCA 2021, LNAI 12733, pp. 175–191, 2021.
https://doi.org/10.1007/978-3-030-77867-5_11

The attribute exploration algorithm asks these questions in an optimized manner such that the expert has to answer as few questions as possible until the validity of every conceivable implication can be inferred from the answers given by the expert. This is the case when every implication either follows from the set of implications accepted as valid or is contradicted by one of the examples given by the expert.

The basic version of attribute exploration requires an all-knowing expert of the domain, i.e. an expert who can answer any question about the domain correctly. It was introduced by Ganter in [3]. Since then, many variants and extensions of attribute exploration have been studied. A good overview can be found in the book *Conceptual Exploration* by Ganter and Obiedkov [6]. These extensions and variants notably include: Attribute exploration with background knowledge and exceptions [4,18], where the idea is to support the exploration with prior knowledge about some of the relations between attributes, for example if one attribute is the negation of another; attribute exploration with partial information [11–13], where the expert is not required to be all-knowing and is also allowed to answer *I do not know* in addition to confirming or refuting a question. Further, the expert is not required to fully specify a counterexample as long as the specified parts contradict the implication in question; and a sketch of how to explore triadic formal contexts [5,6], where the idea of attribute exploration is transferred to triadic concept analysis (an extension of FCA with conditions [17]). We elaborate further on this in Sect. 3.

However, most of the extensions and variants of attribute exploration that have been studied are based on the idea of a single expert answering the questions. As far as we know, there exist only a few papers that mention exploration with multiple experts, notably: Paper [16] deals with how to perform exploration in parallel and potentially offers a way to speed up the exploration with multiple experts; [10] addresses collaborative conceptual exploration based on the notions of local experts for subdomains of a given knowledge domain; and [2] studies attribute exploration in a collaborative exploration setting with multiple experts who share the same view on the domain but only have partial knowledge thereof.

When we explore a domain with multiple experts, one of the fundamental problems we face is that different views on a domain, for example different opinions whether an object has an attribute or not, or whether an implication is valid or not in a domain, are impossible to resolve by combining different pieces of information into one. Either, because there is no clear *right* or *wrong*, e.g. in case of opinions, or simply because we can not know which information to trust most. And, even if we used methods such as majority-voting on information, there is a reasonable chance that the result is not always correct. Combined with the inherent non-robustness of implication theories, i.e., small changes in the underlying data can lead to a very different theory, this suggests that merging different views on a domain is a bad idea for attribute exploration. If we take a closer look at the publications mentioned before, we see that all three avoid this issue in their own way. In [16] the experts all have the same complete knowledge about the domain; in [10] the local experts have partial knowledge about the same consistent domain knowledge; and, in [2] the problem was also avoided by defining expert knowledge as partial knowledge of some consistent domain knowledge.

Attribute exploration where multiple experts can have truly different and even opposing views on the domain has to the best of our knowledge not yet been studied. To this end we develop *triadic exploration* based on ideas presented by Ganter and Obiedkov in [5]. We then adapt triadic exploration to the setting of multiple experts with different views on a domain and thus provide a step in the direction of attribute exploration with multiple experts.

The paper is structured as follows: We begin by giving a brief introduction to the problem in Sect. 1. We recollect some fundamentals of Formal and Triadic Concept Analysis in Sect. 2, in particular *formal* and *triadic contexts*, *attribute implications*, the *relative canonical base* and *attribute exploration*. In Sect. 3, we discuss implications in the triadic setting, in particular, we focus on *conditional attribute implications*. Subsequently, we formulate *triadic exploration*. In Sect. 4, we discuss how to adapt *triadic exploration* to model attribute exploration with multiple experts with different views. Finally, Sect. 5 contains conclusion and outlook. Note that for this paper we do not provide a separate section for related work, instead we address related work throughout the paper whenever appropriate.

## 2 Dyadic and Triadic Formal Contexts

In this section we recollect the fundamentals of (dyadic) Formal Concept Analysis and Triadic Formal Concept Analysis (TCA). We mostly rely on [8,19] for FCA and on [17,20] for TCA. We begin with the definition of *formal contexts* and associated notions. We then introduce *triadic contexts* and give an example which will serve as our running example for the remainder of this paper. Afterwards, we briefly cover *attribute implications*, the *relative canonical base* and *attribute exploration*. This serves as a foundation for Sect. 3, where we look at *implications in the triadic setting* and subsequently develop *triadic exploration*.

### 2.1 Formal Concept Analysis

Formal Concept Analysis was introduced by Wille in [19]. As the theory matured, Ganter and Wille compiled the mathematical foundations of the theory in [8]. A *formal context* $\mathbb{K} = (G, M, I)$ consists of a set $G$ of objects, a set $M$ of attributes and an incidence relation $I \subseteq G \times M$ with $(g, m) \in I$ meaning *object $g$ has attribute $m$*. We define two derivation operators $(\cdot)' : \mathcal{P}(M) \to \mathcal{P}(G)$ and $(\cdot)' : \mathcal{P}(G) \to \mathcal{P}(M)$ in the following way: For a set of objects $A \subseteq G$, the set of *attributes common to the objects in $A$* is provided by $A' := \{m \in M \mid \forall g \in A : (g, m) \in I\}$. Analogously, for a set of attributes $B \subseteq M$, the set of *objects that have all the attributes from $B$* is provided by $B' := \{g \in G \mid \forall m \in B : (g, m) \in I\}$. A *formal concept* of a formal context $\mathbb{K} = (G, M, I)$ is a pair $(A, B)$ with $A \subseteq G$ and $B \subseteq M$ such that $A' = B$ and $A = B'$. We call $A$ the *extent* and $B$ the *intent* of the formal concept $(A, B)$. The set of all formal concepts of a context $\mathbb{K}$ is denoted by $\mathfrak{B}(\mathbb{K})$. Note that for any set $A \subseteq G$ the set $A'$ is the intent of a concept and for any set $B \subseteq M$ the set $B'$ is the extent of a concept.

The subconcept-superconcept relation on $\mathcal{B}(\mathbb{K})$ is formalized by: $(A_1, B_1) \leq (A_2, B_2) :\Leftrightarrow A_1 \subseteq A_2 (\Leftrightarrow B_1 \supseteq B_2)$. The set of concepts together with this order relation $(\mathfrak{B}(\mathbb{K}), \leq)$ forms a complete lattice, the *concept lattice*. The vertical combination of two formal contexts $\mathbb{K}_i = (G_i, M, I_i), i \in \{1, 2\}$ with the same set of attributes $M$ is called the *subposition* of $\mathbb{K}_1$ and $\mathbb{K}_2$. Formally, it is defined as $(\dot{G}_1 \cup \dot{G}_2, M, \dot{I}_1 \cup \dot{I}_2)$, where $\dot{G}_i := \{i\} \times G$ and $\dot{I}_i := \{((i, g), m) | (g, m) \in I_i\}$ for $i \in \{1, 2\}$. The *subposition* of a set of contexts on the same set of attributes is defined analogously and we denote this by $subpos(\cdot)$.

## 2.2   Triadic Concept Analysis

Triadic Concept Analysis (TCA) was introduced by Lehmann and Wille in [17] as an extension to Formal Concept Analysis with conditions. In particular they introduced the notion of *triadic concepts* for which Wille proceeded to show the basic theorem of triadic concept analysis in [20] – clarifying the connection between triadic concepts and complete tri-lattices, analogous to the dyadic case.

The basic structure in TCA is a *triadic context* which is similar to the formal context in FCA. A *triadic context* is defined as a quadruple $\mathbb{T} = (G, M, B, Y)$, where $G, M$ and $B$ are sets and $Y \subseteq G \times M \times B$ is a ternary relation on these sets. The elements of $G, M$ and $B$ are called objects, attributes and conditions respectively. For $g \in G$, $m \in M$ and $b \in B$ with $(g, m, b) \in Y$ we say that *object g has attribute m under condition b*. The conditions are understood in a broad sense, cf. [20]: They comprise, amongst others, relations, interpretations, meanings, purposes and reasons concerning the connections of objects and attributes.

*Example 1.* The following example[1] will serve as our running example throughout the paper. It shows the situation of public transport at the train station Bf. Wilhelmshöhe with direction to the city center in Kassel. From Bf. Wilhelmshöhe you can travel by one of four bus lines (52, 55, 100 and 500), four tram lines (1, 3, 4 and 7), one night tram (N3) and one regional tram (RT5) to the city center. These are the objects $G_{\text{ex.}}$ of our context. The buses and trams leave the station at different times throughout the day. The attributes $M_{\text{ex.}}$ of our context are the aggregated leave-times, more specifically, we have split each day in five distinct time-slots: early morning (4:00 to 7:00), working hours (7:00 to 19:00), evening (19:00 to 21:00), late evening (21:00 to 24:00) and night (0:00 to 4:00). The conditions $B_{\text{ex.}}$ of our context are the days of the week. A bus or tram line is related to a time-slot on a day if a bus or tram of this line leaves the station at least once during the time-slot on the day. This describes the ternary relation $Y \subseteq G_{\text{ex.}} \times M_{\text{ex.}} \times B_{\text{ex.}}$. We have aggregated Monday to Friday into a single condition, because the schedule is the same for these days. Thus, we obtain the context $\mathbb{T}_{\text{ex.}} = (G_{\text{ex.}}, M_{\text{ex.}}, B_{\text{ex.}}, Y)$. The resulting triadic context can be found in Fig. 1.

Naturally, we can view the triadic context as a family of formal contexts, where each context represents one condition, basically slicing the triadic context

---

[1] The example is similar to the one given in [6], which inspired it.

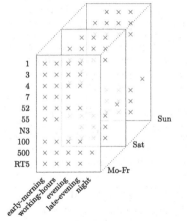

**Fig. 1.** Triadic context $\mathbb{T}_{ex}$. of Example 1

| Mo-Fr | early-morning | working-hours | evening | late-evening | night |
|---|---|---|---|---|---|
| 1 | × | × | × | × | |
| 3 | × | × | × | × | |
| 4 | × | × | × | × | |
| 7 | × | × | × | | |
| 52 | × | × | × | × | |
| 55 | × | × | | | |
| N3 | | | | | |
| 100 | × | × | × | × | |
| 500 | × | × | × | × | × |
| RT5 | × | × | × | × | |

| Sat | early-morning | working-hours | evening | late-evening | night |
|---|---|---|---|---|---|
| 1 | | × | × | × | |
| 3 | × | × | × | × | |
| 4 | × | × | × | × | |
| 7 | | × | | | |
| 52 | × | × | | | |
| 55 | | | | | |
| N3 | | | | | × |
| 100 | × | × | × | × | |
| 500 | | × | × | × | |
| RT5 | × | × | × | × | |

| Sun | early-morning | working-hours | evening | late-evening | night |
|---|---|---|---|---|---|
| 1 | | × | × | × | |
| 3 | × | × | × | × | |
| 4 | × | × | × | × | |
| 7 | | | | | |
| 52 | | | | | |
| 55 | | | | | |
| N3 | | | | | × |
| 100 | × | × | × | × | |
| 500 | | × | × | × | |
| RT5 | | × | × | × | |

**Fig. 2.** The triadic context $\mathbb{T}_{ex}$. from Example 1 represented as context family of the condition contexts $\mathbb{K}_{\text{Mo-Fr}}$, $\mathbb{K}_{\text{Sat}}$ and $\mathbb{K}_{\text{Sun}}$

vertically along the conditions. In Fig. 2 we provide the resulting context family of our running example.

Formally such a family of contexts representing a triadic context $\mathbb{T} = (G, M, B, Y)$ is a set of contexts $\mathbb{K}_b$, $b \in B$ where $\mathbb{K}_b := (G, M, I_b)$ with $(g, m) \in I_b :\Leftrightarrow (g, m, b) \in Y$. We will refer to the contexts $\mathbb{K}_b$ as *condition contexts* of the triadic context $\mathbb{T}$; for our example these are $\mathbb{K}_{\text{Mo-Fr}}$, $\mathbb{K}_{\text{Sat}}$ and $\mathbb{K}_{\text{Sun}}$.

## 2.3 Attribute Implications

Attribute implications are used to describe dependencies between attributes in a formal context. In the following we give a brief introduction. Let $M$ be a set of attributes. (For a start, we do not require it to be related to a specific context.) An *attribute implication* over $M$ is a pair of subsets $A, B \subseteq M$ of $M$. We denote this by $A \to B$. We call $A$ the *premise* and $B$ the *conclusion* of the implication $A \to B$.

We denote the set of all implications over a set $M$ by $\text{Imp}_M = \{A \to B | A, B \subseteq M\}$.

A subset $T \subseteq M$ *respects* an attribute implication $A \to B$ over $M$ if $A \not\subseteq T$ or $B \subseteq T$. We then also call $T$ a *model* of the implication. $T$ *respects a set* $\mathcal{L}$ of implications if $T$ respects all implications in $\mathcal{L}$. An implication $A \to B$ *holds* in a set of subsets of $M$ if each of these subsets respects the implication.

For a formal context $\mathbb{K} = (G, M, I)$ we say that an implication $A \to B$ over $M$ *holds in the context* if for every object $g \in G$ the object intent $g'$ respects the implication. We then also call $A \to B$ a *valid implication* of $\mathbb{K}$. An implication $A \to B$ holds in $\mathbb{K}$ if and only if every object $g \in G$ that has all attributes in $A$ also has all attributes in $B$. Further, an implication $A \to B$ holds in $\mathbb{K}$ if and only if $B \subseteq A''$, or equivalently $A' \subseteq B'$. An implication $A \to B$ *follows* from a set $\mathcal{L}$ of implications over $M$ if each subset of $M$ respecting $\mathcal{L}$ also respects $A \to B$. A family of implications is called *closed* if every implication following from $\mathcal{L}$ is already contained in $\mathcal{L}$. Closed sets of implications are also called *implication theories*.

**Relative Canonical Base.** The set of all implications that hold in a given context $\mathbb{K}$ have a canonical irredundant representation which is called the *canonical base*, cf. [8,9]. Stumme has generalized this representation to the case where some (background) implications are known [18], i.e. attribute implications that are known to hold based on prior knowledge.

Given a formal context $\mathbb{K} = (G, M, I)$ and a set of (background) implications $\mathcal{L}_0$ on $M$ that hold in the context $\mathbb{K}$, a *pseudo-intent* of $\mathbb{K}$ *relative to* $\mathcal{L}_0$ is a set $P \subseteq M$ where $P$ respects $\mathcal{L}_0$, $P \neq P''$ and if $Q \subseteq P$, $Q \neq P$, is a relative pseudo-intent of $\mathbb{K}$ then $Q'' \subseteq P$. The set $\mathcal{L}_{\mathbb{K},\mathcal{L}_0} := \{P \to P'' | P$ relative pseudo-intent of $\mathbb{K}\}$ is called the *canonical base* of $\mathbb{K}$ *relative to* $\mathcal{L}_0$, or simply the *relative canonical base*. All implications in $\mathcal{L}_{\mathbb{K},\mathcal{L}_0}$ hold in $\mathbb{K}$.

**Theorem 1. (see [5,18]).** *If all implications of $\mathcal{L}_0$ hold in $\mathbb{K}$, then*

1. *each implication that holds in $\mathbb{K}$ follows from $\mathcal{L}_{\mathbb{K},\mathcal{L}_0} \cup \mathcal{L}_0$, and*
2. *$\mathcal{L}_{\mathbb{K},\mathcal{L}_0}$ is irredundant w.r.t. 1.*

The notion of a relative canonical base combined with Theorem 1 allows us to reduce the amount of questions that need to be posed during a triadic exploration.

## 2.4 Attribute Exploration

Attribute exploration ([3], cf. also [6,8]) is a knowledge acquisition method based on a question-answer scheme to obtain the implication theory of a domain.

Let us consider a domain (a formal context) $(G, M, I)$ that we do not know completely and that we want to explore and a domain expert for this domain. We start with a (possibly empty) set of known (background) implications $\mathcal{L}$ and a (possibly empty) set $G_E \subseteq G$ of known objects, represented as (possibly empty) formal context $\mathbb{E} = (G_E, M, I_E)$. In every step of the attribute exploration we have a set of already accepted implications $\mathcal{L}$ and a context of already provided counterexamples $\mathbb{E}$. The attribute exploration algorithm picks the next implication $A \to B$ that does not follow from $\mathcal{L}$ and that holds in $\mathbb{E}$. It then asks the expert whether the implication truly holds in the domain. The expert can either confirm that the implication holds or they can refute its validity by providing a counterexample, i.e., an object $g \in G$ whose intent does not respect the implication. If the expert confirms the implication's validity in the domain, it is added to the set $\mathcal{L}$, otherwise the provided counterexample is added to the context of counterexamples $\mathbb{E}$. This process is repeated until there is no implication left to be asked.

After performing the attribute exploration we have the (relative) canonical base of implications from which (combined with the background implications) every valid implication in the domain follows. Furthermore, for every implication that is not valid, the set of examples contains a counterexample.

## 3    Triadic Exploration

In this section we look at *implications* in the triadic setting, in particular, we formally introduce *conditional attribute implications*, and develop a *triadic exploration* for Triadic Concept Analysis as proposed by Ganter and Obiedkov in [5,6].

### 3.1    Conditional Attribute Implications

In formal contexts (of type $(G, M, I)$) the matter of *implications* is fairly straightforward: There are *attribute implications* to describe dependencies between attributes (and dually there are *object implications*). In triadic contexts, the notion of implication is not as simple. This manifests in a multitude of types of implications that have been proposed: The earliest suggestion for a *triadic implication* came from Biedermann [1], where he suggested the study of implications of the form $(R \to S)_C$ which is interpreted as: *If an object has all attributes from R under all conditions from C, then it also has all attributes from S under all conditions from C.*

In [5], Ganter and Obiedkov studied some other types of implications for the triadic setting. They introduced a stronger version of the triadic implication called *conditional attribute implications* to describe dependencies that hold for some conditions. The symmetry arising from the arbitrary choice of objects, attributes and conditions in a triadic context results in five more types of implications. Further, they introduced another generalization of Biedermann's triadic implication called *attribute × condition implication* to express dependencies between combinations of attributes and conditions. For the remainder of this paper we will focus on *conditional attribute implications*, because they best serve our goal of developing attribute exploration with multiple experts.

Given a triadic context $\mathbb{T} = (G, M, B, Y)$, a *conditional attribute implication* is an expression of the form $R \xrightarrow{C} S$ where $R, S \subseteq M$, $C \subseteq B$, which reads as: $R$ *implies* $S$ *under all conditions from* $C$. A conditional attribute implication $R \xrightarrow{C} S$ *holds* in a triadic context $\mathbb{T}$ iff for each condition $c \in C$ it holds that if an object $g \in G$ has all the attributes in $R$ it also has all the attributes in $S$. This is the case if the implication $R \to S$ holds in every conditional context $\mathbb{K}_c$ for $c \in C$.

**Proposition 1.** *Let* $\mathbb{T} = (G, M, B, Y)$ *and* $\mathbb{K}_c$, $c \in B$, *be its respective condition contexts. For a conditional implication* $R \xrightarrow{C} S$ *with* $R, S \subseteq M$ *and* $C \subseteq B$, *the following statements are equivalent:*

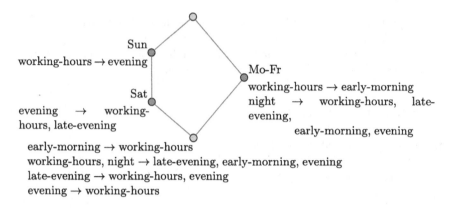

**Fig. 3.** The lattice of conditional implications of the running example $\mathbb{T}_{\text{ex}}$. with simplified labels, which consist of the relative canonical base with respect to the implications in all nodes below. We omit the top label of implications as the extent of this concept is always $\text{Imp}_M$.

1. $R \xrightarrow{C} S$ holds in $\mathbb{T}$
2. $R \rightarrow S$ holds in $\mathbb{K}_c$ for every $c \in C$
3. $R \rightarrow S$ holds in $subpos(\{\mathbb{K}_c | c \in C\})$

*Proof.* 1. $\Leftrightarrow$ 2. follows directly from the definitions of *holds* in the triadic and dyadic setting. 2. $\Leftrightarrow$ 3. follows from the definition of subposition and that an implication $R \rightarrow S$ holds in a context if and only if for every object $g$ the object intent respects the implication.                                                                    □

*Example 2.* In the context family of Example 1 in Fig. 2 we observe that the implication $early - morning \rightarrow working - hours$ holds in all three condition contexts $\mathbb{K}_{\text{Mo-Fr}}$, $\mathbb{K}_{\text{Sat}}$ and $\mathbb{K}_{\text{Sun}}$, hence, $early - morning \xrightarrow{\text{Mo-Fr,Sat,Sun}} working - hours$ holds in $\mathbb{T}_{\text{ex}}$.. In contrast, the implication $working - hours \rightarrow evening$ only holds in the condition context $\mathbb{K}_{\text{Sun}}$ because tram line 7 is a counterexample in $\mathbb{K}_{\text{Sat}}$ and bus line 55 is a counterexample in $\mathbb{K}_{\text{Mo-Fr}}$ and thus $working - hours \xrightarrow{\text{Sun}} evening$ holds in $\mathbb{T}_{\text{ex}}$., but $working - hours \xrightarrow{\text{Mo-Fr,Sat,Sun}} evening$ does not.

Clearly, if a conditional implication $R \xrightarrow{C} S$ holds in a triadic context $\mathbb{T}$ then all conditional implications $R \xrightarrow{D} S$ with $D \subseteq C$ hold as well. Further, for every subset $C \subseteq B$ there is a set of conditional implications $R \xrightarrow{C} S$ that hold in $\mathbb{T}$. This set of conditional implications for a fixed set of conditions $C$ is the implication theory of the subposition of condition contexts $subpos(\{\mathbb{K}_c | c \in C\})$.

**Context of Conditional Implications.** A nice way to structure the conditional implications that hold in a triadic context $\mathbb{T}$ is to use the approach suggested by Ganter and Obiedkov, cf. [5], and to introduce a *context of conditional implications*: Given a triadic context $\mathbb{T}$, we construct a formal context $\mathcal{C}_{imp}(\mathbb{T}) := (\text{Imp}_M, B, I)$, where the set of all possible implications on $M$ is the object set , the set of conditions $B$ of the triadic context $\mathbb{T}$ is the set of attributes and the incidence relation $I$ is determined by

$$(R \to S)Ic :\Leftrightarrow R \xrightarrow{c} S \text{ holds in } \mathbb{T}.$$

The formal concepts of $\mathcal{C}_{imp}(\mathbb{T})$ are pairs $(\mathcal{L}, C)$, where $\mathcal{L}$ is a set of implications and $C$ is a set of conditions, such that $\mathcal{L}$ is the set of all implications $R \to S$ for which $R \xrightarrow{C} S$ holds, and $C$ is the largest set of conditions for which this is the case. These concepts structure the set of conditional implications in a lattice ordered by the conditions for which they hold. Their extents form a system of implication theories.

*Example 3.* For our running example we present the concept lattice of $\mathcal{C}_{imp}(\mathbb{T}_{\text{ex}}.)$ with simplified labels in Fig. 3: The extent of the top node always contains the implications that hold under the empty set of conditions, i.e., the whole set $\text{Imp}_M$. We omit this label. For the other nodes we give the relative canonical base with respect to set of implications from all nodes below. Looking at the implications from Example 2, we find the implication *early − morning →  working − hours* at the *bottom node*, because it holds for all three conditions, whereas we find the implication *working − hours → evening* at the node for *Sunday*, because that is the only condition for which it holds.

## 3.2   Triadic Exploration

Now, we develop *Triadic Exploration* to explore the conditional implications of a triadic domain.

Previously, we have structured the conditional implications of a triadic domain $\mathbb{T}$ as a system of implication theories by utilizing the context of conditional implications $\mathcal{C}_{imp}(\mathbb{T})$. This was possible because we had complete information about the domains implications in the context $\mathbb{T}$. However, it is easy to imagine a situation where we can access the information about a domain only indirectly through a domain expert and where an attribute exploration might be useful. For our running example, imagine someone with a bus and train schedule where the information can be looked up but is not fully available at once. Now the question is: How to explore the complete system of conditional implications?

A naive approach is to explore the implication theory for each fixed subset of the conditions, essentially exploring each node of the system of implication theories independently. But, this is clearly not a good idea; it means answering many questions multiple times for each condition.

A better approach might be to only explore the implication theory for every condition, each providing one column in the context of conditional implications $\mathcal{C}_{imp}$. Then we can compute the concept lattice without any further interactions with the expert.

However, there are some points to consider that suggest a different approach, cf. [6]: First, to stay in the triadic setting, a complete counterexample to a question should describe the new object by the attributes it has *for each of the conditions*, and not only for the one, that is currently under consideration. And second, some implications may hold for several conditions and the domain expert might want to confirm each of them for multiple conditions at once.

Thus, we come back to the context of conditional implications. Ganter and Obiedkov suggested to explore the triadic domain by exploring the nodes in the lattice of conditional implications from the bottom up; using the already known valid implications as background knowledge. Hence, as we explore the system of conditional implications, we successively fill the context of conditional implications.

In the following we describe the nested process of exploring the nodes of the concept lattice of conditional implications with the help of two algorithms: Algorithm 1 for the exploration of the conditional implications for a fixed set of conditions and Algorithm 2 that uses this algorithm as a subroutine to explore all conditional implications of the triadic domain.

**Explore Conditional Implications for a Fixed Set of Conditions.** For a fixed set of conditions $D \subseteq B$ in a triadic domain $\mathbb{T} = (G, M, B, Y)$, the exploration algorithm is an adapted version of the algorithm for attribute exploration with background implications and exceptions, see [6,18]. In Algorithm 1 we present an implementation for the exploration in pseudo-code.

The algorithm starts with some background knowledge, in particular: A triadic context $\mathbb{E} = (G_E, M, B, Y_E)$, that contains some examples from the domain $\mathbb{T}$, and a set of implications $\mathcal{L}_0$ that are known to hold for all conditions in $D$; both of these can be empty. The rest of the domain can only be accessed by the algorithm through interaction with the domain expert. In each step, the algorithm determines the next implication $A \to A''$ to ask the expert. To determine the next question $A \to A''$ the algorithm uses both the information from the examples in $\mathbb{E}$ and the known valid implications in $\mathcal{L}$. It automatically skips questions that follow from the implications in $\mathcal{L}$ or for which $\mathbb{E}$ already contains a counterexample. More precisely, $A$ is the next *relative pseudo-intent* in $subpos(\{\mathbb{K}_d | d \in D\})$, i.e., the lectically smallest set $A$ closed under the set of known valid implications and background implications $\mathcal{L}$ that is not already closed in the subposition context of examples for the conditions in $D$.

Essentially, this algorithm is an attribute exploration with background implications on the subposition of the condition contexts. Additionally, it tracks which implications hold for which conditions in $D$. This enables us to reduce the amount of interaction required from the expert in subsequent explorations by preventing to ask the same question multiple times for different subposition contexts. The proof of correctness for Algorithm 1 is a straightforward adaption of the proof of [18, Theorem 6] and we therefore omit the details.

Note that we chose to collect all implications that are asked about and the subset of conditions of $D$ for which they hold in Line 13 instead of only adding

---

**Algorithm 1.** explore-conditions

---

**Input:** a set of conditions $D \subseteq B$, a triadic context $\mathbb{E} = (G_E, M, B, Y_E)$ of examples (possibly empty) and a set $\mathcal{L}_0$ of background implications known to hold for all conditions in $D$ (also possibly empty)

**Interactive Input:** ($\star$) The expert confirms or rejects an implication to hold for the set of conditions $D$. Upon rejection the expert provides a counterexample $g$ from the domain together with its relation to all conditions and all attributes, i.e., the context $\mathbb{K}_g := (M, B, I)$ where $(m, b) \in I \Leftrightarrow g$ has $m$ under the condition $b$ in the domain.

**Output:** the relative canonical base $\mathcal{L} \setminus \mathcal{L}_0$ of implications that hold for all conditions in $D$ with respect to $\mathcal{L}_0$, a possibly enlarged triadic context of counterexamples $\mathbb{E}$ and the formal context $\mathbb{C}$ of asked implications and the conditions for which they hold.

1   $\mathcal{L} := \mathcal{L}_0$
2   $A := \emptyset$
3   $\mathbb{C} := (\emptyset, B, \emptyset)$
4   **while** $A \neq M$ **do**
5      **while** $A \neq A''$ in $S$ where
6      $S := (G_S, M, J) = $ subposition of $\mathbb{K}_d$ for $d \in D$ with
7      $K_d := (G_E, M, I_d)$ where $(g, m) \in I_d \Leftrightarrow (g, m, d) \in Y_E$
8      **do**
9          Ask the expert if $A \to A''$ holds for all conditions $d \in D$     ($\star$)
10          **if** $A \to A''$ *holds* **then** $\mathcal{L} := \mathcal{L} \cup \{A \to A''\}$
11          **else** extend $\mathbb{E}$ with the counterexample provided by the expert     ($\star$)
12
13          extend $\mathbb{C}$ with the object $A \to A''$ and its relation to all conditions $d \in D$    ($\star$)
14      **end**
15      $A := \text{NextClosure}(A, M, \mathcal{L})$     /* computes the next closure of $A$ in $M$ with respect to the implications in $\mathcal{L}$; see for example [6,8] */
16   **end**
17   **return** $\mathcal{L} \setminus \mathcal{L}_0$, $\mathbb{E}$ and $\mathbb{C}$

---

the implications that hold for all conditions in the context $\mathbb{C}$. Hence, if there is a counterexample, i.e., the implication does not hold for $D$, we track for which subset of $D$ (if any) the implication does hold. This further reduces the number of questions posed in later explorations. The trade-off is that the background knowledge we have is not just of nodes below the currently explored one in the lattice but may also contain implications that first hold for the conditions of the current node. This has no effect on the implication theory of the node but somewhat complicates the labeling of the node – we cannot simply use the relative canonical base with respect to the knowledge we have. In contrast, if we only added the implications that hold for all conditions in the current exploration then the labels are exactly the implications of the relative canonical base, but, we might have to ask some questions multiple times for some of the conditions. For our running example this approach further reduces the number of questions posed to the expert from fifteen to twelve, cf. Example 5 in Sect. 3.3.

**The Order of Explorations.** To determine the sequence in which the nodes of the lattice of conditional implications are explored, Ganter and Obiedkov further suggested to follow a linear extension of the lattice of conditional implications, see [5], and later specified this to follow the *NextExtent-Algorithm*, i.e., *NextClosure* on the extents, on the context of conditional attribute implications, see [6].

However, in our setting the *NextExtent-Algorithm* does not fit. The problem is that we may not have the necessary information to correctly determine the next node to explore.[2] This is because the questions that are asked during the exploration of a node are not guaranteed to discriminate between the conditions that are being explored. Questions that would discriminate between conditions are not asked if there already exists a counterexample for any one of the conditions. This might result in not exploring all nodes of the lattice.

*Example 4.* Let us illustrate the problem with a small example: Take a look at the domain given by the triadic context $\mathbb{T}_1$ in Fig. 4. If we explore this context and begin with the bottom node, i.e., the implications that hold for all conditions without any background knowledge, then the first question posed to the expert is $\emptyset \to ab$?, which the expert refutes with a counterexample – object 1 with all its attributes under all conditions. It substantiates that implication holds for neither of the conditions. The second question that is posed to the expert is $b \to a$? which the expert confirms. This concludes the exploration of the bottom node in this example. If we now compute the next extent in the resulting context of conditional implications $\mathbb{C}$ in order to determine which node to explore next, we obtain NextExtent$(\emptyset) = \{b \to a\}$ with intent $\{d_1, d_2\}$ which we just explored and then NextExtent$(\{b \to a\}) = G_\mathbb{C}$ with intent $\emptyset$ which concludes the exploration. However, clearly the implication $\emptyset \to a$ holds in $d_1$ but not in $d_2$ and is missing in $\mathbb{C}$. The question $\emptyset \to a$? was not posed to the expert because there already existed a counterexample for condition $d_2$ after the first question. Similarly, the implication $a \to b$ holds in $d_2$ but not in $d_1$ and is also missing. In Fig. 4, we present both the lattice of $\mathbb{C}$ and the lattice of $\mathcal{C}_{imp}(\mathbb{T}_1)$. Hence, an exploration that uses the *NextExtent-Algorithm* to determine which nodes of the conditional implications lattice to explore next does not necessarily explore all nodes of the lattice.

To circumvent this problem, we use the suggested strategy of exploring the lattice node by node from the bottom up with the already known valid conditional implications in $\mathcal{C}_{imp}$ as background knowledge. But, instead of using the *NextExtent-Algorithm* to incrementally determine the next combination of conditions to explore, we simply follow a linear extension of $(\mathcal{P}(B) \setminus \emptyset, \supseteq)$. Which means, we walk through all subsets of $B$ sorted by their cardinality from biggest to smallest and stop when we have explored all subsets of cardinality one. At first glance this might look as if we explore more nodes than necessary, because

---

[2] For the same reason, the nested application of NextClosure for computing all concepts of a triadic context, as described in [14,15], cannot serve as a base for the triadic exploration.

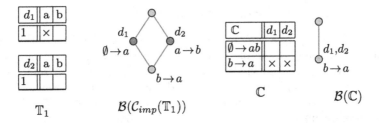

$$\mathbb{T}_1 \qquad\qquad \mathcal{B}(\mathcal{C}_{imp}(\mathbb{T}_1))$$

**Fig. 4.** A triadic context $\mathbb{T}_1$, the lattice of conditional implications of $\mathbb{T}_1$, the context $\mathbb{C}$ after exploring the conditional implications of $\mathbb{T}_1$ using the *NextExtent-Algorithm* to determine the next conditions to explore, and the lattice of $\mathbb{C}$

---

**Algorithm 2.** triadic-exploration

**Input:** a triadic context $\mathbb{E} = (G_E, M, B, Y_E)$ of examples (possibly empty) and a context $\mathbb{C} = (G_\mathbb{C}, B, I_\mathbb{C})$ of implications known to hold for some conditions

**Output:** a triadic context of counterexamples $\mathbb{E}$ and the context of conditional implications $\mathbb{C}$, from which all valid conditional implications can be inferred

1 **for** $D$ *in linear extension of* $(\mathcal{P}(B) \setminus \emptyset, \supseteq)$ **do**
2     $\mathcal{L} := D'$ (in $\mathbb{C}$)
3     $\mathcal{L}_D, \mathbb{E}_D, \mathbb{C}_D := \text{explore-conditions}(D, \mathbb{E}, \mathcal{L})$
4     $\mathbb{E} := \mathbb{E}_D$
5     $\mathbb{C} := \mathbb{C} \cup \mathbb{C}_D = (G_\mathbb{C} \cup G_{\mathbb{C}_D}, B, I_\mathbb{C} \cup I_{\mathbb{C}_D})$
6 **end**
7 **return** $\mathbb{E}$ and $\mathbb{C}$

---

the implication theory of a condition might be included in another one and thus is explored at least twice – once in combination and once alone. But, because we only ask questions about implications that are unknown with respect to the knowledge we already have when the condition is explored alone, these questions won't be asked again.

In Algorithm 2 we present the algorithm for *triadic exploration* in pseudo-code: We walk through $(\mathcal{P}(B) \setminus \emptyset, \supseteq)$, i.e. the subsets of conditions, in Line 1. For each set of conditions $D \subseteq B$ we determine the implications $\mathcal{L}$ that are known to hold for all conditions in $D$ in Line 2. We compute the canonical base relative to $\mathcal{L}$ in the subposition of condition context of $D$ Line 3. Then, update the known examples $E$ and the known implications in Lines 4 and 5.

### 3.3   An Example for Triadic Exploration

*Example 5.* We now give a brief example for a triadic exploration of the domain of our running example (Example 1): Let us assume we only have a triadic expert for this domain and not the whole domain information – imagine someone with access to a search interface for the bus and train schedule of Fig. 2. In Fig. 5 we have listed all interactions with the expert. Each row shows one interaction and the order of interactions is from top to bottom. The resulting lattice of conditional implications is exactly the lattice shown in Fig. 3. The extent of each concept of this lattice is a generating set for the implication theory of implications that hold for all conditions of the intent which follows from Theorem 1, and, because we iteratively computed relative canonical bases. Thus, we know that, for each concept, the implications in its extent are complete, but – as a union of "stacked" relative canonical bases – not necessarily irredundant.

## 4   Application for Exploration with Multiple Experts

In this section we discuss how to adapt triadic exploration to a setting where we have multiple experts with different views on a domain (i.e., a set of attributes). In Sect. 1, we have briefly discussed the problem of exploration with multiple experts with different views and concluded that combining answers from different experts is not a good strategy for attribute exploration in general. We have further established that all previous methods for multi-expert exploration avoided this problem by assuming that the experts' knowledge is derived from some consistent domain knowledge.

| Conditions | Question – does the implication hold? | Question holds for | Answer |
|---|---|---|---|
| Mo-Fr, Sat, Sun | $\emptyset \rightarrow$ working-hours, late-evening, early-morning, night, evening | $\emptyset$ | RT5 |
| Mo-Fr, Sat, Sun | $\emptyset \rightarrow$ working-hours, late-evening, evening | $\emptyset$ | 52 |
| Mo-Fr, Sat, Sun | evening $\rightarrow$ working-hours, late-evening | Sat, Sun | 7 |
| Mo-Fr, Sat, Sun | evening $\rightarrow$ working-hours | Mo-Fr, Sat, Sun | true |
| Mo-Fr, Sat, Sun | night $\rightarrow$ working-hours, late-evening, early-morning, evening | Mo-Fr | N3 |
| Mo-Fr, Sat, Sun | early-morning $\rightarrow$ working-hours | Mo-Fr, Sat, Sun | true |
| Mo-Fr, Sat, Sun | late-evening $\rightarrow$ working-hours, evening | Mo-Fr, Sat, Sun | true |
| Mo-Fr, Sat, Sun | working-hours, night $\rightarrow$ late-evening, early-morning, evening | Mo-Fr, Sat, Sun | true |
| Mo-Fr, Sun | working-hours $\rightarrow$ evening | Sun | 55 |
| Mo-Fr, Sat | working-hours, evening $\rightarrow$ early-morning | Mo-Fr | 500 |
| Sun | working-hours, late-evening, early-morning, evening $\rightarrow$ night | $\emptyset$ | 4 |
| Mo-Fr | working-hours $\rightarrow$ early-morning | Mo-Fr | true |

**Fig. 5.** Triadic Exploration of the running example. Each row represents one interaction with the expert. It comprises the set of conditions that is explored, the question posed in form of an implication, the conditions for which the implication holds, and, the answer given by the expert.

Here, we suggest a different approach that allows for a group of experts with different, opposing views on a domain. The basic idea is to accept all answers equally and look for the subset of knowledge that all experts agree on. To explore the domain we then explore the agreed-upon knowledge of different subsets of the expert group.

If all experts know about the same objects of the domain we can regard the group of experts as a triadic domain where each experts view is expressed as one condition. In our running example, imagine that there are three experts for the bus and train schedule: One for Monday-Friday, one for Saturday and one for Sunday. The three experts will have different opinions about the implication theory of the time slots.

To explore the dependencies of attributes in this triadic multi-expert domain, we utilize triadic exploration. To ask about a conditional implication then means to ask all experts if the implication holds in their view. However, a simple translation back to the triadic case means that each time an expert gives a counterexample to a question, all experts must be consulted about their view on the counterexample to stay within the triadic setting (because we need the full slice of the triadic context). This is not ideal, but we can adapt the triadic exploration to avoid this issue: Since we do not rely on any specific properties of the triadic context other than being able to form the subposition of the condition contexts $\mathbb{K}_d$, we can simply leave the triadic setting behind and transfer the idea of conditional attribute implications to a setting where we replace the triadic context with a context family on the same set of attributes (but not necessarily the same set of objects), i.e., a context family $\{\mathbb{K}_e = (G_e, M, I_e)|e \in E\}$ for a group of experts $E$.

Note that we could also explore the implication theory for each context in such a context family independently and combine the results afterwards, as initially suggested in Sect. 3. It is not obvious how this approach compares to the triadic one. However, the triadic approach also allows to only explore a subset of the system of implication theories.

A real world example for such a context family can be found in the *BSI-IT-Grundschutzkatalog*[3], a publication by the German *Federal Office for Information Security*, which contains security recommendations on a wide variety of IT topics. There, a general set of elementary threats is defined and for topics where these threats are present (for example organizational, infrastructure and personnel) a set of measures is defined where each measure combats one or multiple of the threats. Hence, if we regard the elementary threats as attributes, the topics as conditions/experts and the measures as objects, we have a context family on the same set of attributes but with different object sets.

To explore the domain of such a context family, where the set $G$ varies for the different conditions, we have to slightly alter Algorithms 1 and 2. In particular, we need to replace the triadic contexts with context families and the conditions with the experts. The triadic context of counterexamples becomes a context family of counterexamples where each expert has their own context of counterexamples and the objects between them can differ. Hence, in Algorithm 1, $A''$ is computed on the subposition of the respective contexts of counterexamples and asking about the implication $A \to A''$ means asking each of the experts.

---

[3]  https://www.bsi.bund.de/EN/Topics/ITGrundschutz/itgrundschutz_node.html.

Now, counterexamples from one expert can be accepted without having to ask all other experts about their view on the example.

If we explore a triadic domain in this more abstract setting of context families on the same set of attributes, the trade-off is that we obtain less complete information about the counterexamples. However, we still obtain the same knowledge in terms of conditional attribute implications that hold in the domain.

In addition, we gain the ability to explore context families that do not fit the triadic setting or only do so after some modifications, as for example, the context family of the *BSI-IT-Grundschutzkatalog*. Another example can be derived from the running example: If we look at $\mathbb{K}_{\text{Mo-Fr}}$ in Fig. 2, imagine that instead of one context (and thus one expert) of bus and tram lines we had one for bus lines and one for tram lines. Clearly, this family of two contexts could be transformed into a triadic context, however, to do so we would have to add bus lines to the tram lines context and vice versa – mixing domains that might be perceived as different.

## 5   Conclusion and Outlook

In this paper, we addressed the problem of multi-expert attribute exploration in Formal Concept Analysis. To this end, we developed *triadic exploration* – an analogue to attribute exploration – for Triadic Concept Analyis, which extends Formal Concept Analysis with the notion of conditions. Triadic exploration helps a triadic domain expert to explore the structure of the conditional attribute implications of the domain.

We adapted triadic exploration to a multi-expert setting by considering the experts' views of a domain as conditions in a triadic setting. We discussed the ramifications of this approach and subsequently suggested to adapt triadic exploration to the more general setting of context families on the same set of attributes.

This paper is a step towards multi-expert exploration where experts can have different views on a domain. In contrast to the few prior works on this subject, here the experts can have opposing views. A next step is the combination of this approach with the notion of partial expert knowledge and a more in depth study of context families as a foundation for multi-expert explorations.

## References

1. Biedermann, K.: A foundation of the theory of trilattices. Dissertation, shaker, TU Darmstadt, Aachen (1998)
2. Felde, M., Stumme, G.: Interactive collaborative exploration using incomplete contexts. CoRR abs/1908.08740 (2019). http://arxiv.org/abs/1908.08740
3. Ganter, B.: Two basic algorithms in concept analysis. In: Kwuida, L., Sertkaya, B. (eds.) ICFCA 2010. LNCS (LNAI), vol. 5986, pp. 312–340. Springer, Heidelberg (2010). https://doi.org/10.1007/978-3-642-11928-6_22
4. Ganter, B.: Attribute exploration with background knowledge. Theor. Comput. Sci. **217**(2), 215–233 (1999)

5. Ganter, B., Obiedkov, S.: Implications in triadic formal contexts. In: Wolff, K.E., Pfeiffer, H.D., Delugach, H.S. (eds.) ICCS-ConceptStruct 2004. LNCS (LNAI), vol. 3127, pp. 186–195. Springer, Heidelberg (2004). https://doi.org/10.1007/978-3-540-27769-9_12

6. Ganter, B., Obiedkov, S.: More expressive variants of exploration. In: Conceptual Exploration, pp. 237–292. Springer, Heidelberg (2016). https://doi.org/10.1007/978-3-662-49291-8_6

7. Ganter, B., Wille, R.: Conceptual scaling. In: Roberts, F. (ed.) Applications of Combinatorics and Graph Theory to the Biological and Social Sciences, pp. 139–167. Springer-Verlag (1989)

8. Ganter, B., Wille, R.: Formal Concept Analysis: Mathematical Foundations. Springer-Verlag, Berlin/Heidelberg (1999)

9. Guigues, J.L., Duquenne, V.: Familles minimales d'implications informatives résultant d'un tableau de données binaires. Mathématiques et Sci. Humaines **95**, 5–18 (1986)

10. Hanika, T., Zumbrägel, J.: Towards collaborative conceptual exploration. In: Chapman, P., Endres, D., Pernelle, N. (eds.) ICCS 2018. LNCS (LNAI), vol. 10872, pp. 120–134. Springer, Cham (2018). https://doi.org/10.1007/978-3-319-91379-7_10

11. Holzer, R.: Methoden der formalen Begriffsanalyse bei der Behandlung unvollständigen Wissens. Dissertation, shaker, TU Darmstadt (2001)

12. Holzer, R.: Knowledge acquisition under incomplete knowledge using methods from formal concept analysis: Part i. Fundam. Informaticae **63**(1), 17–39 (2004)

13. Holzer, R.: Knowledge acquisition under incomplete knowledge using methods from formal concept analysis: Part ii. Fundam. Informaticae **63**(1), 41–63 (2004)

14. Jäschke, R., Hotho, A., Schmitz, C., Ganter, B., Stumme, G.: Discovering shared conceptualizations in folksonomies. Web Semant. **6**(1), 38–53 (2008). https://doi.org/10.1016/j.websem.2007.11.004

15. Jäschke, R., Hotho, A., Schmitz, C., Ganter, B., Stumme, G.: Trias - an algorithm for mining iceberg tri-lattices. In: Proceedings of 6th ICDM conference. Hong Kong, December 2006. https://doi.org/10.1109/ICDM.2006.162

16. Kriegel, F.: Parallel attribute exploration. In: Haemmerlé, O., Stapleton, G., Faron Zucker, C. (eds.) ICCS 2016. LNCS (LNAI), vol. 9717, pp. 91–106. Springer, Cham (2016). https://doi.org/10.1007/978-3-319-40985-6_8

17. Lehmann, F., Wille, R.: A triadic approach to formal concept analysis. In: Ellis, G., Levinson, R., Rich, W., Sowa, J.F. (eds.) ICCS-ConceptStruct 1995. LNCS, vol. 954, pp. 32–43. Springer, Heidelberg (1995). https://doi.org/10.1007/3-540-60161-9_27

18. Stumme, G.: Attribute exploration with background implications and exceptions. In: Bock, H.H., Polasek, W. (eds.) Data Analysis and Information Systems. Statistical and Conceptual approaches. Proceedings of GfKl 1995. Studies in Classification, Data Analysis, and Knowledge Organization 7, pp. 457–469. Springer, Heidelberg (1996)

19. Wille, R.: Restructuring lattice theory: an approach based on hierarchies of concepts. In: Rival, I. (ed.) Ordered Sets, pp. 445–470. Reidel, Dordrecht-Boston (1982)

20. Wille, R.: The basic theorem of triadic concept analysis. Order **12**(2), 149–158 (1995). https://doi.org/10.1007/BF01108624

# Towards Interactive Transition from AOC Poset to Concept Lattice

Tim Pattison$^{(\boxtimes)}$ and Aaron Ceglar

Defence Science and Technology Group, Edinburgh, SA, Australia
{tim.pattison,aaron.ceglar}@dst.defence.gov.au

**Abstract.** Efficient algorithms exist for constructing the attribute-object concept (AOC) partially-ordered set (poset) from a formal context. The atoms and co-atoms of the corresponding concept lattice can be determined from this AOC poset and horizontally ordered so as to reduce arc crossings in a layered drawing of the AOC poset initially, and ultimately of the concept lattice digraph. The remaining, abstract concepts must then be computed and progressively inserted into the AOC poset to construct the lattice digraph. This paper describes the preparation of a formal context for efficient computation of these abstract concepts, and the consequent localisation in the AOC poset digraph of any resultant insertions. In particular, it provides simple screening tests for identifying bigraph edges, and hence also any attributes and objects, which do not contribute to abstract concepts. Elimination of these bigraph elements reduces the size of the context and paves the way for dividing and conquering the enumeration of the abstract concepts. These screening tests are also used to determine *ab initio* which arcs in the AOC poset digraph will not be subject to subsequent transitive reduction. These arcs are visually distinguished in the line diagram to focus attention on the remaining digraph arcs where the insertion of additional concepts may yet occur, and where the graphical interpretation of meets and joins is unsafe.

**Keywords:** Formal concept analysis · Abstract concepts · Lattice drawing · Divide and conquer

## 1 Introduction

### 1.1 Formal Concept Analysis and Scalability

A formal context is a bipartite graph – henceforth *bigraph* – whose vertices are partitioned into objects and attributes, and whose edges are specified by a binary relation $I \subseteq G \times M$ between the sets $G$ of objects and $M$ of attributes. Formal Concept Analysis transforms this bigraph into a partially-ordered set – henceforth *poset* – of formal concepts. Each formal concept consists of a maximal set of objects, called its *extent*, and a maximal set of attributes, called its *intent*, such that each object in the extent is adjacent in the bigraph to each attribute

© Crown 2021
A. Braud et al. (Eds.): ICFCA 2021, LNAI 12733, pp. 192–207, 2021.
https://doi.org/10.1007/978-3-030-77867-5_12

in the intent, and vice versa. The set $g' \subseteq M$ of attributes adjacent to $g \in G$ is the intent of the corresponding *object concept*, and the set $m' \subseteq G$ of objects adjacent to $m \in M$ is the extent of the corresponding *attribute concept*. We refer to object and attribute concepts collectively as *concrete* concepts, and to the remainder as *abstract* concepts.

The set of formal concepts, partially ordered by extent set inclusion, forms a complete lattice. This concept lattice can be represented as an acyclic directed graph – henceforth *digraph* – whose vertices are formal concepts and whose directed edges correspond to the cover relation – the transitive reduction of the ordering relation – between concepts. A line diagram is an upward drawing of this digraph in which each object and attribute concept is labelled with the corresponding object(s) and attribute(s) respectively. Abstract concepts can be readily recognised from this diagram as those which are not labelled.

A formal context may give rise to as many as $2^{\min(|G|,|M|)}$ formal concepts, of which at most $|G| + |M|$ are concrete. Abstract concepts therefore differ quantitatively from concrete ones in that they are potentially far more numerous, and actually so in pathological cases such as the contranominal scale [1]. The potential combinatorial explosion of concepts with increasing size of the formal context poses challenges for the computation, layout and visualisation of, as well as interaction with, the lattice digraph. Contexts of even moderate size can produce a large number of resultant vertices and arcs, which compete for limited screen real estate and challenge user comprehension. On-demand construction and layout of the entire lattice digraph cannot be achieved in interactive timescales for large lattices, so either prior or user-guided construction and layout is required to support responsive interaction.

## 1.2   AOC Poset

Some analytic objectives, such as identifying the upper neighbours of the infimum and lower neighbours of the supremum, can be achieved without enumeration of the full concept lattice. The Attribute-Object Concept (AOC) poset [2] of a formal context consists of only the concrete concepts, once again ordered by extent set inclusion. The AOC poset therefore has at most $|G| + |M|$ elements, allowing its elements and cover relation to be computed in less time [2], and its line diagram presented using less screen real estate, than the concept lattice. For convenience, we include the supremum and infimum of the concept lattice in the AOC poset and corresponding line diagram, regardless of whether they are concrete concepts.

Applying FCA to the domain of object-oriented software engineering, Godin and Mili [5] used the term "abstract" to describe concepts lacking an object label. The analogy between abstract concepts in FCA and abstract classes in object-oriented software engineering is straighforward – an abstract class is one from which an object cannot be directly instantiated, and in this sense (only) abstracts from the properties of the classes which inherit it. Godin and Mili [5] noted the existence of concepts which have neither object nor attribute labels – for which we have reserved the term "abstract" – and described the benefits of including them in their analysis, both for domain understanding and object-oriented design. The greatest lower and least upper bounds – *meet* and *join* respectively – exist for any subset of concepts in the concept lattice, and can be "read" from the line diagram as the concepts at which downward and upward paths, respectively, from those concepts converge. Due to the absence of abstract concepts, however, these bounds are not guaranteed to exist in the AOC poset, and hence the corresponding interpretation of the line diagram is unsafe. The AOC poset is consequently insufficient for analytical tasks which rely on the existence of these bounds, such as deriving a basis for attribute implications.

### 1.3   Morphing AOC Poset into Concept Lattice

Pattison and Ceglar [9] therefore proposed a hybrid approach, which exploits the computational and graph drawing benefits of the AOC poset to rapidly present its line diagram to the user for familiarisation while they await computation of the abstract concepts. Progressive insertion of the abstract concepts then morphs this line diagram into that for the concept lattice, which contains all abstract concepts and delivers the attendant benefits noted by Godin and Mili [5].

The lower neighbours of the supremum and upper neighbours of the infimum in the concept lattice are referred to as *atoms* and *co-atoms* respectively. Already present in the AOC poset, these can be horizontally ordered so as to reduce arc crossings in – and hence improve the readability of – a layered drawing of the AOC poset initially, and ultimately of the concept lattice digraph [10]. The horizontal positions of the remaining concepts are derived from those of their atomic descendants and co-atomic ancestors [9]. The AOC poset with atoms and co-atoms so ordered therefore provides a suitable substrate for the subsequent progressive insertion of the remaining abstract concepts. This progressive approach to construction of the lattice digraph allows users to familiarise themselves with the line diagram of the AOC poset while the abstract concepts are being computed, and ideally to preserve their resultant mental model throughout subsequent concept insertions.

### 1.4   Generating only Abstract Concepts

Pattison and Ceglar [9] did not specify how, having already identified the concrete concepts and constructed the line diagram for the AOC poset, an FCA algorithm should thereafter efficiently and promptly produce only the remaining, abstract concepts. Once the AOC poset has been constructed, a conventional FCA algorithm could obviously be modified to simply discard any concepts it generates which are not abstract. However re-generating, identifying and discarding each concrete concept is not only inefficient, but also delays production of the abstract concepts awaited by the user. A second alternative is that an existing FCA algorithm might be modified to first produce the AOC poset, followed by (only) the remaining, abstract, concepts. A third alternative is that the AOC poset and its line diagram are generated by an existing algorithm such as Hermes [2], and a novel algorithm generates only the abstract concepts for subsequent insertion into the line diagram. This third option, and in particular exposition of the novel algorithm, is the focus of this paper.

The generation of abstract concepts proceeds in three stages: pre-processing of the clarified formal context to remove edges and vertices which do not satisfy necessary conditions for their participation in abstract concepts; conventional FCA of the pre-processed context; and efficient elimination of any resultant concepts which are either not valid or not abstract in the original context. The pre-processing step removes bigraph elements, and thereby *ablates* the formal context, while preserving all abstract concepts. Established precedents for context ablation include clarification and reduction [3], which remove selected vertices and adjacent edges from a formal context while preserving the structure of the lattice digraph. In addition to simplifying and expediting the subsequent process of Formal Concept Analysis, context ablation has the beneficial side-effect of reducing the cardinalities $|G|$ and $|M|$ of the context bigraph vertex sets, and thereby lowering the *a priori* exponential bound on the number of concepts.

In [9], the user could either await the insertion of additional abstract concepts potentially anywhere in the evolving poset digraph, or prioritise their generation by selecting existing concepts and requesting their meet or join. Thus the user may waste time waiting for, or trying to prioritise, the enumeration of abstract concepts which do not exist. If *immutable* arcs in the AOC poset digraph – i.e. those which will not be subject to subsequent transitive reduction – could be identified *ab initio*, the remainder could be visually distinguished to focus user attention on areas where the insertion of abstract concepts may yet occur, and hence where the interpretation of meets and joins is unsafe. We argue that these immutable digraph arcs include those corresponding to bigraph edges removed during our context ablation step, and demonstrate empirically that many such arcs can be identified as a by-product of context ablation.

## 1.5  Organisation

This paper is organised as follows. Section 2 describes the ablation of a formal context to reduce the number of bigraph elements while preserving all abstract concepts. It outlines the strategy, and details supporting theory, for the elimination of bigraph elements, illustrating their application using a worked example. Those familiar with FCA theory can start from Definition 11. Section 3 describes analysis of the ablated context to generate only abstract concepts for progressive insertion into the line diagram of the AOC poset. As the line diagram is thereby morphed into that for the concept lattice, the user's attention is directed to where such insertions may still occur. Section 4 summarises the contribution.

# 2  Context Ablation

## 2.1  Preliminaries

**Definition 1.** *A formal context* $\mathbb{K} = (G, M, I)$ *is a labelled bipartite graph, or* bigraph, *with object vertex set* $G$, *attribute vertex set* $M$, *and undirected edge set* $I \subseteq G \times M$.

Each vertex has a unique label which derives from the domain of application. For bibliographic analysis, for example, the objects may represent publications labelled by their title and the attributes may represent authors labelled by their full name.

**Definition 2.** *A sub-context* $\underline{\mathbb{K}} = (\underline{G}, \underline{M}, \underline{I})$ *of a formal context* $\mathbb{K} = (G, M, I)$ *is a formal context for which* $\underline{G} \subseteq G$, $\underline{M} \subseteq M$ *and* $\underline{I} \subseteq I \cap (\underline{G} \times \underline{M})$.

**Definition 3.** $\underline{\mathbb{K}} \leq \mathbb{K}$ *iff* $\underline{\mathbb{K}}$ *is a sub-context of* $\mathbb{K}$.

**Definition 4.** *A biclique of the formal context* $\mathbb{K}$ *is a sub-context* $\underline{\mathbb{K}} = (\underline{G}, \underline{M}, \underline{I})$ *satisfying* $\underline{I} = \underline{G} \times \underline{M} \subseteq I$.

**Definition 5.** *A biclique* $(\underline{G}, \underline{M}, \underline{G} \times \underline{M})$ *is proper if* $\underline{G} \neq \emptyset$ *and* $\underline{M} \neq \emptyset$.

**Definition 6.** *A biclique* $(\mathcal{E}, \mathcal{I}, \mathcal{E} \times \mathcal{I})$ *of the formal context* $\mathbb{K}$ *is* maximal *if no proper superset* $\overline{\mathcal{E}} : \mathcal{E} \subset \overline{\mathcal{E}} \subseteq G$ *satisfies* $\overline{\mathcal{E}} \times \mathcal{I} \subseteq I$ *and no proper superset* $\overline{\mathcal{I}} : \mathcal{I} \subset \overline{\mathcal{I}} \subseteq M$ *satisfies* $\mathcal{E} \times \overline{\mathcal{I}} \subseteq I$.

**Definition 7.** *A formal concept of the formal context* $\mathbb{K}$ *is an ordered pair* $(\mathcal{E}, \mathcal{I})$ *consisting of the object set* $\mathcal{E} \subseteq G$ *and attribute set* $\mathcal{I} \subseteq M$ *of a maximal biclique.*

The set $\mathcal{E}$ is called the *extent* of the formal concept, and the set $\mathcal{I}$ is called the *intent*. A formal concept may have empty intent or extent, and hence need not correspond to a *proper* biclique [4].

**Definition 8.** *The* intent operator *maps any set* $\mathcal{A} \subseteq G$ *of object vertices to the maximal set* $\mathcal{A}' \subseteq M$ *of attribute neighbours satisfying* $\mathcal{A} \times \mathcal{A}' \subseteq I$. *The* extent operator *maps any set* $\mathcal{B} \subseteq M$ *of attribute vertices to the maximal set* $\mathcal{B}' \subseteq G$ *of object neighbours satisfying* $\mathcal{B}' \times \mathcal{B} \subseteq I$.

Since it is obvious from the context which of these two operators is intended, the same symbol $'$ usually suffices for both.

**Observation 1** ([3]). *The pair* $(\mathcal{E}, \mathcal{I})$ *with* $\mathcal{E} \subseteq G$ *and* $\mathcal{I} \subseteq M$ *is a formal concept of* $\mathbb{K}$ *iff* $\mathcal{I} = \mathcal{E}'$ *and* $\mathcal{E} = \mathcal{I}'$.

The intent $\mathcal{I} = \mathcal{E}'$ and extent $\mathcal{E} = \mathcal{I}'$ of a concept are closed under the composition $''$ of these two operators, since $\mathcal{I} = \mathcal{E}' = \mathcal{I}''$ and $\mathcal{E} = \mathcal{I}' = \mathcal{E}''$.

**Definition 9.** *Formal concepts are partially ordered such that*

$$(\mathcal{E}_1, \mathcal{I}_1) < (\mathcal{E}_2, \mathcal{I}_2) \iff \mathcal{E}_1 \subset \mathcal{E}_2 \text{ and } \mathcal{I}_1 \supset \mathcal{I}_2$$

The set $\mathfrak{B}$ of formal concepts of the formal context $\mathbb{K} = (G, M, I)$, partially ordered as per Definition 9, constitutes a complete lattice. The least upper bound, or *supremum*, and the greatest lower bound, or *infimum*, of $\mathfrak{B}$ are referred to collectively as the *extrema*.

**Definition 10.** *The* object concept *for object* $i$ *is* $(i'', i')$, *and the* attribute concept *for attribute* $j$ *is* $(j', j'')$.

**Observation 2.** *A formal concept $(\mathcal{E}, \mathcal{I})$ satisfies*

$$(j', j'') \geq (\mathcal{E}, \mathcal{I}) \geq (i'', i') \qquad \forall (i, j) \in \mathcal{E} \times \mathcal{I} \tag{1}$$

**Definition 11.** *A formal concept $(\mathcal{E}, \mathcal{I})$ is abstract if $\mathcal{E} \times \mathcal{I} \neq \emptyset$ and*

$$(j', j'') > (\mathcal{E}, \mathcal{I}) > (i'', i') \qquad \forall (i, j) \in \mathcal{E} \times \mathcal{I} \tag{2}$$

Definition 11 excludes attribute and object concepts, to which we refer collectively as *concrete* concepts. It also excludes the supremum, since: the inequality is impossible – and the supremum an attribute concept – for any universal attribute $j \in \mathcal{I}$; and if instead $\mathcal{I} = \emptyset$, then $\mathcal{E} \times \mathcal{I} = \emptyset$. Definition 11 similarly excludes the infimum. $\mathfrak{B}$ can therefore be partitioned into the extrema, (other) concrete concepts and the set of abstract concepts.

**Corollary 1.** *A formal concept $(\mathcal{E}, \mathcal{I}) : \mathcal{E} \times \mathcal{I} \neq \emptyset$ is abstract iff*

$$|\mathcal{E}| > k_{\mathcal{E}} \geq 1 \tag{3a}$$
$$|\mathcal{I}| > k_{\mathcal{I}} \geq 1 \tag{3b}$$

*where*

$$k_{\mathcal{E}} = \max_{i \in \mathcal{E}} |i''| \tag{4a}$$
$$k_{\mathcal{I}} = \max_{j \in \mathcal{I}} |j''| \tag{4b}$$

*Proof.* This result follows from Observation 2 and Definitions 11 and 9. The constraint $\mathcal{E} \times \mathcal{I} \neq \emptyset$ is required to avoid maximisation over an empty set in Eq. 4.

We denote by $\mathfrak{B}^*$ the set of abstract concepts of $\mathbb{K}$, and refer to the corresponding proper maximal bicliques as *abstract bicliques*. We denote by $\mathbb{K}^* = (G^*, M^*, I^*)$ the sub-context of $\mathbb{K}$ consisting of the union of all abstract bicliques.

**Observation 3.** $(\mathcal{E}, \mathcal{I}) \in \mathfrak{B}^*$ *is a formal concept of any $\mathbb{K}' : \mathbb{K}^* \leq \mathbb{K}' \leq \mathbb{K}$.*

*Proof.* By our premise, $(\mathcal{E}, \mathcal{I})$ is a proper maximal biclique in $\mathbb{K}$. It remains a proper biclique in $\mathbb{K}' \geq \mathbb{K}^*$ because all of its constituent edges and vertices are present in $\mathbb{K}^*$, and it remains maximal because all edges in $\mathbb{K}' \leq \mathbb{K}$ are also in $\mathbb{K}$.

## 2.2 Strategy

In order to enumerate only abstract concepts, we seek a procedure which identifies and removes edges and vertices in $\mathbb{K} \setminus \mathbb{K}^* = (G \setminus G^*, M \setminus M^*, I \setminus I^*)$. Such a procedure would ideally terminate when, and only when, $\mathbb{K}$ has been transformed into $\mathbb{K}^*$. The remaining subgraph $\mathbb{K}' : \mathbb{K}^* \leq \mathbb{K}' \leq \mathbb{K}$ would then be divided into its connected components, the components subjected to independent FCA, and those resultant concepts which are both valid in $\mathbb{K}$ and abstract as per Corollary 1, progressively inserted into the AOC poset. Given that the elements of the AOC poset have already been enumerated, the desired procedure can use as input the properties of concrete concepts, such as intent or extent set cardinality.

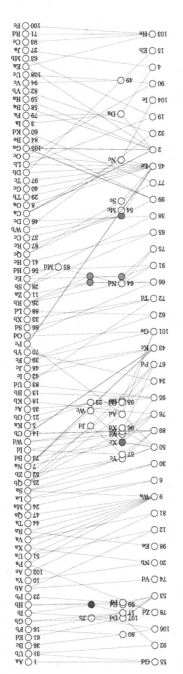

**Fig. 1.** Line diagram of concept lattice for `InfoVis 151` omitting the extrema. Atoms (bottom) and co-atoms (top) are ordered using resistance distance, and the remaining concepts placed at the horizontal barycenter of their co-atomic ancestors and atomic descendants.

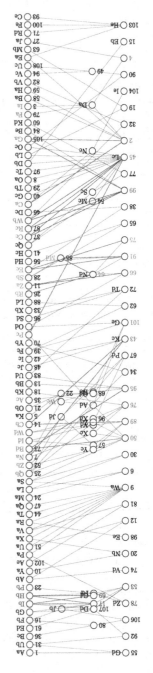

**Fig. 2.** Line diagram of AOC poset for `InfoVis 151` omitting the extrema. Non-black object and attribute labels are colour-coded for membership of the connected components of the context bigraph remaining after the ablation described in Example 4.

Iteration over the edges of $\mathbb{K} = (G, M, I)$ to eliminate those not involved in abstract concepts is viable provided the context is sparse. By "sparse" we mean that the cardinalities $|I|, |G|, |M|$ of the context relation $I$, object set $G$ and attribute set $M$ satisfy $|I| \ll |G||M|$. We use as a running empirical example a sub-context of the InfoVis 2004 bibliographic data set [12] having 151 proper maximal bicliques. This clarified sub-context has 108 objects, 113 attributes and $273 \ll 108 \times 113$ edges. Hereafter, we refer to this context as InfoVis 151. A drawing of the lattice digraph for this context, omitting the extrema, is shown in Fig. 1. Abstract concepts are shown with coloured fill.

## 2.3   Elimination of Bigraph Elements

**Observation 4.** *Let bigraph edge $(i, j)$ participate in an abstract concept $(\mathcal{E}, \mathcal{I})$ of $\mathbb{K}$. Then*

$$|j'| > |\mathcal{E}| > |i''| \tag{5a}$$
$$|i'| > |\mathcal{I}| > |j''| \tag{5b}$$

Observation 4 follows from Eq. 2 of Definition 11, and necessarily precludes $(\mathcal{E}, \mathcal{I})$ from being the object concept for any $g \in i''$ or the attribute concept for any $m \in j''$. However, it might still be the object concept for some $g \in j' \setminus i''$ or the attribute concept for some $m \in i' \setminus j''$. The following corollary therefore provides a necessary but not sufficient condition for any concepts intervening between $(i'', i')$ and $(j', j'')$ to be abstract.

**Corollary 2.** *Let edge $(i, j)$ participate in an abstract concept of $\mathbb{K}$. Then*

$$|i'| \geq |j''| + 2 \tag{6a}$$
$$|j'| \geq |i''| + 2 \tag{6b}$$

If $(i, j) \in I$ does not satisfy Eq. 6, then $(i, j) \in I \setminus I^*$ and can be safely eliminated without compromising any abstract maximal bicliques of $\mathbb{K}$. Eliminated bigraph edges for which $(j', j'') > (i'', i')$ correspond to arcs in the concept lattice digraph. Such arcs are already present in the line diagram of the AOC poset, and are immutable in the sense that they will not be interrupted by the subsequent insertion of the remaining (abstract) concepts.

*Example 1.* Of the 11 edges in the simple context bigraph depicted in Fig. 3a, Corollary 2 eliminates the 7 shown black. Of these, the 2 solid edges satisfy $(j', j'') > (i'', i')$, and hence correspond to lattice arcs already present in the line diagram of the AOC poset shown in Fig. 3b; the remaining 5 dashed edges do not. Solid bigraph edges correspond to mutable (red) and immutable (black) arcs in the AOC poset line diagram; dashed edges have no corresponding poset arc. The mutable (red) arcs in Fig. 3b are interrupted by the subsequent insertion of the abstract concept $(\{1, 2\}, \{\mathtt{a}, \mathtt{b}\})$ shown grey in Fig. 3c.

**Corollary 3.** *Let bigraph vertex $\alpha$ participate in an abstract concept of $\mathbb{K}$. Then in $\mathbb{K}' : \mathbb{K}^* \leq \mathbb{K}' \leq \mathbb{K}$, $|\alpha'| \geq 2$.*

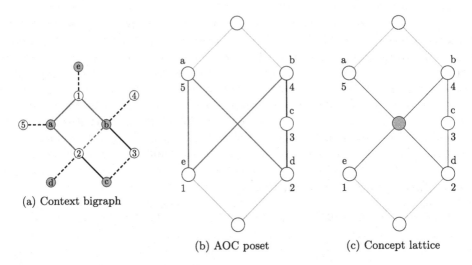

(a) Context bigraph

(b) AOC poset          (c) Concept lattice

**Fig. 3.** Simple context bigraph (a) and line diagrams for the corresponding AOC poset (b) and concept lattice (c). Corollary 2 eliminates the black bigraph edges. Solid edges in (a) correspond to mutable (red) and immutable (black) arcs in (b). The surviving red edges and adjacent vertices in (a) correspond to the abstract (grey) concept in (c), whose insertion interrupts the mutable arcs. (Color figure online)

*Proof.* Any vertex in a biclique has at least as many neighbours as the biclique contains vertices of the opposite type. Equation 3 requires that $|\mathcal{E}| \geq 2$ and $|\mathcal{I}| \geq 2$ for an abstract biclique, so that $|\alpha'| \geq 2$ in $\mathbb{K}$. Observation 3 and Corollary 2 ensure that no edge within an abstract biclique is deleted, so that $|\alpha'| \geq 2$ also holds in $\mathbb{K}'$.

Vertices of degree less than two in $\mathbb{K}'$ can be safely deleted along with their adjacent edges, since neither can participate in an abstract concept of $\mathbb{K}$. By removing the unrealistic requirement for apriori knowledge of the elements of $\mathcal{E}$ and $\mathcal{I}$, Corollary 3 clears the way for practical application of Corollary 1 to context ablation in preparation for FCA.

*Example 2.* Following application of Corollary 2 to the example context in Fig. 3a, Corollary 3 eliminates objects $\{3, 4, 5\}$ and attributes $\{c, d, e\}$, leaving only the red edges and adjacent vertices. These constitute the abstract biclique $(\{1, 2\}, \{a, b\})$ corresponding to the grey vertex in Fig. 3c.

Since Corollaries 2 and 3 impose conditions on bigraph elements which are necessary but not sufficient for participation in abstract bicliques, the ablated formal context may more generally still contain superfluous elements.

The deletion of vertices and their adjacent edges from $\mathbb{K}'$ according to Corollary 3 produces a new context $\mathbb{K}'' : \mathbb{K}^* \leq \mathbb{K}'' \leq \mathbb{K}' \leq \mathbb{K}$ to which Corollary 3 also applies. Corollary 3 should be applied iteratively, since the deletion of a vertex having a single neighbour can cause the (formerly) adjacent vertex to subsequently fail the neighbour cardinality test. In the worst case, one more iteration may be required than the longest chain of vertices of degree 2.

*Example 3.* Following application of Corollary 2 to InfoVis 151, 2 iterations of Corollary 3 eliminate 9 of the remaining 29 objects, 12 of 41 attributes and 20 incident edges, leaving three connected components of sizes $12 \times 20$, $4 \times 5$ and $4 \times 4$. Additional iterations do not eliminate any further bigraph elements. FCA can be applied independently to these three components.

Whilst Corollaries 2 and 3 can be applied in either order, the cardinalities $|i'|$, $|j'|$, $|i''|$ and $|j''|$ required by the former must be calculated on the original context $\mathbb{K}$. Edge deletion may have decreased the vertex degrees $|i'|$ and $|j'|$, and, by removing the structural distinction between some vertices, increased the closure cardinalities $|j''|$ and $|i''|$. The vertex degrees $|i'|$ and $|j'|$ cannot have increased, since no edges were added to $\mathbb{K}$. Furthermore $|j''|$ and $|i''|$ cannot have decreased, since vertices which were structurally equivalent in $\mathbb{K}$ remain so in $\mathbb{K}'$.

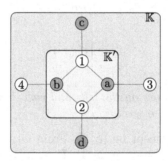

**Fig. 4.** Example illustrating that closure cardinality may be higher in $\mathbb{K}'$ than $\mathbb{K}$.

The example context in Fig. 4 contains the maximal biclique $(\{1, 2\}, \{a, b\})$. Each vertex of this biclique has an additional distinct neighbour so that the biclique edges satisfy Corollary 2. The edges $(1, c)$, $(2, d)$, $(3, a)$ and $(4, b)$ do not satisfy Corollary 2, and hence are not present in $\mathbb{K}'$. The degrees of objects 1 and 2 and attributes a and b have each decreased by 1. Object 1 is thereby rendered indistinguishable from object 2, and attribute a from b. Thus for example $|1''|$ is 1 in $\mathbb{K}$ but 2 in $\mathbb{K}'$.

## 2.4   Iterative Edge Elimination

We have seen that significant numbers of edges not involved in abstract concepts of $\mathbb{K}$ can be removed by Corollary 2 to produce $\mathbb{K}' : \mathbb{K}^* \leq \mathbb{K}' \leq \mathbb{K}$. Since Corollary 2 compares the cardinalities $|i'|$, $|j'|$, $|i''|$ and $|j''|$ calculated on $\mathbb{K}$, it no longer applies to $\mathbb{K}'$, and is therefore of no further use in approaching $\mathbb{K}^*$. In Example 3, additional vertices and adjacent edges in $\mathbb{K}' \setminus \mathbb{K}^*$ were then eliminated through the iterative application of Corollary 3 to produce $\mathbb{K}'' \geq \mathbb{K}^*$. We now develop iterative variants of Corollary 2 which can be applied either to $\mathbb{K}'$ or to $\mathbb{K}'' \geq \mathbb{K}^*$, and demonstrate empirically that in the former case they can delete more edges than Corollary 3.

**Observation 5.** *Let edge $(i,j)$ participate in an abstract concept $(\mathcal{E}, \mathcal{I})$ of the formal context $\mathbb{K}$. Then in $\mathbb{K}' : \mathbb{K}^* \leq \mathbb{K}' \leq \mathbb{K}$, $(j', j'') \geq (\mathcal{E}, \mathcal{I}) \geq (i'', i')$.*

*Proof.* Denote by $\mathfrak{B}'$ the set of formal concepts of $\mathbb{K}'$. Observation 3 ensures that $(\mathcal{E}, \mathcal{I}) \in \mathfrak{B}'$, and the result follows from Observation 2. Equalities are possible – and hence $(\mathcal{E}, \mathcal{I})$ may not be abstract in $\mathbb{K}'$ – because, as a consequence of edge deletions, it may now be the object [attribute] concept for $i$ [$j$].

**Corollary 4.** *Let bigraph edge $(i,j)$ participate in an abstract concept of the formal context $\mathbb{K}$. Then in $\mathbb{K}' : \mathbb{K}^* \leq \mathbb{K}' \leq \mathbb{K}$*

$$|i'| \geq |j''| \tag{7a}$$
$$|j'| \geq |i''| \tag{7b}$$

**Observation 6.** *Let bigraph edge $(i,j)$ participate in an abstract concept of the formal context $\mathbb{K}$. Then*

$$|i'| \geq |j''| + 1 \tag{8a}$$
$$|j'| \geq |i''| + 1 \tag{8b}$$

*where the derivation operators on the left- and right-hand sides are with respect to $\mathbb{K}' : \mathbb{K}^* \leq \mathbb{K}' \leq \mathbb{K}$ and $\mathbb{K}$ respectively.*

*Proof.* Let the abstract concept be $(\mathcal{E}, \mathcal{I})$. From Observation 5, $|i'| \geq |\mathcal{I}|$ and $|j'| \geq |\mathcal{E}|$ in $\mathbb{K}'$ and from Observation 4, $|\mathcal{I}| \geq |j''| + 1$ and $|\mathcal{E}| \geq |i''| + 1$ in $\mathbb{K}$. Pairing these inequalities and noting that $\mathcal{E}$ and $\mathcal{I}$ – and hence also their cardinalities – are the same in both contexts yields the result.

Corollary 4 and Observation 6 both provide lower bounds on $|i'|$ and $|j'|$ in $\mathbb{K}'$. The greater of these two lower bounds should be applied when determining whether any additional bigraph edges can be eliminated.

In contrast with Corollary 2, which can be applied only once to $\mathbb{K}$, Corollaries 3 and 4 and Observation 6 can be applied iteratively. The removal of a non-compliant bigraph edge $(i,j)$ in one iteration reduces $|i'|$ and $|j'|$, and can thereby cause other edges adjacent to object $i$ or attribute $j$ to fail Eq. 8 in the next. Since only edges in $\mathbb{K} \setminus \mathbb{K}^*$ are deleted, the iteration must halt when no further edges can be deleted, leaving at least the abstract concepts intact.

*Example 4.* Following application of Corollary 2 to `InfoVis` 151, 4 iterations of Observation 6 eliminated 29 additional bigraph edges, isolating an additional 11 objects and 16 attributes. No further edges were eliminated by the fifth iteration. Three connected components remained, as for Example 3, but now with sizes $10 \times 17$, $4 \times 5$ and $4 \times 3$. The labels of the objects and attributes belonging to these connected components, and the abstract concepts to which these components give rise, are coloured in Figs. 2 and 1 respectively.

The overall context ablation procedure is listed in Algorithm 1. Corollary 2, Observation 6 and Corollary 3 are implemented in turn by the loops commencing at lines 8 and 13 and the assignments commencing at line 26 respectively. Since Corollary 2 and Observation 6 both eliminate edges adjacent to vertices of degree one, the subsequent application of Corollary 3 amounts to deleting vertices of degree zero. Their more timely deletion would confer no computational benefit, since Corollary 2 and Observation 6 both cycle through edges vice vertices. The cardinalities $|i'|, |i''| \ \forall i \in G$ and $|j'|, |j''| \ \forall j \in M$ can either be passed to ABLATE() following preparation of the AOC poset, or re-calculated from the context at line 3. The values for $|i''|$ and $|j''|$ are re-used throughout Algorithm 1. Vertex $k$ currently has Neighbours(k) neighbours, and Dirty(k) is true if one or more of its adjacent edges has been deleted during the most recent pass through the set of edges. Unless no adjacent edges remain, this flag is cleared after one complete pass through the set of edges (adjacent to "dirty" vertices) fails to delete any further adjacent edges.

# 3    Formal Concept Analysis of Ablated Context

Section 2 described the elimination of elements of the context bigraph $\mathbb{K}$ which cannot participate in abstract formal concepts. Examples 3 and 4 demonstrated that the resultant formal context can have considerably fewer graph elements, thereby significantly constraining the potential combinatorial explosion of abstract concepts. In this section, the remaining context $\mathbb{K}'$ is subjected to conventional FCA, and the resultant concepts screened to eliminate any which are either not valid in $\mathbb{K}$ or are not abstract. For reviews of FCA algorithms applicable to the analysis of $\mathbb{K}'$, the interested reader is referred to [7,8,13].

## 3.1    Finding Valid Abstract Concepts

For the Infovis 151 context processed as per Example 4, $\mathbb{K}'$ contains a total of 35 concepts, compared with the 151 concepts for $\mathbb{K}$. Definition 11 and Corollary 1 justify the immediate elimination of any resultant formal concepts which have extent or intent cardinality less than 2, which for Infovis 151 leaves only 14 concepts as candidate abstract concepts of $\mathbb{K}$. The more stringent form of Corollary 1 involving Eq. 4 eliminates any concepts which are valid and concrete in $\mathbb{K}$; in this instance it eliminates a further 2 concepts.

Observation 3 applies only to the abstract concepts of $\mathbb{K}$, since at least some concrete concepts must be affected by the removal of edges or vertices to form $\mathbb{K}'$. Indeed, as we have seen in the case of Infovis 151, many of the objects and attributes whose corresponding attribute and object concepts are members of the AOC poset are no longer present in $\mathbb{K}'$. For those which remain, their corresponding object and attribute concepts in $\mathbb{K}'$ may not be valid concepts in $\mathbb{K}$. Screening the concepts generated by conventional FCA of $\mathbb{K}'$ for validity in $\mathbb{K}$ according to Observation 1 can therefore eliminate some concepts which are not viable candidates for abstract concepts of $\mathbb{K}$, while preserving all abstract

---

**Algorithm 1.** Implement Corollaries 2 and 3 and Observation 6

---

**Require:** Formal context $(G, M, I)$, preferrably clarified
**Ensure:** $\mathfrak{B}'(G', M', I')$ contains all abstract concepts of $(G, M, I)$
1: **procedure** ABLATE(G,M,I)
2:     **for all** $i \in G, j \in M$ **do**
3:         Calculate $|i'|, |i''|, |j'|, |j''|$
4:         Neighbours$(i) \leftarrow |i'|$, Dirty$(i) \leftarrow$ False, NowDirty$(i) \leftarrow$ False
5:         Neighbours$(j) \leftarrow |j'|$, Dirty$(j) \leftarrow$ False, NowDirty$(j) \leftarrow$ False
6:     **end for**
7:     $I' \leftarrow I$
8:     **for all** $(i, j) \in I : |i'| < |j''| + 2$ or $|j'| < |i''| + 2$ **do**           ▷ Corollary 2
9:         $I' \leftarrow I' \setminus (i, j)$
10:         Neighbours$(i)--$, Dirty$(i) \leftarrow$ True
11:         Neighbours$(j)--$, Dirty$(i) \leftarrow$ True
12:     **end for**
13:     **repeat**                                                    ▷ Observation 6
14:         **for all** $(i, j) \in I' :$ Dirty$(i)$ or Dirty$(j)$ **do**
15:             **if** Neighbours$(i) < |j''| + 1$ or Neighbours$(j) < |i''| + 1$ **then**
16:                 $I' \leftarrow I' \setminus (i, j)$
17:                 Neighbours$(i)--$, NowDirty$(i) \leftarrow$ True
18:                 Neighbours$(j)--$, NowDirty$(j) \leftarrow$ True
19:             **end if**
20:         **end for**
21:         **for all** $(i, j) \in I'$ **do**
22:             Dirty$(i) \leftarrow$NowDirty$(i)$, NowDirty$(i) \leftarrow$ False
23:             Dirty$(j) \leftarrow$NowDirty$(j)$, NowDirty$(j) \leftarrow$ False
24:         **end for**
25:     **until** $\nexists (i, j) \in I' :$ Dirty$(i)$ or Dirty$(j)$
26:     $G' \leftarrow G \setminus \{i :$ Neighbours$(i) = 0\}$                    ▷ Corollary 3
27:     $M' \leftarrow M \setminus \{j :$ Neighbours$(j) = 0\}$
28:     **return** $(G', M', I')$
29: **end procedure**

---

concepts according to Observation 3. Of the 12 remaining candidate concepts for Infovis 151, only 6 are valid in $\mathbb{K}$. In contrast, naïve application of FCA to the original Infovis 151 generates and tests the novelty of at least 151 concepts.

## 3.2    Dividing and Conquering

A beneficial side-effect of the context ablation described in Sect. 2 is that it allows us to divide and conquer the process of concept generation. The InfoVis 151 bigraph $\mathbb{K}$ is connected by virtue of pre-processing by the CARVE algorithm [11]. However, as Examples 3 and 4 have demonstrated, deletion of some bigraph elements from $\mathbb{K}$ to form $\mathbb{K}'$ can cause the latter to be disconnected. In this case, enumeration of its formal concepts can be divided and conquered through independent FCA of each of its connected components. Pattison et al. [11] described how the lattice digraph can be constructed from the digraphs for each of these connected components. As per Example 4 for the `InfoVis 151` context, which contains 108 objects and 113 attributes, three connected components with a total of 18 objects and 25 attributes can be identified following application of Corollary 2 and Observation 6. The divide and conquer approach also extends to checking of Corollary 1; for this purpose, each attribute and object should be accompanied by the cardinality of its closure from the original AOC poset.

### 3.3   Anticipating and Localising Change in the AOC Poset

Pattison and Ceglar [9] described an incremental approach to drawing a concept lattice whereby the elements of the attribute-object concept (AOC) poset were first identified and positioned horizontally, and then the remaining concepts progressively inserted into a two-dimensional drawing of the poset. This paper has described an efficient, divide and conquer approach for producing those remaining abstract concepts. As the line diagram for the AOC poset is morphed into that for the concept lattice, user attention can be directed to or from regions of the line diagram where the insertion of abstract concepts is still possible, and hence where the graphical interpretation of meets and joins is unsafe.

Figure 2 shows the line diagram for the AOC poset of the `Infovis 151` context. The arcs shown red correspond to edges in the context bigraph which satisfy the necessary conditions in Corollary 2 and Observation 6 for participation in an abstract concept. For comparison, the line diagram for the concept lattice digraph is shown in Fig. 1, with abstract concepts shown coloured. Many of the arcs highlighted in Fig. 2 have subsequently been removed as a result of transitive reduction. However, some of these remain in Fig. 1, indicating that there may be opportunity to further refine the necessary conditions in Sect. 2 to reduce the number of such false alarms. The highlighting in Fig. 1 is for illustrative purposes only: once a connected component has been processed, and its abstract concepts inserted into the line diagram, highlighting can be removed from any of its arcs which remain highlighted.

## 4   Discussion and Summary

The context ablation procedure described in Sects. 2.3 and 2.4 does not require the context to have been clarified. In an unclarified context, however, some bigraph edges corresponding to immutable lattice arcs may pass the test in Eq. 6, and thereby escape elimination. This would result in the generation of additional concepts which must then be explicitly eliminated as per Sect. 3.1.

The technique described in [10] for horizontally ordering the lattice atoms and co-atoms in the AOC poset requires that these concept sets are disjoint. FCA of a formal context which does not meet this requirement can be recursively divided using the CARVE algorithm [11] into sub-contexts which do. Only the resultant connected bigraphs would then be subjected to the techniques described in this paper. These may be disconnected by the context ablation in Sects. 2.3 and 2.4, and thereby further divided and conquered.

A subgraph of the context bigraph is *biconnected* if it remains connected upon removal of any one of its vertices, or *2-connected* if a single vertex is not considered "connected". A *bridge* is a biconnected component of the context bigraph containing exactly two vertices. An abstract maximal biclique of $\mathbb{K}$ is a 2-connected, bridgeless subgraph of $\mathbb{K}' : \mathbb{K}^* \leq \mathbb{K}' \leq \mathbb{K}$ and hence is contained within a single biconnected component of $\mathbb{K}'$. Efficient algorithms exist for finding the biconnected components of a simple graph [6], and should be investigated for simultaneous ablation and partitioning of $\mathbb{K}$. Excluding any bridges from further analysis effectively ablates the constituent edges from, and thereby disconnects, the context bigraph.

Enumeration and visualisation of the AOC poset, vice concept lattice, scales to larger formal contexts, since the former has at most $|G| + |M|$ elements and the latter $|\mathfrak{B}| \leq 2^{\min(|G|,|M|)}$. It therefore constitutes a more reliable first step for interactive, on-demand FCA. For sparse contexts, mutable poset arcs can be efficiently identified using context ablation, and the quantity

$$\min(|j'| - |i''|, |i'| - |j''|) - 1$$

calculated for each as an upper bound on the number of abstract concepts between the object concept for $i$ and the attribute concept for $j$. Aggregating these bounds constrains the number of abstract concepts, from which the feasibility of interactive visualisation of the concept lattice can be assessed. The ablation and bounding steps are of course unneccessary if the results of prior, as opposed to on-demand, computation and layout of the concept lattice digraph are available. Visualisation of the AOC poset may still be appropriate in this case if the digraph size challenges user comprehension. Selection of mutable arcs might then drive on-demand insertion of any intervening abstract concepts.

Instead of morphing the line diagram for the AOC poset into that for the concept lattice, the latter could be presented to the user as a separate view. This would clearly distinguish between the line diagrams in which meets and joins are and are not guaranteed to exist. This approach would transform the challenge of maintaining the user's mental model throughout the morph into one of acquiring the association between the two views. A comparative evaluation of the user experience will be required to decide between these two approaches.

This paper has described the ablation of a formal context to eliminate many of the concrete concepts, dividing and conquering FCA of the remaining sub-context, and screening the remaining formal concepts to produce only the required abstract concepts. These are progressively inserted into the line digram for the AOC poset, morphing it into that for the concept lattice. The user's attention is directed to where such insertions of missing meets or joins may still occur. The potential utility of this approach has been demonstrated using a single real-world context. Empirical studies will be required to confirm and qualify its wider applicability to sparse contexts.

# References

1. Albano, A.: Upper bound for the number of concepts of contra nominal-scale free contexts. In: Glodeanu, C.V., Kaytoue, M., Sacarea, C. (eds.) Formal Concept Analysis, pp. 44–53. Springer (2014). DOI: https://doi.org/10.1007/978-3-319-07248-7_4

2. Anne, B., Alain, G., Marianne, H., Amedeo, N., Alain, S.:Hermes: a simple and efficient algorithm for building the AOC-post of a binary relation. Ann. Math. Artif. Intell. 45–71 (2014). https://doi.org/10.1007/s10472-014-9418-6

3. Formal Concept Analysis. Springer, Heidelberg (1999). https://doi.org/10.1007/978-3-642-59830-2

4. Gaume, B., Navarro, E., Prade, H.: A parallel between extended formal concept analysis and bipartite graphs analysis. In: Hüllermeier, E., Kruse, R., Hoffmann, F. (eds.) Computational Intelligence for Knowledge-Based System Design. LNCS, vol. 6178, pp. 270–280. Springer (2010). DOI: https://doi.org/10.1007/978-3-642-14049-5_28

5. Godin, R., Mili, H.: Building and maintaining analysis-level class hierarchies using Galois lattices. In: OOPSLA '93: Proceedings of the 8th Annual Conference Object-Oriented Programming Systems, Languages, and Applications. pp. 394–410 (1993). 10.1145/165854.165931

6. Hopcroft, J., Tarjan, R.: Algorithm 447: efficient algorithms for graph manipulation. Commun. ACM **16**(6), 372–378 (1973). https://doi.org/10.1145/362248.362272

7. Krajca, P., Outrata, J., Vychodil, V.: Advances in algorithms based on CbO. In: Kryszkiewicz, M., Obiedkov, S.A. (eds.) Proceedings of the 7th International Conference. CLA. vol. 672, pp. 325–327. CEUR-WS.org (2010)

8. Kuznetsov, S.O., Obiedkov, S.A.: Algorithms for the construction of concept lattices and their diagram graphs. In: Proceeding of the 5th European Conf. Principles of Data Mining and Knowledge Discovery (PKDD 2001), LNCS, vol. 2168, pp. 289–300. Springer (2001). 10.1007/3-540-44794-6_24

9. Pattison, T., Ceglar, A.: Interaction challenges for the dynamic construction of partially-ordered sets. In: Bertet, K., Rudolph, S. (eds.) Proceedings of the 11th International Conference CLA, vol. 1252, pp. 23–34. CEUR-WS.org (2014)

10. Pattison, T., Ceglar, A.: Simultaneous, polynomial-time layout of context bigraph and lattice digraph. In: Cristea, D., Le Ber, F., Sertkaya, B. (eds.) Proceedings of the 15th International Conference FCA. LNCS: Artificial Intelligence, vol. 11511, pp. 223–240 (2019). 10.1007/978-3-030-21462-3_15

11. Pattison, T., Ceglar, A., Weber, D.: Efficient formal concept analysis through recursive context partitioning. In: Ignatov, D.I., Nourine, L. (eds.) Proceedings of the 14th International Conference CLA, vol. 2123, pp. 219–230. CEUR-WS.org (2018)

12. Plaisant, C., Fekete, J.D., Grinstein, G.: Promoting insight-based evaluation of visualizations: from contest to benchmark repository. IEEE Trans. Vis. Comp. Graph. **14**(1), 120–134 (2008). https://doi.org/10.1109/TVCG.2007.70412

13. Priss, U.: Formal concept analysis in information science. Ann. Rev. Inf. Sci. Tech. **40**, 521–543 (2006). https://doi.org/10.1002/aris.1440400120

# Visualization of Statistical Information in Concept Lattice Diagrams

Jana Klimpke$^{(\boxtimes)}$ (ID) and Sebastian Rudolph (ID)

Computational Logic Group, TU Dresden, Dresden, Germany

**Abstract.** We propose a method of visualizing statistical information in concept lattice diagrams. To this end, we examine the characteristics of support, confidence, and lift, which are parameters used in association analysis. Based on our findings, we develop the notion of *cascading line diagrams*, a visualization method that combines the properties of additive line diagrams with association analysis. In such diagrams, one can read the size of a concept's extent from the height of the corresponding node in the diagram and, at the same time, the geometry of the formed quadrangles illustrates whether two attributes are statistically independent or dependent and whether they are negatively or positively correlated. In order to demonstrate this visualization method, we have developed a program generating such diagrams.

## 1 Introduction

*Formal concept analysis* (FCA) is a mathematical approach for analyzing conceptual hierarchies arising from relationships between objects and attributes. By means of an order relation, hierarchically grouped sets of entities can be sorted by set inclusion and later visualized by means of *line diagrams*, from which *qualitative*, crisp dependencies between the examined attributes, called *implications*, can be read off easily.

*Association analysis* is a data mining technique, used to discover and evaluate *quantitative* relationships and dependencies in a data set. It offers ways of characterizing the strength of these relationships, using the statistical measures *support*, *confidence* and *lift*.

The use of association analysis to extract "imperfect implications" – referred to as *association rules* – has been explored widely in data mining, but also specifically in FCA [11,8,9,3]. Unfortunately, the same cannot be said about the visualization part, i.e., representation of statistical information in line diagrams. One 2004 paper introduced a method of lattice drawing where concepts were placed at positions that were related to their support [10]. However, the authors stated that this method sometimes created nearly horizontal lines in the diagram. To address this problem, they introduced a spring-based lattice drawing method in a follow-up paper [7]. Additionally, the authors proposed to generate a lattice diagram in $\mathbb{R}^3$ and then allow the user to find a "best" projection into $\mathbb{R}^2$ by rotating the lattice around a central axis.

© Springer Nature Switzerland AG 2021
A. Braud et al. (Eds.): ICFCA 2021, LNAI 12733, pp. 208–223, 2021.
https://doi.org/10.1007/978-3-030-77867-5_13

Our work aims at an approach for visualizing concept lattice diagrams in a way that, on top of displaying all perfect relationships between attributes, also reflects the frequencies of the depicted concepts as well as correlation strength between attributes. To this end, we propose *cascading line diagrams*, realized by means of a positioning rule that is inspired by the notion of additive line diagrams but adjusts the height of the concept nodes according to their extent's cardinality. By choosing a logarithmic scale for the latter, statistic independence between attributes manifests itself in perfect parallelograms, whereas positive and negative correlations lead to obtuse or acute deviations from this parallelogram shape. We present an open-source prototypical implementation for drawing cascading line diagrams, which also allows for an intuitive interactive adjustment of its parameters along the remaining degrees of freedom.

## 2   Preliminaries

We start by briefly introducing the basic notions of FCA [6]. A *formal context* is a triple $(G, M, I)$, consisting of a set $G$ of *objects*, a set $M$ of *attributes* as well as a binary *incidence relation* $I \subseteq G \times M$ between $G$ and $M$. As usual, the fact that $I$ relates an object $g$ to an attribute $m$ will be written as $(g, m) \in I$.

**Table 1.** Cross table displaying the relation between objects and attributes of a given formal context.

| $G \setminus M$ | a | b | c | d | e |
|---|---|---|---|---|---|
| T1 | | × | | × | |
| T2 | | × | | | × |
| T3 | | | × | | |
| T4 | × | × | × | | |
| T5 | | | | × | |
| T6 | | × | × | | |
| T7 | | | | | × |

The set of all object-attribute relationships of a formal context can be written down by means of an incidence matrix. An example of such a matrix, also known as cross table, can be seen in Table 1. The set of attributes, shared by a set $A \subseteq G$ of objects can be derived by

$$A' := \{m \in M \mid \forall\, g \in A : (g, m) \in I\}. \tag{1}$$

Dually, the set of all objects that have each of the attributes in a set $B \subseteq M$ can be obtained by

$$B' := \{g \in G \mid \forall\, m \in B : (g, m) \in I\}. \tag{2}$$

A pair $(A, B)$, with $A \subseteq G$ and $B \subseteq M$ will be called a *formal concept* of a context $(G, M, I)$ if $A' = B$ and $B' = A$. The set $A$ will be called the *extent* and the set $B$ the *intent* of the concept. The set of all formal concepts of a formal context $(G, M, I)$ is denoted by $\mathfrak{B}(G, M, I)$.

Concepts can be ordered hierarchically by using the order relation $\leq$. For the concepts $(A_1, B_1)$ and $(A_2, B_2)$ one lets $(A_1, B_1) \leq (A_2, B_2)$ iff $A_1 \subseteq A_2$. We call $(A_1, B_1)$ a *lower neighbor* of $(A_2, B_2)$ – and write $(A_1, B_1) \prec (A_2, B_2)$ – whenever both $(A_1, B_1) \leq (A_2, B_2)$ and there is no "intermediate concept" $(\dot{A}, \dot{B})$ satisfying $(A_1, B_1) \leq (\dot{A}, \dot{B}) \leq (A_2, B_2)$. It turns out the set of a context's formal concepts together with $\leq$ is not only an ordered set, but even a complete lattice, called the *concept lattice* of the context. As for partial orders in general, it is common to visualize concept lattices by means of *line diagrams*, where each node represents one concept of the concept lattice and, for all $(A_1, B_1) \prec (A_2, B_2)$ there is an ascending straight line connecting the node representing $(A_1, B_1)$ to the node representing $(A_2, B_2)$. While this requirement puts some constraints on the vertical positioning of the nodes in a lattice diagram, there is still a lot of leeway and it is a non-trivial question how to arrive at a "good" diagram.

One approach to obtain particularly well-readable diagrams, called *additive line diagrams*, will be explained in the following.

## 2.1   Additive Line Diagrams

An attribute $m \in M$ is called *irreducible* if there is no set $X \subseteq M$ of attributes with $m \notin X$ so that $\{m\}' = X'$. The set of all irreducible attributes is denoted by $M_{irr}$. The set $irr(A, B)$ of all irreducible attributes of a concept is defined by $B \cap M_{irr}$.

An additive line diagram is obtained based on a function mapping each irreducible attribute to a two-dimensional vector according to Formula (3). The position in the plane at which the node representing concept $(A, B)$ should be drawn is determined according to Formula (4) [6].

$$vec : M_{irr} \to \mathbb{R} \times \mathbb{R}_{<0} \qquad (3)$$

$$pos(A, B) := \sum_{m \in irr(A,B)} vec(m) \qquad (4)$$

The resulting diagrams are characterized by many parallel lines, which makes them easier to read than most other diagrams. An example is displayed in Fig. 1.

An *implication* is a rule of the form $X \to Y$, where $X$ is the body and $Y$ is the head of the rule. $X$ and $Y$ are sets of attributes. Given a formal context an implication $X \to Y$ is valid iff $Y \subseteq X''$. It is not too difficult and a standard exercise in FCA to directly read implications from a concept lattice diagram.

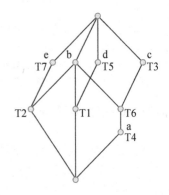

**Fig. 1.** Example of an additive line diagram based on the context given in Table 1.

## 2.2  Association Analysis

Association rule learning is a data mining method used to discover and evaluate relationships and dependencies in databases. It is used, among others, in shopping cart analysis. Put into FCA terminology, the goal of association analysis is to find connections between individual attributes in order to be able to make statements about which of them often co-occur together in objects. For this purpose, one uses the statistical characteristics *support*, *confidence* and *lift* [4,5], which are explained in this section together with their relationship to FCA.

The basis of association analysis are *association rules*, which have the form $X \Rightarrow Y$. $X$ and $Y$ represent disjoint and real subsets of the set of all attributes $M$. An object fulfills such a rule if it has all the attributes, which occur in $X$ and $Y$.

**Support.** The support of an attribute set $X$ describes the relative frequency of their joint occurrence in the data (that is: the context) and is calculated by

$$supp(X) := \frac{|\{g \in G \mid X \subseteq \{g\}'\}|}{|G|} = \frac{|X'|}{|G|}. \tag{5}$$

In the same way, the support of an association rule $X \Rightarrow Y$ describes the frequency with which the union $X \cup Y$ occurs in the data, that is

$$supp(X \Rightarrow Y) := supp(X \cup Y) = \frac{|(X \cup Y)'|}{|G|}. \tag{6}$$

**Confidence.** The confidence is a measure of how often a rule is fulfilled in relation to $X$. It is calculated by

$$conf(X \Rightarrow Y) := \frac{supp(X \Rightarrow Y)}{supp(X)}. \tag{7}$$

**Lift.** Since the confidence does not refer to the frequency with which the head of the formula occurs "normally", it cannot be used to make a statement about how strong the body of an association rule really "promotes" the head. In order to account for this fact, the lift is used, which is calculated by

$$lift(X \Rightarrow Y) := \frac{conf(X \Rightarrow Y)}{supp(Y)} = \frac{supp(X \cup Y)}{supp(X) \cdot supp(Y)}. \tag{8}$$

As is obvious from that formula, $lift(X \Rightarrow Y) = lift(Y \Rightarrow X)$ always holds and the corresponding value provides information regarding the correlation between attribute occurrences $X$ and $Y$. The following correspondences apply:

- $lift(X \Rightarrow Y) > 1$ : $X$ and $Y$ positively correlated
- $lift(X \Rightarrow Y) = 1$ : $X$ and $Y$ not correlated
- $lift(X \Rightarrow Y) < 1$ : $X$ and $Y$ negatively correlated

## 3  Weight-Dependent Positioning

This section describes how support, confidence, and lift can be read from a line diagram where each concept node's y-coordinate is chosen according to the size of its extent. Refining this idea, we then introduce the diagram type *cascading (additive) line diagram*, where dependencies between attributes, can be read using the *parallelogram method*, which is also presented.

### 3.1  Weighted Formal Contexts

Often, formal contexts representing large real-world data sets contain many objects that coincide in terms of their attributes. To represent such data in a succinct but statistically faithful[1] manner, we endow formal contexts with weights. A *weighted formal context* is a quadruple $(G, M, I, mult)$ extending a formal context $(G, M, I)$ by a mapping $mult$, which assigns a *weight* (or *multiplicity*)

$$mult : G \to \mathbb{N}^+ \tag{9}$$

**Table 2.** Example of a weighted context.

|    | $mult$ | a | b | c | d |
|----|--------|---|---|---|---|
| O1 | 10     | × | × |   |   |
| O2 | 10     |   |   | × | × |
| O3 | 1      | × |   |   |   |
| O4 | 1      |   | × |   |   |
| O5 | 1      |   |   | × |   |
| O6 | 1      |   |   |   | × |

to every object. In our setting, $mult(g) = n$ means that object $g$ occurs $n$ times in our data set.[2] In order to reflect this in cross tables, we extend them by a column $mult$, which contains the weight of each object (cf. Table 2, where O1 and O2 are taken to appear 10-fold).

Based on this, the weight associated to a concept $(A, B)$ is defined as the sum of the weights of all objects contained in its extent $A$. We define:

$$wgt : \mathfrak{P}(G) \to \mathbb{N} \tag{10}$$

$$wgt(A) := \sum_{g \in A} mult(g) \tag{11}$$

### 3.2  Linear Vertical Positioning

In this paper, the nodes in concept lattice diagrams are positioned in a 2-dimensional Cartesian coordinate system. Thereby, in order to implement our goal that statistical information be readable from the diagram, we first investigate the approach where we let the y-position of a node be defined by its weight. In the lattice diagram of a given weighted formal context $(G, M, I, mult)$ defined this way, the highest node has the y-coordinate $wgt(G)$. We note that such a positioning always creates admissible lattice diagrams, since the function defined in Formula (11) is a monotonic mapping from $(\mathfrak{P}(G), \subset)$ to $(\mathbb{N}, <)$.

---

[1] clarifying the context would prune duplicates but distort the statistical information

[2] Note, however, that the notion easily generalizes to settings where the weight expresses other qualities that justify to assign more statistical importance to certain objects (in which case one might rather choose $\mathbb{Q}$ or $\mathbb{R}$ as codomain).

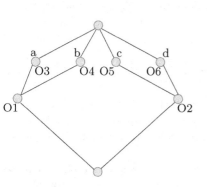

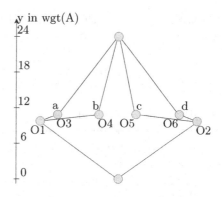

**Fig. 2.** Additive line diagram.          **Fig. 3.** Y-weighted, additive line diagram.

In additive line diagrams, the position of a node is determined by summing up the vectors associated to all irreducible nodes that define it and adding them to a normalization vector. The lattice diagrams represented in this way are characterized by many parallel lines, which increases readability. By simply adjusting the y-coordinate in the way described above, this advantage is lost. To illustrate this, consider the context from Table 2.

Figure 2 shows a possible additive line diagram which can be derived from the context. Fixing the y-coordinate of each node with its weight results in Fig. 3. The parallel sides of the diagram are lost and some of the lines have become almost horizontal, which is obstructing readability. In this work, therefore, the x-coordinate is not determined as an unweighted sum of the irreducible nodes, as is the case with additive line diagrams. Instead, the x-position is determined by compressing or stretching the vector resulting from the summation of the vectors of all upper neighbor nodes. Figure 4 illustrates this approach. Figure 5 shows the resulting line diagram without auxiliary lines.

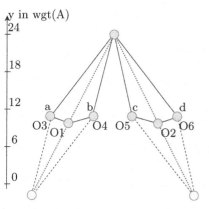

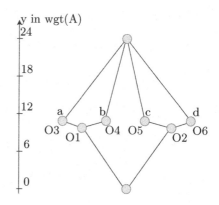

**Fig. 4.** Weighted line diagram with auxiliary lines.

**Fig. 5.** X-Y-weighted, additive line diagram.

**Support.** Having in mind the meaning of multiplicities, the support of a weighted context's concept $(A, B)$ is defined analogous to Formula (5) by its weight as

$$supp(B) := \frac{wgt(B')}{wgt(G)} = \frac{wgt(A)}{wgt(G)} =: supp(A) \tag{12}$$

Since $wgt(G)$ is constant and $wgt(A)$ was used as the weight of the concept to define the y-coordinate of the corresponding node, the support can also be read from the previously defined lattice diagram. This only requires an adjustment of the labeling of the y-axis.

In order to illustrate this, we introduce a new example, which will be used in the following to explain how to read off the confidence, the lift, and the correlation. The used weighted context is shown in Table 3. It reflects the statistical distribution by height and gender in Germany in 2006, differentiated by height <175 and ≥175 centimeters. The *mult* column shows the percentage of the respective gender for each height.

Figure 6 shows a line diagram that can be derived from the given context. On the y-axis, in addition to the weight $wgt(A)$, the support $supp(A)$ is shown. The

**Table 3.** Distribution by height and gender in Germany 2006 [2].

|     | *mult* | male | female | <175 | ≥175 |
|-----|--------|------|--------|------|------|
| M1  | 31     | ×    |        | ×    |      |
| M2  | 69     | ×    |        |      | ×    |
| F1  | 91     |      | ×      | ×    |      |
| F2  | 9      |      | ×      |      | ×    |

support is calculated by Formula (12). This line diagram allows for reading the weight as well as the support of the displayed concepts from the y-axis.

The natural way of defining the support of an implication in a weighted context is by the weight of all objects that fulfill this implication, relative to the weight of the set of all objects $G$. It can be formalized in two ways using the intents $X$ and $Y$ or the extents $X'$ and $Y'$.

$$supp(X \rightarrow Y) := \frac{supp(X' \cap Y')}{supp(G)} = \frac{supp(X \cup Y)}{supp(G)} \tag{13}$$

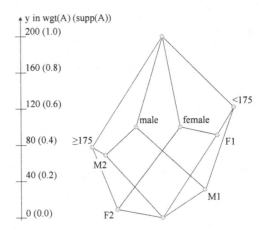

**Fig. 6.** Gender and size distribution with representation of $wgt(A)$ and $supp(A)$.

**Confidence.** As explained in Section 2.1, implications can be read off from the concept lattice diagram. For example, the line diagram displayed in Fig. 6 can be used to find the implication $\{male\} \rightarrow \{\geq 175\}$.

According to Section 2.2, the confidence of an association rule can be determined using Formula (7). For an implication $X \rightarrow Y$, the confidence is defined analogously as:

$$conf(X \rightarrow Y) := \frac{supp(X \rightarrow Y)}{supp(X)} \tag{14}$$

## 3.3   Logarithmic Vertical Positioning

Since the confidence is defined as a fraction, it is difficult to read it from the previously defined representation of the concept lattice diagram. With the help of the logarithmic law $\log_a (x/y) = \log_a(x) - \log_a(y)$, Formula (14) can be transformed and displayed as subtraction.

$$\log_a(conf(X \rightarrow Y)) := \log_a \left( \frac{supp(X \rightarrow Y)}{supp(X)} \right) \tag{15}$$

$$= \log_a(supp(X \rightarrow Y)) - \log_a(supp(X)) \tag{16}$$

The logarithmized confidence could be easily read from a line diagram, where the logarithmic support is shown on the y-axis. There are several possibilities for the choice of the base $a$ of the used logarithm, but they only lead to a linear vertical scaling. In the following, the base 2 was chosen, since $\log_2(0.5) = -1.0$ applies and thus the results can be easily estimated. The concept lattice with logarithmically scaled y-axis is shown in Fig. 7.

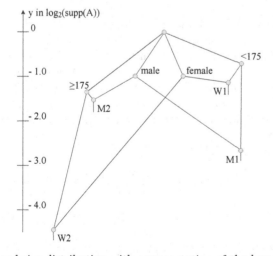

**Fig. 7.** Gender and size distribution with representation of the logarithmized support of each node on the y-axis.

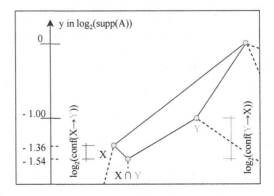

**Fig. 8.** Example for reading the logarithmic confidence for the implications $X \to Y$ and $Y \to X$.

Since nodes with weight zero would have the value $-\infty$ with this scaling, they are not displayed. To be able to see if a bottom element with weight 0 exists, short vertical auxiliary lines are attached to those nodes, that share an edge with it. Those can be seen in Fig. 7. Figure 8 illustrates the process of reading the logarithmic confidence for the implications $X \to Y$ and $Y \to X$. It shows an excerpt of Figure 7.

**Reading correlation.** As described in Section 2.2, it is possible to derive from the lift of an association rule $X \Rightarrow Y$ how $X$ and $Y$ are correlated. Analogous to Formula (8) for association rules, the lift for implications $X \to Y$ in weighted contexts is defined by

$$lift(X \to Y) := \frac{conf(X \to Y)}{supp(Y)}. \tag{17}$$

Since statements about the polarity and approximate strength of the correlation are usually more important than the correlation coefficient's exact value, it is not necessary to calculate the lift exactly. In many cases, a quantitative estimation is sufficient. If the lift is greater than one, $X$ and $Y$ are positively correlated. If it is smaller than one, they are negatively correlated. As the lift is defined as a fraction, it is advisable to apply the logarithm, similar to the case of the confidence. This way, we obtain the following characterization for the case of positive correlation:

$$\frac{conf(X \to Y)}{supp(Y)} > 1.0 \tag{18}$$

$$\log_2\left(\frac{conf(X \to Y)}{supp(Y)}\right) > \log_2(1.0) \tag{19}$$

$$\log_2(conf(X \to Y)) - \log_2(supp(Y)) > 0.0 \tag{20}$$

$$\log_2(conf(X \to Y)) > \log_2(supp(Y)) \tag{21}$$

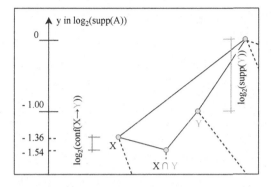

**Fig. 9.** Estimating the dependence of X and Y by reading off confidence and support.

To estimate whether $X$ and $Y$ are positively correlated, it is therefore sufficient to compare the logarithmic values of the implication's confidence and the head's support. This can be done purely graphically.

Figure 9 shows the values to be read for the estimation of the lift for the implication $X \to Y$. It shows a cutout of Fig. 7, with $X$ being $\geq 175$ and $Y$ being $male$. It can therefore be concluded that there is a positive lift for the implication $\{\geq 175\} \to \{male\}$, since $\log_2(conf(\{\geq 175\} \to \{male\}))$ is $-0.18$ and $\log_2(supp(male))$ is $-1.0$. Comparing the two values shows

$$- 0.18 > -1.0 \ \to \ \text{positively correlated} \tag{22}$$

A second possibility to read the correlation between $X$ and $Y$ in the logarithmic lattice diagram can be derived from the definition of the lift. This is due to the fact that the lift for the implication $X \to Y$ mathematically equals the lift for the implication $Y \to X$.

Another way to determine whether $X$ and $Y$ are statistically dependent or independent is the *parallelogram method* presented below. As already described, $X$ and $Y$ are statistically independent exactly if

$$\log_2(conf(X \to Y)) = \log_2(supp(Y)). \tag{23}$$

If the independence of $X$ and $Y$ can be derived from $X \to Y$, then follows

$$\log_2(conf(Y \to X)) = \log_2(supp(X)). \tag{24}$$

If $X$ and $Y$ are independent, i.e. Formulas (23) and (24) are satisfied, a parallelogram is formed in the diagram. However, if they are dependent, a one-sided distortion of the parallelogram along the diagonal occurs. A simple, convex square is formed. If the downward-pointing half of the parallelogram is compressed, $X$ and $Y$ are positively correlated. If it is stretched, they are negatively correlated. Table 4 shows all three possible correlation types together with an example.

On the basis of the deviation of the calculated parallelogram the dependence of $X$ and $Y$ can be read. For this purpose, it may be helpful to mark the result-

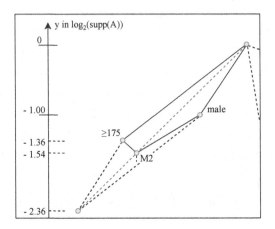

**Fig. 10.** Illustrating the parallelogram method. The downward-pointing half of the quadrangle is compressed, indicating that $\geq 175$ and *male* are positively correlated.

ing parallelogram, which symbolizes independence, with auxiliary lines in the diagram. Figure 10 shows such an example.

### 3.4   Cascading Additive Line Diagrams

In order to achieve that the parallelogram method can also be used in situations where the nodes representing $X$ and $Y$ do not share an edge with the top node, we refine our idea leading to the definition of *cascading additive line diagrams*[3], detailed below.

The y-coordinate of each node in a cascading additive line diagram is defined by the logarithm of the support of the concept associated with the corresponding node, just as it was described in Section 3.1. The x-coordinate, on the other hand, is defined taking into account the positions of all upper neighbor nodes, leading to a recursive definition – hence the name 'cascading'. The upper neighbors (also called direct predecessors) of the node associated to the concept $(A, B)$ are determined according to the order relation as defined in Section 2 and Formula (25).

$$pre(A, B) := \{(C, D) \in \mathfrak{B}(G, M, I) \mid (A, B) \prec (C, D)\} \tag{25}$$

Remember that in the diagram, these are just the nodes that have a higher y-position and share an edge with the node in question. The *unscaled position* $upos(A, B)$ of each concept node is defined by adding the positions of all predecessors:

$$upos(A, B) := \sum_{(C,D)\in pre(A,B)} pos(C, D) + \sum_{m\in M_{irr},\, A=\{m\}'} vec(m) \tag{26}$$

---

[3] short: cascading line diagrams

**Table 4.** Minimal examples showcasing the three different types of correlation. The diagrams have been created with the help of the program described in Section 4.

| negatively correlated | independent | positively correlated |
|---|---|---|

| | *mult* | a | b |
|---|---|---|---|
| T | 10 | | |
| A | 30 | × | |
| B | 30 | | × |
| AB | 10 | × | × |

| | *mult* | a | b |
|---|---|---|---|
| T | 20 | | |
| A | 20 | × | |
| B | 20 | | × |
| AB | 20 | × | × |

| | *mult* | a | b |
|---|---|---|---|
| T | 30 | | |
| A | 10 | × | |
| B | 10 | | × |
| AB | 30 | × | × |

$$lift(\{a\} \to \{b\}) = \frac{1/8}{1/2 \cdot 1/2} = 1/2 < 1.0$$

$$lift(\{a\} \to \{b\}) = \frac{1/4}{1/2 \cdot 1/2} = 1.0$$

$$lift(\{a\} \to \{b\}) = \frac{3/8}{1/2 \cdot 1/2} = 3/2 > 1.0$$

supp(X) | log2(supp(X)) | wgt(X)

1.00 | 0.00 | 80

0.50 | -1.00 | 40

0.13 | -3.00 | 10

supp(X) | log2(supp(X)) | wgt(X)

1.00 | 0.00 | 80

0.50 | -1.00 | 40

0.25 | -2.00 | 20

supp(X) | log2(supp(X)) | wgt(X)

1.00 | 0.00 | 80

0.50 | -1.00 | 40

0.38 | -1.42 | 30

| stretched | well-formed | compressed |
|---|---|---|

By scaling the unscaled position with respect to the y-coordinate, the final position of the node is determined:

$$pos(A, B) := \frac{\log_2(supp(B))}{upos_y(A, B)} \cdot upos(A, B) \tag{27}$$

To illustrate this presentation method in more detail, we use the context shown in Table 5. This context is characterized by the independence of $\{k\}$ and $\{m\}$, which is proved mathematically in Formula (28) by calculating the lift of $\{k\} \to \{m\}$.

$$lift(\{k\} \to \{m\}) := \frac{supp(\{k\} \to \{m\})}{supp(\{k\}) \cdot supp(\{m\})}$$
$$= \frac{0.05}{0.25 \cdot 0.2} = 1.0 \tag{28}$$

**Table 5.** Example context.

|        | *mult* | g | h | k | m |
|--------|--------|---|---|---|---|
| G      | 25     | × |   |   |   |
| H      | 30     |   | × |   |   |
| GH     | 5      | × | × |   |   |
| GHK    | 20     | × | × | × |   |
| GHM    | 15     | × | × |   | × |
| GHKM   | 5      | × | × | × | × |

Figure 11 displays the corresponding concept lattice as a cascading additive line diagram. It can be seen that the top-element forms a parallelogram with the nodes labeled with "GHK", "GHM" and "GHKM", thus displaying the independence of $\{k\}$ and $\{m\}$. If the node labeled "GHKM" were higher, meaning it had a larger y-coordinate, then $\{k\}$ and $\{m\}$ would be positively correlated. If the y-coordinate were smaller, the parallelogram would be stretched on the lower side. This would correspond to a negative correlation.

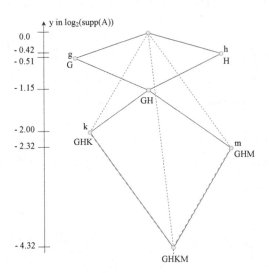

**Fig. 11.** Cascading line diagram of the context shown in Table 5.

The cascading additive line diagram allows to read at each node that has exactly two predecessors, the (in)dependence of these predecessors using the parallelogram method. Furthermore, in all cases, it is possible to read the dependence by means of the y-coordinates, as discussed in Section 3.1.

## 4  Prototype Implementation

To demonstrate the operation of the cascading additive line diagram, we implemented a visualisation prototype. The source code is freely available online [1]. The tool creates concept lattice diagrams for given formal contexts. It is able to generate both additive and cascading line diagrams.

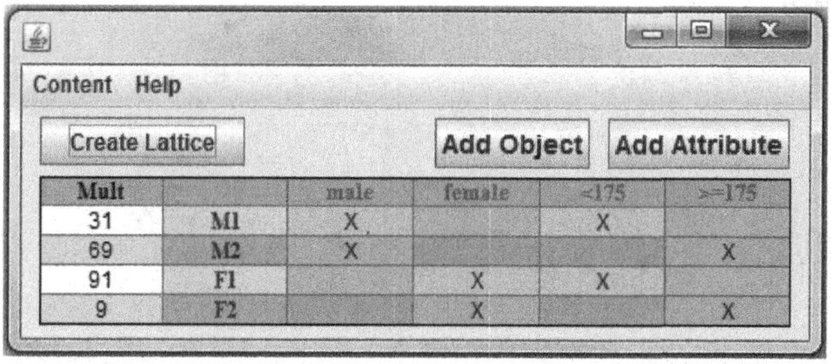

**Fig. 12.** Display of a cross table in the program.

Figure 12 shows the user interface for entering the formal context. In the screenshot, the example from Table 3 has been entered. Clicking on 'Create Lattice' triggers the creation of the concept lattice diagram. The additive line diagram is created as described in Section 2.1. The cascading line diagram is created as described in Section 3.4.

Upon displaying the diagram, the irreducible nodes are marked by green squares. In the cascading view, the independence of two attributes is expressed by the parallelogram formed with the top element. By left-clicking on a reducible node with two predecessors, the parallelogram that would result if the predecessors were independent is shown in red. This is shown in Fig. 13, using the context from Table 3, which was considered in detail in Section 3.2. Figure 13 also shows an information window, which opens upon right-clicking on a node.

## 5  Conclusion

The goal of this paper was to develop a representation for concept lattice diagrams that – in addition to the attribute-logical relationships – allows statistical relationships between attributes to be read off the diagram. Toward a more succinct representation of the input data, the cross table defining the formal context was extended by a multiplier column, where the positive natural number in this

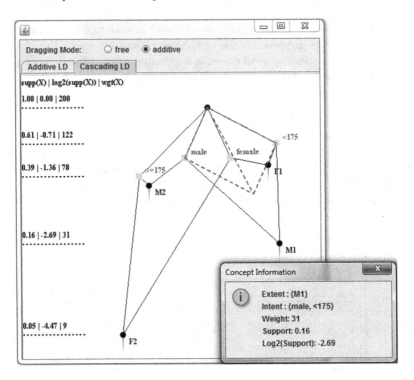

**Fig. 13.** Display of a cascading additive line diagram in the program with the parallelogram with the top element shown and the additional window opened.

cell indicates the weight or multiplicity of the associated object. This can be understood to represent an aggregation of several objects with the same intent from a normal cross table.

Inspired by the display form of additive line diagrams, a new display form was then developed, which allows for reading statistical and associative measures such as support, confidence, and lift from the concept lattice diagram. Furthermore, the statistical (in)dependence of two attributes can be derived from the diagram. This was achieved by a vertical positioning following the logarithmized support of the associated concept. Through additive scaling, it was also possible to develop a form of representation, called cascading additive line diagram, in which the dependence or independence of two attributes can be intuitively grasped from their formed parallelograms (or, rather, the deviation from the parallelogram shape).

We presented a prototype implementation that allows to input a formal context as a cross table and, from this, to derive, display, and adjust both additive and a cascading line diagrams.

It has been shown in this paper that a formal context's concept lattice can be represented by a line diagram from which, beyond the classical logical dependencies, statistical relationships between attributes can be read rather directly and intuitively.

As one avenue of future work, the proposed representation paradigm could be coupled with existing optimization approaches toward a beneficial choice of the vector assignment *vec* to the irreducible attributes, with the goal of avoiding (near-)overlap of nodes, nodes being positioned on (or close to) edges they are not incident with, and reducing the overall number of edge crossings.

Finally, our novel visualization approach will have to be tested empirically: determining if this way of presenting statistical information is indeed useful to human users and how well this paradigm works for larger concept lattices can only be found out by means of comprehensive user studies.

# References

1. https://github.com/Klimpke/Cascading-Line-Diagrams-Visualizer
2. Körpergröße in Deutschland nach Geschlecht 2006 [Body height in Germany by gender 2006]. https://de.statista.com/statistik/daten/studie/1825/umfrage/koerpergroesse-nach-geschlecht/. Accessed 23 Sept 2020
3. Abdullah, Z., Saman, M.Y.M., Karim, B., Herawan, T., Deris, M.M., Hamdan, A.R.: FCA-ARMM: a model for mining association rules from formal concept analysis. In: Herawan, T., Ghazali, R., Nawi, N.M., Deris, M.M. (eds.) SCDM 2016. AISC, vol. 549, pp. 213–223. Springer, Cham (2017). https://doi.org/10.1007/978-3-319-51281-5_22
4. Agrawal, R., Imieliński, T., Swami, A.: Mining association rules between sets of items in large databases. In: Proceedings International Conference on Management of Data (SIGMOD), pp. 207–216. ACM (1993)
5. Brin, S., Motwani, R., Ullman, J.D., Tsur, S.: Dynamic itemset counting and implication rules for market basket data. SIGMOD Rec. **26**(2), 255–264 (1997)
6. Ganter, B., Wille, R.: Formal Concept Analysis – Mathematical Foundations. Springer-Verlag (1999). https://doi.org/10.1007/978-3-642-59830-2
7. Hannan, T., Pogel, A.: Spring-based lattice drawing highlighting conceptual similarity. In: Missaoui, R., Schmidt, J. (eds.) ICFCA 2006. LNCS (LNAI), vol. 3874, pp. 264–279. Springer, Heidelberg (2006). https://doi.org/10.1007/11671404_18
8. Lakhal, L., Stumme, G.: Efficient mining of association rules based on formal concept analysis. In: Ganter, B., Stumme, G., Wille, R. (eds.) Formal Concept Analysis. LNCS (LNAI), vol. 3626, pp. 180–195. Springer, Heidelberg (2005). https://doi.org/10.1007/11528784_10
9. Mondal, K.C., Pasquier, N., Mukhopadhyay, A., Maulik, U., Bandhopadyay, S.: A new approach for association rule mining and bi-clustering using formal concept analysis. In: Perner, P. (ed.) MLDM 2012. LNCS (LNAI), vol. 7376, pp. 86–101. Springer, Heidelberg (2012). https://doi.org/10.1007/978-3-642-31537-4_8
10. Pogel, A., Hannan, T., Miller, L.: Visualization of concept lattices using weight functions. In: Wolff, K.E., Pfeiffer, H.D., Delugach, H.S. (eds.) Supp. Proceedings 12th International Conference on Conceptual Structures (ICCS), pp. 1–14. Shaker (2004)
11. Stumme, G., Taouil, R., Bastide, Y., Pasquier, N., Lakhal, L.: Intelligent structuring and reducing of association rules with formal concept analysis. In: Baader, F., Brewka, G., Eiter, T. (eds.) KI 2001. LNCS (LNAI), vol. 2174, pp. 335–350. Springer, Heidelberg (2001). https://doi.org/10.1007/3-540-45422-5_24

# Force-Directed Layout of Order Diagrams Using Dimensional Reduction

Dominik Dürrschnabel[1,2(✉)] ⓘ and Gerd Stumme[1,2] ⓘ

[1] Knowledge and Data Engineering Group, University of Kassel, Kassel, Germany
[2] Interdisciplinary Research Center for Information System Design,
University of Kassel, Kassel, Germany
{duerrschnabel,stumme}@cs.uni-kassel.de

**Abstract.** Order diagrams allow human analysts to understand and analyze structural properties of ordered data. While an expert can create easily readable order diagrams, the automatic generation of those remains a hard task. In this work, we adapt force-directed approaches, which are known to generate aesthetically-pleasing drawings of graphs, to the realm of order diagrams. Our algorithm `ReDraw` thereby embeds the order in a high dimension and then iteratively reduces the dimension until a two-dimensional drawing is achieved. To improve aesthetics, this reduction is equipped with two force-directed steps where one step optimizes the distances of nodes and the other one the distances of lines in order to satisfy a set of a priori fixed conditions. By respecting an invariant about the vertical position of the elements in each step of our algorithm we ensure that the resulting drawings satisfy all necessary properties of order diagrams. Finally, we present the results of a user study to demonstrate that our algorithm outperforms comparable approaches on drawings of lattices with a high degree of distributivity.

**Keywords:** Ordered sets · Order diagram drawing · Lattice drawing · Force-directed algorithms · Dimensional reduction · Graph drawing

## 1 Introduction

*Order diagrams*, also called *line diagrams*, *Hasse diagrams* (or simply *diagrams*) are a graphical tool to represent ordered sets. In the context of *ordinal data analysis*, i.e., data analysis investigating ordered sets, they provide a way for a human reader to explore and analyze complex connections. Every element of the ordered set is thereby visualized by a node and two elements are connected by a straight line if one is lesser than the other and there is no element "in between".

The general structure of such an order diagram is therefore fixed by these conditions. Nonetheless, finding good coordinates for the nodes representing the elements such that the drawing is perceived as "readable" by humans is not a trivial task. An expert with enough practice can create such a drawing; however, this is a time-consuming and thus uneconomical task and therefore rather

© Springer Nature Switzerland AG 2021
A. Braud et al. (Eds.): ICFCA 2021, LNAI 12733, pp. 224–240, 2021.
https://doi.org/10.1007/978-3-030-77867-5_14

uncommon. Still, the availability of such visualizations of order diagrams is an integral requirement for developing ordinal data science and semi-automated data exploration into a mature instrument; thus, making the automatic generation of such order diagrams an important problem. An example of a research field that is especially dependent on the availability of such diagrams is Formal Concept Analysis, a field that generates a conceptual lattices from binary datasets.

The generated drawings have to satisfy a set of *hard constraints* in order to guarantee that the drawing accurately represents the ordered set. First of all, for comparable elements the node representing the greater element has to have a larger $y$-coordinate then the lesser node. Secondly, no two nodes are allowed to be positioned on the same coordinates. Finally, nodes are not allowed to touch non-adjacent lines. Beside those hard criteria, there is a set of commonly accepted *soft criteria* that are generally considered to make a drawing more readable, as noted in [16]. Those include maximizing the distances between element nodes and lines, minimizing the number of crossing lines, maximizing the angles of crossing lines, minimizing the number of different line directions or organizing the nodes in a limited number of layers. While it is not obvious how to develop an algorithm that balances the soft criteria and simultaneously guarantees that the hard criteria are satisfied, such an algorithm might not even yield readable results as prior works such as [3] suggests. Furthermore, not every human reader might perceive the same aspects of an order diagram as "readable"; conversely, it seems likely that every human perceives different aspects of a good drawing as important. It is thus almost impossible to develop with a good fitness function for readable graphs. Those reasons combined make the automatic generation of readable graph drawing - and even their evaluation - a surprisingly hard task.

There are some algorithms today that can produce readable drawings to some extent; however, none of them is able to compete with the drawings that are manually drawn by an expert. In this paper we address this problem by proposing our new algorithm `ReDraw` that adapts the force-directed approach of graph drawing to the realm of order diagram drawing. Thereby, a physical simulation is performed in order to optimize a drawing by moving it to a state of minimal stress. Thus, the algorithm proposed in this paper provides a way to compute sufficiently readable drawings of order diagrams. We compare our approach to prior algorithms and show that our drawings are more readable under certain conditions or benefits from lesser computational costs. We provide the source code[1] so that other researchers can conduct their own experiments and extend it.

## 2    Related Work

Order diagram drawing can be considered to be a special version of the graph drawing problem, where a graph is given as a set of vertices and a set of edges and a readable drawing of this graph is desired. Thereby, each vertex is once again

---

[1] https://github.com/domduerr/redraw

represented by a node and two adjacent vertices are connected by a straight line. The graph drawing problem suffers from a lot of the same challenges as order diagram drawing and thus a lot of algorithms that were developed for graph drawing can be adapted for diagram drawing. For a graph it can be checked in linear time whether it is planar [11]), i.e., whether it has a drawing that has no crossing edges. In this case a drawing only consisting of straight lines without bends or curves can always be computed [12] and should thus be preferred. For a directed graph with a unique maximum and minimum, like for example a lattice, it can be checked in linear time whether an upward planar drawing exists. Then such a drawing can be computed in linear time [2]. The work of Battista et al. [2] provides an algorithm to compute straight-line drawings for "serial parallel graphs", which is a special family of planar, acyclic graphs. As symmetries are often preferred by readers, the algorithm was extended [10] to reflect them in the drawings based on the automorphism group of the graph. However, lattices that are derived from real world data using methods like Formal Concept Analysis rarely satisfy the planarity property [1]. The work of Sugiyama et al. [14], usually referred to as Sugiyama's framework, from 1981 introduces an algorithm to compute layered drawings of directed acyclic graphs and can thus be used for drawing order diagram. Force-directed algorithms were introduced in [5] and further refined in [7]. They are a class of graph drawing algorithms that are inspired by physical simulations of a system consisting of springs.

The most successful approaches for order diagram drawing are a work of Sugiyama et al. [14] which is usually referred to as Sugiyama's framework and a work of Freese [6]. Those algorithms both use the structure of the ordered set to decide on the height of the element nodes; however, the approach choosing the horizontal coordinates of a node differ significantly. While Sugiyama's framework minimizes the number of crossing lines between different vertical layers, Freese's layout adapts a force-directed algorithm to compute a three-dimensional drawing of the ordered set. DimDraw [4] on the other hand is not an adapted graph drawing algorithm, but tries to emphasize the dimensional structure that is encapsulated in the ordered set itself. Even though this approach is shown to outperform Freese's and Sugiyama's approach in [4], it is not feasible for larger ordered sets because of its exponential nature. In [4] a method is proposed to draw order diagrams based on structural properties of the ordered set. In doing so, two maximal differing linear extensions of the ordered set are computed. The work in [8] emphasizes additive order diagrams of lattices. Another force-directed approach that is based on minimizing a "conflict distance" is suggested in [17].

In this work we propose the force-directed graph drawing algorithm ReDraw that, similarly to Freese's approach, operates not only in two but in higher dimensions. Compared to Freese's layout, our algorithm however starts in an arbitrarily high dimension and improves it then by reducing the number of dimensions in an iterative process. Thus, it minimizes the probability to stop the algorithm early with a less pleasing drawing. Furthermore, our approach gets rid of the ranking function to determine the vertical position of the elements and instead uses the force-directed approach for the vertical position of nodes as well. We achieve this by defining a vertical invariant which is respected in each step of the algorithm. This invariant guarantees that the resulting drawing will respect the hard condition of placing greater elements higher than lesser elements.

# 3   Fundamentals and Basics

This section recalls fundamentals and lays the foundations to understand the design choices of our algorithm. This includes recalling mathematical notation and definitions as well as introducing the concept of force-directed graph drawing.

## 3.1   Mathematical Notations and Definitions

We start by recalling some standard notations that are used throughout this work. An *ordered set* is a pair $(X, \leq)$ with $\leq \,\subseteq (X \times X)$ that is reflexive $((a, a) \in \leq$ for all $a \in X)$, antisymmetric (if $(a, b) \in \leq$ and $(b, a) \in \leq$, then $a = b$) and transitive (if $(a, b) \in \leq$ and $(b, c) \in \leq$, then $(a, c) \in \leq$). The notation $(a, b) \in \leq$ is used interchangeably with $a \leq b$ and $b \geq a$. We call a pair of elements $a, b \in X$ *comparable* if $a \leq b$ or $b \leq a$, otherwise we call them *incomparable*. A subset of $X$ where all elements are pairwise comparable is called a *chain*. An element $a$ is called *strictly less than* an element $b$ if $a \leq b$ and $a \neq b$ and is denoted by $a < b$, the element $b$ is then called *strictly greater than* $a$. For an ordered set $(X, \leq)$, the associated covering relation $\prec \,\subseteq <$ is given by all pairs $(a, c)$ with $a < c$ for which no element $b$ with $a < b < c$ exists. A *graph* is a pair $(V, E)$ with $E \subseteq \binom{V}{2}$. The set $V$ is called the set of *vertices* and the set $E$ is called the set of *edges*, two vertices $a$ and $b$ are called *adjacent* if $\{a, b\} \in E$.

From here out we give some notations in a way that is not necessarily standard but will be used throughout our work. A $d$-dimensional *order diagram* or *drawing* of an ordered set $(X, \leq)$ is denoted by $(\vec{p}_a)_{a \in X} \subseteq \mathbb{R}^d$ whereby $\vec{p}_a = (x_{a,1}, \ldots, x_{a,d-1}, y_a)$ for each $a \in X$ and for all $a \prec b$ it holds that $y_a < y_b$. Similarly, a $d$-dimensional *graph drawing* of a graph $(V, E)$ is denoted by $(\vec{p}_a)_{a \in V}$ with $\vec{p}_a = (x_{a,1}, \ldots, x_{a,d-1}, y_a)$ for each $a \in V$. If the dimension of a order diagram or a graph drawing is not qualified, the two-dimensional case is assumed. In this case an order diagram can be depicted in the plane by visualizing the elements as a node or *nodes* and connecting element pairs in the covering relation by a straight line. In the case of a graph, vertices are depicted by a node and adjacent vertices are connected by a straight line. We call $y_a$ the *vertical component* and $x_{a,1}, \ldots, x_{a,d-1}$ the *horizontal components* of of $\vec{p}_a$ and denote $(\vec{p}_a)_x = (x_{a,1}, \ldots, x_{a,d-1}, 0)$. The forces operating on the vertical component are called the *vertical force* and the forces operating on the horizontal components the *horizontal forces*. The Euclidean distance between the representation of $a$ and $b$ is denoted by $d(\vec{p}_a, \vec{p}_b) = |\vec{v}_a - \vec{v}_b|$, while the distance between the vertical components is denoted by $d_y(\vec{p}_a, \vec{p}_b)$ and the distance in the horizontal components is denoted by $d_x(\vec{p}_a, \vec{p}_b) = d((\vec{p}_a)_x, (\vec{p}_b)_x)$. The unit vector from $\vec{p}_a$ to $\vec{p}_b$ is denoted by $\vec{u}(\vec{p}_a, \vec{p}_b)$, the unit vector operating in the horizontal dimensions is denoted by

$\vec{u}_x(\vec{p}_a, \vec{p}_b)$. Finally, the cosine-distance between two vector pairs $(\vec{a}, \vec{b})$ and $(\vec{c}, \vec{d})$ with $\vec{a}, \vec{b}, \vec{c}, \vec{d} \in \mathbb{R}^d$ is given by $d_{\cos}((\vec{a}, b), (\vec{c}, \vec{d})) := 1 - \frac{\sum_{i=1}^{d}(b_i - a_i)\cdot(d_i - c_i)}{d(a,b)\cdot d(c,d)}$.

### 3.2  Force-Directed Graph Drawing

The general idea of force-directed algorithms is to represent the graph as a physical model consisting of steel rings each representing a vertex. For every pair of adjacent vertices, their respective rings are connected by identical springs. Using a physical simulation, this system is then moved into a state of minimal stress, which can in turn be used as the drawing. Many modifications to this general approach, that are not necessarily based on springs, were proposed in order to encourage additional conditions in the resulting drawings.

The idea of force-directed algorithms was first suggested by Eades [5]. His algorithmic realization of this principle is done using an iterative approach where in each step of the simulation the forces that operate on each vertex are computed and summed up (cf. Algorithm 1). Based on the sum of the forces operating on each vertex, they are then moved. This is repeated for either a limited number of rounds or until there is no stress left in the physical model. While a system consisting of realistic springs would result in linear forces between the vertices, Eades claims that those are performing poorly and thus introduces an artificial spring force. This force operates on each vertex $a$ for adjacent pairs $\{a, b\} \in E$ and is given as $f_{\text{spring}}(\vec{p}_a, \vec{p}_b) = -c_{\text{spring}} \cdot \log\left(\frac{d(\vec{p}_a, \vec{p}_b)}{l}\right) \cdot \vec{u}(\vec{p}_a, \vec{p}_b)$, whereby $c_{\text{spring}}$ is the spring constant and $l$ is the equilibrium length of the spring. The spring force repels two vertices if they are closer then this optimal distance $l$ while it operates as an attracting force if two vertices have a distance greater then $l$, see Fig. 1. To enforce that non-connected vertices are not placed too close to each other, he additionally introduces the repelling force that operates between non-adjacent vertex pairs as $f_{\text{rep}}(\vec{p}_a, \vec{p}_b) = \frac{c_{\text{rep}}}{d(\vec{p}_a, \vec{p}_b)^2} \cdot \vec{u}(\vec{p}_a, \vec{p}_b)$. The value for $c_{\text{rep}}$ is once again constant. In a realistic system, even a slightest movement of a vertex changes the forces that are applied to its respective ring. To depict this realistically a damping factor $\delta$ is introduced in order to approximate the realistic system. The smaller

---

**Algorithm 1.** Force-Directed Algorithm by Eades

| **Input:** | Graph: $(V, E)$ | **Constants:** $K \in \mathbb{N},\ \varepsilon > 0,\ \delta > 0$ |
|---|---|---|

**Input:**  Graph: $(V, E)$     **Constants:** $K \in \mathbb{N}, \varepsilon > 0, \delta > 0$
  Initial drawing: $p = (\vec{p}_a)_{a \in V} \subseteq \mathbb{R}^2$
**Output:** Drawing: $p = (\vec{p}_a)_{a \in V} \subseteq \mathbb{R}^2$

---

$t = 1$
**while** $t < K$ **and** $\max_{a \in V} \|F_a(t)\| > \varepsilon$ :
  **for** $a \in V$ :
    $F_a(t) := \sum_{\{a,b\} \notin E} f_{\text{rep}}(\vec{p}_a, \vec{p}_b) + \sum_{\{a,b\} \in E} f_{\text{spring}}(\vec{p}_a, \vec{p}_b)$
  **for** $a \in V$ :
    $\vec{p}_a := \vec{p}_a + \delta \cdot F_a(t)$
  $t = t + 1$

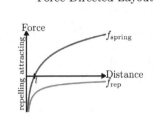 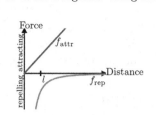 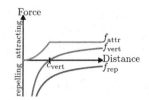

**Fig. 1.** The forces for graphs as introduced by Eades in 1984. The $f_{spring}$ force operates between adjacent vertices and has an equilibrium at $l$, the force $f_{rep}$ is always a repelling force and operates on non-adjacent pairs.

**Fig. 2.** Horizontal forces for drawing order diagrams introduced by Freese in 2004. The force $f_{attr}$ operates between comparable pairs, the force $f_{rep}$ between incomparable pairs. There is no vertical force.

**Fig. 3.** Our forces for drawing order diagrams. $f_{vert}$ operates vertically between node pairs in the covering relation, the force $f_{attr}$ between comparable pairs and the force $f_{rep}$ between incomparable pairs.

this damping factor is chosen, the closer the system is to a real physical system. However, a smaller damping factor results in higher computational costs. In some instances this damping factor is replaced by a cooling function $\delta(t)$ to guarantee convergence. The physical simulation stops if the total stress of the system falls below a constant $\varepsilon$. Building on this approach, a modification is proposed in the work of Fruchterman and Reingold [7] from 1991. In their algorithm, the force $f_{attr}(\vec{p}_a, \vec{p}_b) = -\frac{d(\vec{p}_a, \vec{p}_b)^2}{l} \cdot \vec{u}(\vec{p}_a, \vec{p}_b)$ is operating between every pair of connected vertices. Compared to the spring-force in Eades' approach, this force is always an attracting force. Additionally the force $f_{rep}(\vec{p}_a, \vec{p}_b) = \frac{l^2}{d(\vec{p}_a, \vec{p}_b)} \cdot \vec{u}(\vec{p}_a, \vec{p}_b)$ repels every vertex pair. Thus, the resulting force that is operating on adjacent vertices is given by $f_{spring}(\vec{p}_a, \vec{p}_b) = f_{attr}(\vec{p}_a, \vec{p}_b) + f_{rep}(\vec{p}_a, \vec{p}_b)$ and has once again its equilibrium at length $l$. These forces are commonly considered to achieve better drawings than Eades' approach and are thus usually preferred.

While the graph drawing algorithms described above lead to sufficient results for undirected graphs, they are not suited for order diagram drawings as they do not take the direction of an edge into consideration. Therefore, they will not satisfy the hard condition that greater elements have higher $y$-coordinates. Freese [6] thus proposed an algorithm for lattice drawing that operates in three dimensions, where the ranking function $rank(a) = height(a) - depth(a)$ fixes the vertical component. The function $height(a)$ thereby evaluates to the length of the longest chain between $a$ and the minimal element and the function $depth(a)$ to the length of the longest chain to the maximal element. While this ranking function guarantees that lesser elements are always positioned below greater elements, the horizontal coordinates are computed using a force-directed approach. Freese introduces an attracting force between comparable elements that is given by $f_{attr}(\vec{p}_a, \vec{p}_b) = -c_{attr} \cdot d_x(\vec{p}_a, \vec{p}_b) \cdot \vec{u}_x(\vec{p}_a, \vec{p}_b)$, and a repelling force that is given by $f_{rep}(\vec{p}_a, \vec{p}_b) = c_{rep} \cdot \frac{d_x(\vec{p}_a, \vec{p}_b)}{|y_b - y_a|^3 + |x_{b,1} - x_{a,1}|^3 + |x_{b,2} - x_{a,2}|^3} \cdot \vec{u}_x(\vec{p}_a, \vec{p}_b)$ operating on incomparable pairs only, (cf. Fig. 2). The values for $c_{attr}$ and $c_{rep}$ are constants. A parallel projection is either done by hand or chosen automatically in order to compute a two-dimensional depiction of the three-dimensional drawing.

## 4    The ReDraw Algorithm

Our algorithm ReDraw uses a force-directed approach similar to the one that is used in Freese's approach. Compared to Freese's algorithm, we however do not use a static ranking function to compute the vertical positions in the drawing. Instead, we use forces which allow us to incorporate additional properties like the horizontal distance of vertex pairs, into the vertical distance. By respecting a vertical invariant, that we will describe later, the vertical movement of the vertices is restricted so that the hard constraint on the $y$-coordinates of comparable nodes can be always guaranteed. However, the algorithm is thus more likely to get stuck in a local minimum. We address this problem by computing the first drawing in a high dimension and then iteratively reducing the dimension of this drawing until a two-dimensional drawing is achieved. As additional degrees of freedom allow the drawing to move less restricted in higher dimensions it thus reduces the probability for the system to get stuck in a local minimum.

Our algorithm framework (cf. Algorithm 2) consists of three individual algorithmic steps that are iteratively repeated. We call one repetition of all three steps a *cycle*. In each cycle the algorithm is initialized with the $d$-dimensional drawing and returns a $(d-1)$-dimensional drawing. The first step of the cycle, which we refer to as the *node step*, improves the $d$-dimensional drawing by optimizing the proximity of nodes in order to achieve a better representation of the ordered set. In the second step, which we call the *line step*, the force-directed approach is applied to improve distances between different lines as well as between lines and nodes. The resulting drawing thereby achieves a better satisfaction of soft criteria and thus improves the readability for a human reader. Finally, in the *reduction step* the dimension of the drawing is reduced to $(d-1)$ by using a parallel projection into a subspace that preserves the vertical dimension. In the last (two-dimensional) cycle, the dimension reduction step is omitted.

---

**Algorithm 2.** ReDraw Algorithm

---

| **Input:** | Ordered set: $O = (X, \leq)$ | **Constants:** $K \in \mathbb{N}$, $\varepsilon > 0$, $\delta > 0$, |
|---|---|---|
| | Initial dimension: $d$ | $c_{vert} > 0$, $c_{hor} > 0$, |
| **Output:** | Drawing: $p = (\vec{p}_a)_{a \in V} \subseteq \mathbb{R}^2$ | $c_{par} > 0$, $c_{ang} > 0$, $c_{dist} > 0$ |

---

```
p = initial_drawing (O)
while d ≥ 2 :
    node_step (O, p, d, K, ε, δ, c_vert, c_hor )
    line_step (O, p, d, K, ε, δ, c_par, c_ang, c_dist )
    if d > 2 :
        dimension_reduction (O, p, d)
        d = d − 1
```

---

The initial drawing used in the first cycle is randomly generated. The vertical coordinate of each node is given by its position in a randomly chosen linear extension of the ordered set. The horizontal coordinates of each element are set to a random value between $-1$ and $1$. This guarantees that the algorithm does not start in an unstable local minimum. Every further cycle then uses the output of the previous cycle as input to further enhance the resulting drawing.

Compared to the approach taken by Freese we do not fix the vertical component by a ranking function. Instead, we recompute the vertical position of each element in each step using our force-directed approach. To ensure that the resulting drawing is in fact a drawing of the ordered set we guarantee that in every step of the algorithm the following property is satisfied:

**Definition 1.** *Let $(X, \leq)$ be an ordered set with a drawing $(\vec{p}_a)_{a \in X}$. The drawing $(\vec{p}_a)_{a \in X}$ satisfies the vertical constraint, iff $\forall a, b \in X : a < b \Rightarrow y_a < y_b$.*

This vertical invariant is preserved in each step of the algorithm and thus in the final drawing the comparabilities of the order are correctly depicted.

## 4.1  Node Step

The first step of the iteration is called the *node step*, which is used in order to compute a $d$-dimensional representation of the ordered set. It thereby emphasizes the ordinal structure by positioning element pairs in a similar horizontal position, if they are comparable. In this step we define three different forces that operate simultaneously. For each $a \leq b$ on $a$ the vertical force $f_{\text{vert}}(\vec{p}_a, \vec{p}_b) = \left( 0, \ldots, 0, -c_{\text{vert}} \cdot \left( \frac{1 + d_x(\vec{p}_a, \vec{p}_b)}{d_y(\vec{p}_a, \vec{p}_b)} - 1 \right) \right)$ operates while on $b$ the force $-f_{\text{vert}}(\vec{p}_a, \vec{p}_b)$ operates. If two elements have the same horizontal coordinates it has its equilibrium if the vertical distance is at the constant $c_{\text{vert}}$. Then, if two elements are closer then this constant it operates repelling and if they are farther away the force operates as an attracting force. Thus, the constant $c_{\text{vert}}$ is a parameter that can be used to tune the *optimal vertical distance*. By incorporating the horizontal distance into the force, it can be achieved that vertices with a high horizontal distance will also result in a higher vertical distance. Note, that this force only operates on the covering relation instead of all comparable pairs, as Otherwise, chains would be contracted to be positioned close to a single point.

On the other hand there are two different forces that operate in the horizontal direction. Similar to Freese's layout, there is an attracting force between comparable and repelling force between incomparable element pairs; however, the exact forces are different. Between all comparable pairs $a$ and $b$ the force $f_{\text{attr}}(\vec{p}_a, \vec{p}_b) = -\min\left( d_x(\vec{p}_a, \vec{p}_b)^3, c_{\text{hor}} \right) \cdot \vec{u}_x(\vec{p}_a, \vec{p}_b)$ is operating. Note that in contrast to $f_{\text{vert}}$ this force operates not only on the covering but on all comparable pairs and thus encourages chains to be drawn in a single line. Similarly, incomparable elements should not be close to each other and thus the force $f_{\text{rep}}(\vec{p}_a, \vec{p}_b) = \frac{c_{\text{hor}}}{d_x(\vec{p}_a, \vec{p}_b)} \cdot \vec{u}_x(\vec{p}_a, \vec{p}_b)$, repels incomparable pairs horizontally.

We call the case that an element would be placed above a comparable greater element or below a lesser element, *overshooting*. However, to ensure that every

---

**Algorithm 3.** ReDraw - Node step

---

**Input:**    Ordered set: $(X, \leq)$          **Constants:** $K \in \mathbb{N}$, $\varepsilon > 0$, $\delta > 0$,
             Drawing $p = (\vec{p}_a)_{a \in X} \subseteq \mathbb{R}^d$          $c_{\text{vert}} > 0$, $c_{\text{hor}} > 0$
**Output:** Drawing: $p = (\vec{p}_a)_{a \in X} \subseteq \mathbb{R}^d$

---

$t = 1$
**while** $t < K$ **and** $\max_{a \in X} \|F_a(t)\| > \varepsilon$ :
   **for** $a \in X$ :
      $F_a(t) := \sum_{a \prec b} f_{\text{vert}}(\vec{p}_a, \vec{p}_b) - \sum_{b \prec a} f_{\text{vert}}(\vec{p}_a, \vec{p}_b)$
         $+ \sum_{a \leq b} f_{\text{attr}}(\vec{p}_a, \vec{p}_b) + \sum_{a \not\leq b} f_{\text{rep}}(\vec{p}_a, \vec{p}_b)$
   **for** $a \in X$ :
      $\vec{p}_a := \texttt{overshooting\_protection}\,(\vec{p}_a + \delta \cdot F_a(t))$
   $t = t + 1$

---

intermediate drawing that is computed in the node step still satisfies the vertical invariant we have to prohibit overshooting. Therefore, we add overshooting protection to the step in the algorithm where $(\vec{p}_a)_{a \in X}$ is recomputed. This is done by restricting the movement of every element such that it is placed maximally $\frac{c_{\text{vert}}}{10}$ below the lowest positioned greater element, or symmetrically above the greatest lower element. If the damping factor is chosen sufficiently small overshooting is rarely required. This is, because our forces are defined such that the closer two elements are positioned the stronger they repel each other, see Fig. 3.

All three forces are then consolidated into a single routine that is repeated at most $K$ times or until the total stress falls below a constant $\varepsilon$, see Algorithm 3. The general idea of our forces is similar to the forces described in Freese's approach, as comparable elements attract each other and incomparable elements repel each other. However, we are able to get rid of the ranking function that fixes $y$-coordinate and thus have an additional degree of freedom which allows us to include the horizontal distance as a factor to determine the vertical positions. Furthermore, our forces are formulated in a general way such that the drawings can be computed in arbitrary dimensions, while Freese is restricted to three dimensions. This overcomes the problem of getting stuck in local minima and enables us to recompute the drawing in two dimensions in the last cycle.

### 4.2    Line Step

While the goal of the node step is to get a good representation of the internal structure by optimizing on the proximity of nodes, the goal of the line step is to make the resulting drawing more aesthetically pleasing by optimizing distances between lines. Thus, in this step the drawing is optimized on three soft criteria. First, we want to maximize the number of parallel lines. Secondly, we want to achieve large angles between two lines that are connected to the same node. Finally, we want to have a high distance between elements and non-adjacent lines. We achieve a better fit to these criteria by applying a force-directed algorithm with three different forces, each optimizing on one criterion. While the

---

**Algorithm 4.** ReDraw - Line Step

---

**Input:**   Ordered set: $(X, \leq)$          **Constants:** $K \in \mathbb{N},\ \varepsilon > 0,\ \delta > 0,$
              Drawing $p = (\vec{p}_a)_{a \in X} \subseteq \mathbb{R}^d$                    $c_{\mathrm{par}} > 0,\ c_{\mathrm{ang}} > 0,\ c_{\mathrm{dist}} > 0$
**Output:** Drawing: $p = (\vec{p}_a)_{a \in X} \subseteq \mathbb{R}^d$

---

$t = 1$
**while** $t < K$ **and** $\max_{a \in X} \| F_a(t) \| > \varepsilon$ :
$\quad A = \{\{(a,b), (c,d)\} \mid a \prec b, c \prec d, d_{\cos}((\vec{p}_a, \vec{p}_b), (\vec{p}_c, \vec{p}_d)) < c_{\mathrm{par}}\}$
$\quad B = \{\{(a,c), (b,c)\} \mid (a \prec c, b \prec c) \text{ or } (c \prec a, c \prec b), d_{\cos}((\vec{p}_a, \vec{p}_c), (\vec{p}_b, \vec{p}_c)) < c_{\mathrm{ang}}\}$
$\quad C = \{(a, (b,c)) \mid a \in X, b \prec c, d(\vec{p}_a, (\vec{p}_b, \vec{p}_c)) < c_{\mathrm{dist}}\}$
$\quad$ **for** $a \in X$ :
$\quad\quad F_a(t) := \sum_{\{(a,b),(c,d)\} \in A} f_{\mathrm{par}}((\vec{p}_a, \vec{p}_b), (\vec{p}_c, \vec{p}_d)) + \sum_{(a,(b,c)) \in C} f_{\mathrm{dist}}(\vec{p}_a, (\vec{p}_b, \vec{p}_c))$
$\quad\quad\quad - \sum_{\{(b,a),(c,d)\} \in A} f_{\mathrm{par}}((\vec{p}_a, \vec{p}_b), (\vec{p}_c, \vec{p}_d)) - \frac{1}{2} \sum_{(b,(a,c)) \in C} f_{\mathrm{dist}}(\vec{p}_a, (\vec{p}_b, \vec{p}_c))$
$\quad\quad\quad + \sum_{\{(a,c),(b,c)\} \in B} f_{\mathrm{ang}}((\vec{p}_a, \vec{p}_c), (\vec{p}_b, \vec{p}_c))$
$\quad$ **for** $a \in X$ :
$\quad\quad \vec{p}_a := \mathtt{overshooting\_protection}(\vec{p}_a + \delta \cdot F_a(t))$
$\quad t = t + 1$

---

previous step does not directly incorporate the path of the lines, this step incorporates those into its forces. Therefore, we call this step the *line step*.

The first force of the line step operates on lines $(a, b)$ and $(c, d)$ with $a \neq c$ and $b \neq d$ if their cosine distance is below a threshold $c_{\mathrm{par}}$. The horizontal force $f_{\mathrm{par}}((\vec{p}_a, \vec{p}_b), (\vec{p}_c, \vec{p}_d)) = - \left( 1 - \frac{d_{\cos}((\vec{p}_a, \vec{p}_b),(\vec{p}_c, \vec{p}_d))}{c_{\mathrm{par}}} \right) \cdot \left( \frac{(\vec{p}_b - \vec{p}_a)_x}{y_b - y_a} - \frac{(\vec{p}_d - \vec{p}_c)_x}{y_d - y_c} \right)$ operates on $a$ and the force $-f_{\mathrm{par}}((\vec{p}_a, \vec{p}_b), (\vec{p}_c, \vec{p}_d))$ operates to $b$. This result of this force is thus that almost parallel lines are moved to become more parallel. Note, that this force becomes stronger the more parallel the two lines are.

The second force operates on lines that are connected to the same node and have a small angle, i.e., lines with cosine distance below a threshold $c_{\mathrm{ang}}$. Let $(a, c)$ and $(b, c)$ be such a pair then the horizontal force operating on $a$ is given by $f_{\mathrm{ang}}((\vec{p}_a, \vec{p}_c), (\vec{p}_b, \vec{p}_c)) = \left( 1 - \frac{d_{\cos}((\vec{p}_a, \vec{p}_c),(\vec{p}_b, \vec{p}_c))}{c_{\mathrm{ang}}} \right) \cdot \left( \frac{(\vec{p}_c - \vec{p}_a)_x}{y_c - y_a} - \frac{(\vec{p}_c - \vec{p}_b)_x}{y_c - y_b} \right)$. In this case, once again the force is stronger for smaller angles; however, the force is operating in the opposite direction compared to $f_{\mathrm{par}}$ and thus makes the two lines less parallel. Symmetrically, for each pair $(c, a)$ and $(c, b)$ the same force operates on $a$. There are artifacts from $f_{\mathrm{par}}$ that operate against $f_{\mathrm{ang}}$ in opposite direction. This effect should be compensated for by using a much higher threshold constant $c_{\mathrm{ang}}$ than $c_{\mathrm{par}}$, otherwise the benefits of this force are diminishing.

Finally, there is a force that operates on all pairs of nodes $a$ and lines $(b, c)$, for which the distance between the element and the line is closer then $c_{\mathrm{dist}}$. The force $f_{\mathrm{dist}}(\vec{p}_a, (\vec{p}_b, \vec{p}_c)) = \frac{1}{d(\vec{p}_a,(\vec{p}_b, \vec{p}_c))} \cdot \left( (\vec{p}_a - \vec{p}_c) - \frac{(\vec{p}_a - \vec{p}_c) \cdot (\vec{p}_b - \vec{p}_c)}{(\vec{p}_b - \vec{p}_c) \cdot (\vec{p}_b - \vec{p}_c)} (\vec{p}_b - \vec{p}_c) \right)$ is applied to $a$ and $-f_{\mathrm{dist}}(\vec{p}_a, (\vec{p}_b, \vec{p}_c))/2$ is applied to $b$ and $c$. This results in a force whose strength is linearly stronger, the closer the distance $d(\vec{p}_a, (\vec{p}_b, \vec{p}_c))$. It operates perpendicular to the line and repels the node and the line.

Similar to the node step, all three forces are combined into a routine that is repeated until the remaining energy in the physical system drops below a certain stress level $\varepsilon$. Furthermore a maximal number of repetitions $K$ is fixed. We also once again include the overshooting protection as described in the previous section to make sure that the vertical invariant stays satisfied.

The line step that is described in this section is a computational demanding task, as in every repetition of the iterative loop the sets of almost parallel lines, small angles and elements that are close to lines have to be recomputed. To circumvent this problem on weaker hardware, there are a number of possible speedup techniques. First of all, the sets described above do not have to be recomputed every iteration, but can be cached over a small number of iterations. In Algorithm 4 these are the sets $A$, $B$ and $C$. By recomputing those sets only every $k$-th iteration a speedup to almost factor $k$ can be achieved. Another speedup technique that is possible is to only execute the line step in the last round. Both of these techniques however have a trade off for the quality of the final drawing and are thus not further examined in this paper.

### 4.3   Dimension Reduction

In the dimension reduction step, we compute a $(d-1)$-dimensional drawing from the $d$-dimensional drawing with the goal of reflecting the structural details of the original drawing like proximity and angles. Our approach to solve this is to compute a $(d-1)$-dimensional linear subspace of the $d$-dimensional space. By preserving the vertical dimension we can ensure that the vertical invariant stays satisfied. Then a parallel projection into this subspace is performed.

As such a linear subspace always contains the origin, we center our drawing around the origin. Thereby, the whole drawing $(\vec{p}_a)_{a \in X}$ is geometrically translated such that the mean of every coordinate becomes 0. The linear subspace projection is performed as follows: The last coordinate of the linear subspace will be the vertical component of the $d$-dimensional drawing to ensure that the vertical invariant is preserved. For the other $(d-1)$ dimensions of the original space, a principle component analysis [13] is performed to reduce them to a $(d-2)$-dimensional subspace. By combining this projection with the vertical dimension a $(d-1)$-dimensional drawing is achieved, that captures the structure of the original, higher-dimensional drawing and represents its structural properties.

It is easily possible to replace PCA in this step by any other dimension reduction technique. It would thus be thinkable to just remove the first coordinate in each step and hope that the drawing in the resulting subspace has enough information encapsulated in the remaining coordinates. Also other ways of choosing the subspace in which is projected could be considered. Furthermore, non-linear dimension reduction methods could be tried in order to achieve drawings, however our empirical experiments suggest, that PCA hits a sweet spot. The payoff of more sophisticated dimension reduction methods seems to be negligible as each drawing is further improved in lower dimensions. On the other hand we observed local minima if we used simpler dimension reduction methods.

## 5     Evaluation

As we described in the previous sections, it is not a trivial task to evaluate the quality of an order diagram drawing. Drawings that one human evaluator might consider as favorably might not be perceived as readable by others. Therefore, we evaluate our generated drawing with a large quantity of domain experts.

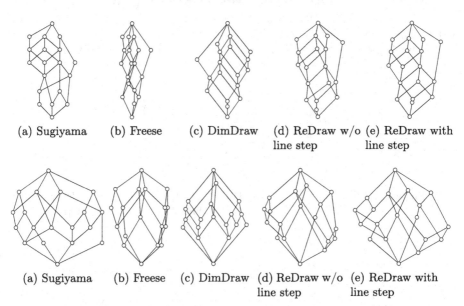

(a) Sugiyama     (b) Freese     (c) DimDraw     (d) ReDraw w/o     (e) ReDraw with
                                                line step           line step

(a) Sugiyama     (b) Freese     (c) DimDraw     (d) ReDraw w/o     (e) ReDraw with
                                                line step           line step

**Fig. 4.** Top: Drawing of the lattices for the formal contexts "forum romanum" (top) and "living beings and water" (bottom) from the test dataset.

### 5.1     Run-Time Complexity

The run-time of the node step is limited by $\mathcal{O}(n^2)$ with $n$ being the number of elements, as the distances between every element pair are computed. The run-time of the line step is limited by $\mathcal{O}(n^4)$, as the number of lines is bounded by $\mathcal{O}(n^2)$. Finally, the run-time of the reduction step is determined by PCA which is known to be bounded by $O(n^3)$. Therefore, the total run-time of the algorithm is polynomial in $\mathcal{O}(n^4)$. This is an advantage compared to DimDraw and Sugiyama's framework, which both solve exponential problems; however, Sugiyama is usually applied with a combination of heuristics to overcome this problem. Freese's layout has by its nature of being a force-directed order diagram drawing algorithm, similar to our approach, polynomial run-time. Thus, for larger diagrams, only **ReDraw**, Freese's algorithm and Sugiyama's framework (the latter with its heuristics) are suitable, while DimDraw is not.

## 5.2   Tested Datasets

Our test dataset consists of 77 different lattices including all classical examples of lattices described in [9]. We enriched these by lattices of randomly generated contexts and some sampled contexts from large binary datasets. An overview of all related formal contexts for these lattices, together with their drawing generated by `ReDraw` is published together with its source code. We restrict the test dataset to lattices, as lattice drawings are of great interest for the formal concept analysis community. This enables us to perform a user study using domain experts for lattices from the FCA community to evaluate the algorithm.

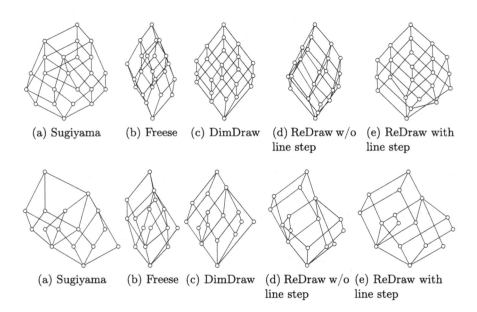

(a) Sugiyama     (b) Freese   (c) DimDraw   (d) ReDraw w/o   (e) ReDraw with
                                            line step        line step

(a) Sugiyama     (b) Freese   (c) DimDraw   (d) ReDraw w/o   (e) ReDraw with
                                            line step        line step

**Fig. 5.** Top: Drawing of the lattices for the formal contexts "therapy" (top) and "ice cream" (bottom) from the test dataset.

## 5.3   Recommended Parametrizations

As it is hardly possible to conduct a user study for every single combination of parameters, our recommendations are based on empirical observations. We used a maximal number of $K = 1000$ algorithm iterations or stopped if the stress in the physical system fell below $\varepsilon = 0.0025$. Our recommended damping factor $\delta = 0.001$. In the node step we set $c_{\mathrm{vert}} = 1$ as the optimal vertical distance and $c_{\mathrm{hor}} = 5$. We used the thresholds $c_{\mathrm{par}} = 0.005$, $c_{\mathrm{ang}} = 0.05$ and $c_{\mathrm{dist}} = 1$ in the line step. The drawing algorithms are started with 5 dimensions as we did not observe any notable improvements with higher dimensional drawings. Finally the resulting drawing is scaled in horizontal direction by a factor of 0.5.

## 5.4   Empirical Evaluation

To demonstrate the quality of our approach we compare the resulting drawings to the drawings generated by a selected number of different algorithms in Fig. 4 and Fig. 5. The different drawings are computed using Sugiyama's framework, Freese's layout, DimDraw and our new approach. Additionally, a drawing of our approach before the line step is presented to show the impact of this line step. In the opinion of the authors of this paper, the approach proposed in this paper achieves satisfying results for these ordered sets. In most cases, we still prefer the output of DimDraw (and sometimes Sugiyama), but ReDraw is able to cope with much larger datasets because of its polynomial nature. Modifications of ReDraw that combine the node step and the line step into a single step were tried by the authors; however, the then resulting algorithm did not produce the anticipated readability, as the node and line forces seem to work against each other.

## 5.5   User Evaluation

To obtain a measurable evaluation we conducted a user study to compare the different drawings generated by our algorithm to two other algorithms. We decided to compare our approach to Freese's and Sugiyama's algorithm, as those two seem to be the two most popular algorithms for lattice drawing at the moment. We decided against including DimDraw into this study as, even though it is known to produce well readable drawings, it struggles with the computational costs for drawings of higher order dimensions due to its exponential nature.

**Experimental Setup.** In each step of the study, all users are presented with three different drawings of one lattice from the dataset in random order and have to decide which one they perceive as "most readable". The term "most readable" was neither further explained nor restricted.

**Results.** The study was conducted with nine experts from the formal concept analysis community to guarantee expertise with order diagrams among the participants. Thus, all ordered sets in this study were lattices. The experts voted 582 times in total; among those votes, 35 were cast for Freese's algorithm, 266 for our approach and 281 for Sugiyama. As a common property of lattices is to contain a high degree of truncated distributivity [15], which makes this property of special interest, we decided to compute the share of distributive triples for each lattice excluding those resulting in the bottom-element. We call the share of such distributive triples of all possible triples the *truncated relative distributivity (RTD)*. Based on the RTD we compared the share of votes for Sugiyama's framework and ReDraw for all order diagrams that are in a specific truncated distributivity range. The results of this comparison are depicted in Fig. 6. The higher the RTD, the better ReDraw performs in comparison. The only exception in the range 0.64–0.68 can be traced back to a small test set with $n = 4$.

**Discussion.** As one can conclude from the user study, our force-directed algorithm performs on a similar level to Sugiyama's framework while outperforming Freese's force-directed layout. In the process of developing `ReDraw` we also conducted a user-study that compared an early version to DimDraw which suggested that `ReDraw` cannot compete with DimDraw. However, DimDraw's exponential run-time makes computing larger order drawings unfeasible. From the comparison of `ReDraw` and Sugiyama's, that takes the RTD into account, we can follow that our algorithm performs better on lattices that have a higher RTD. We observed similar results when we computed the relative normal distributivity. The authors of this paper thus recommend to use `ReDraw` for larger drawings that are highly distributive. Furthermore, the authors observed, that `ReDraw` performs better if there are repeating structures or symmetries in the lattice as each instance of such a repetition tends to be drawn similarly. This makes it the algorithm of choice for ordered sets that are derived from datasets containing high degrees of symmetries. Anyway, the authors of this paper are convinced that there is no single drawing algorithm that can produce readable drawings for all different kinds of order diagrams. It is thus always recommended to use a combination of different algorithms and then decide on the best drawing.

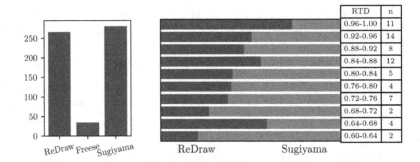

**Fig. 6.** Results of the user study. L: Number of votes for each algorithm. R: Share of votes for ordered sets divided into ranges of different truncated distributivity.

## 6    Conclusion and Outlook

In this work we introduced our novel approach `ReDraw` for drawing diagrams. Thereby we adapted a force-directed algorithm to the realm of diagram drawing. In order to guarantee that the emerging drawing satisfies the hard conditions of order diagrams we introduced a vertical invariant that was satisfied in every step of the algorithm. The algorithm consists of two main ingredients, the first being the node step that optimizes the drawing in order to represent structural properties using the proximity of nodes. The second is the line step that improves the readability for a human reader by optimizing the distances of lines. To avoid local minima, our drawings are first computed in a high dimension and then iterativly reduced into two dimensions. To make the algorithm easily accessible, we published the source code and gave recommendations for parameters. Generated drawings were, in our opinion, suitable to be used for ordinal data analysis. A study using domain experts to evaluate the quality of the drawings confirmed this observation.

Further work in the realm of order diagram drawing could be to modify the line step and combine it with algorithms such as DimDraw. Also modifications that produce additive drawings are of great interest and should be investigated further. Finally, in the opinion of the authors the research fields of ordinal data analysis and graph drawing would benefit significantly from the establishment of a "readability measure" or at least of a decision procedure that, given two visualizations of the same ordered set identifies the more readable one.

# References

1. Albano, A., Chornomaz, B.: Why concept lattices are large: extremal theory for generators, concepts, and vc-dimension. Int. J. Gen. Syst. **46**(5), 440–457 (2017)
2. Battista, G.D., Eades, P., Tamassia, R., Tollis, I.G.: Graph Drawing: Algorithms for the Visualization of Graphs. Prentice-Hall, Upper Saddle River (1999)
3. Demel, A., Dürrschnabel, D., Mchedlidze, T., Radermacher, Ml, Wulf, L.: A greedy heuristic for crossing-angle maximization. In: Biedl, T., Kerren, A. (eds.) GD 2018. LNCS, vol. 11282, pp. 286–299. Springer, Cham (2018). https://doi.org/10.1007/978-3-030-04414-5_20
4. Dürrschnabel, D., Hanika, T., Stumme, G.: Drawing order diagrams through two-dimension extension. CoRR abs/1906.06208 (2019)
5. Eades, P.: A heuristic for graph drawing. Congressus Numer. **42**, 149–160 (1984)
6. Freese, R.: Automated Lattice Drawing. In: Eklund, P. (ed.) ICFCA 2004. LNCS (LNAI), vol. 2961, pp. 112–127. Springer, Heidelberg (2004). https://doi.org/10.1007/978-3-540-24651-0_12
7. Fruchterman, T.M.J., Reingold, E.M.: Graph drawing by force-directed placement. Softw. Pract. Exp. **21**(11), 1129–1164 (1991)
8. Ganter, B.: Conflict avoidance in additive order diagrams. J. Univ. Comput. Sci. **10**(8), 955–966 (2004)
9. Ganter, B., Wille, R.: Formal Concept Analysis. Springer, Heidelberg (1999). https://doi.org/10.1007/978-3-642-59830-2
10. Hong, S., Eades, P., Lee, S.H.: Drawing series parallel digraphs symmetrically. Comput. Geom. **17**(3–4), 165–188 (2000)
11. Hopcroft, J.E., Tarjan, R.E.: Efficient planarity testing. J. ACM **21**(4), 549–568 (1974)
12. Nishizeki, T., Rahman, M.S.: Planar graph drawing. In: Lecture Notes Series on Computing, vol. 12. World Scientific (2004)
13. Pearson, K.: Liii. on lines and planes of closest fit to systems of points in space. London Edinburgh Dublin Philos. Mag. J. Sci. **2**(11), 559–572 (1901)
14. Sugiyama, K., Tagawa, S., Toda, M.: Methods for visual understanding of hierarchical system structures. IEEE Trans. Syst. Man Cybern. **11**(2), 109–125 (1981)

15. Wille, R.: Truncated distributive lattices: conceptual structures of simple-implicational theories. Order **20**(3), 229–238 (2003)
16. Yevtushenko, S.A.: Computing and visualizing concept lattices. Ph.D. thesis, Darmstadt University of Technology, Germany (2004)
17. Zschalig, C.: An FDP-algorithm for drawing lattices. In: Eklund, P.W., Diatta, J., Liquiere, M. (eds.) Proceedings of the Fifth International Conference on Concept Lattices and Their Applications, CLA 2007, Montpellier, France, 24–26, October 2007. CEUR Workshop Proceedings, vol. 331. CEUR-WS.org (2007)

# Short Papers

# Sandwich: An Algorithm for Discovering Relevant Link Keys in an LKPS Concept Lattice

Nacira Abbas[1]([✉]), Alexandre Bazin[1], Jérôme David[2], and Amedeo Napoli[1]

[1] Université de Lorraine, CNRS, Inria, Loria, 54000 Nancy, France
Nacira.Abbas@inria.fr, {Alexandre.Bazin,Amedeo.Napoli}@loria.fr
[2] Université Grenoble Alpes, Inria, CNRS, Grenoble INP, LIG, 38000 Grenoble, France
Jerome.David@inria.fr

**Abstract.** The discovery of link keys between two RDF datasets allows the identification of individuals which share common key characteristics. Actually link keys correspond to closed sets of a specific Galois connection and can be discovered thanks to an FCA-based algorithm. In this paper, given a pattern concept lattice where each concept intent is a link key candidate, we aim at identifying the most relevant candidates w.r.t adapted quality measures. To achieve this task, we introduce the "Sandwich" algorithm which is based on a combination of two dual bottom-up and top-down strategies for traversing the pattern concept lattice. The output of the Sandwich algorithm is a poset of the most relevant link key candidates. We provide details about the quality measures applicable to the selection of link keys, the Sandwich algorithm, and as well a discussion on the benefit of our approach.

## 1 Introduction

Linked data are structured data expressed in the RDF (Resource Description Framework) model where resources are identified by Internationalized Resources Identifiers (IRIs) [7]. Data interlinking is a critical task for ensuring the wide use of linked data. It consists in finding pairs of IRIs representing the same entity among different RDF datasets and returning a set of identity links between these IRIs. Many approaches have been proposed for data interlinking [8–11]. In this paper, we focus on the discovery of *link keys* [2]. Link keys extend the notion of a key as used in databases and allow the inference of identity links between RDF datasets. A link key is based on two sets of pairs of properties and a pair of classes. The pairs of properties express sufficient conditions for two subjects, instances of the classes, to be the identical. The link key

$$k = (\{\langle \texttt{designation},\texttt{titre}\rangle\},\{\langle \texttt{designation},\texttt{titre}\rangle,\langle \texttt{author},\texttt{auteur}\rangle\},\langle \texttt{Book},\texttt{Livre}\rangle)$$

states that whenever an instance $a_1$ from the class Book and an instance $b_1$ from the class Livre have the same values for the property designation and for the property titre, and that $a_1$ and $b_1$ share at least one value for the properties author and auteur, then $a_1$ and $b_1$ denote the same entity. We say that a link key $k$ generates the identity link $\langle a_1,b_1 \rangle$.

A. Napoli—This work is supported by the French ANR Elker Project ANR-17-CE23-0007-01.

A. Braud et al. (Eds.): ICFCA 2021, LNAI 12733, pp. 243–251, 2021.
https://doi.org/10.1007/978-3-030-77867-5_15

Link keys are in general not provided and they have to be discovered in the datasets under study. The discovery of link keys consists in extracting link key candidates from a pair of datasets and then to evaluate their relevance for the interlinking task. The relevance of a given link key is measured w.r.t. two criteria, (i) correctness, and (ii) completeness.

Link key candidates correspond to closed sets of a specific Galois connection. For this reason the question of using Formal Concept Analysis (FCA) [6] to discover link keys was naturally raised [3]. In [4] authors proposed a formal context for the discovery of link key candidates for a given pair of classes. However, when there is no alignment between classes, the choice of the right pair of classes is not necessarily straightforward. To overcome this limitation, a generalization of link key discovery based on Pattern Structures [5] was proposed in [1]. The authors introduced a specific pattern structure for link key candidate discovery over two datasets $D_1$ and $D_2$ called $LK$-pattern structure without requiring an a priori alignment. An LKPS-lattice is the lattice of pattern concepts generated from an $LK$-pattern structure. Each concept intent is a link key candidate and each extent is the link set generated by the link key in the intent.

The size of an LKPS-lattice may be prohibitively large and not all link key candidates are relevant for the interlinking task. Our purpose in this paper is to identify the relevant link keys in the LKPS-lattice and to discard the irrelevant ones. The evaluation criteria of a link key candidate, i.e., completeness and correctness, are based on adapted evaluation measures that can be used for selecting the relevant link keys, i.e., the value taken by a candidate for such a measure is above a given threshold. The evaluation measures should also be monotone, i.e., increasing or decreasing, w.r.t. the order of the LKPS-lattice. Moreover, completeness and correctness verify an "inverse" relationship, as correctness tends to decrease when completeness increases. In this paper we rely on this observation and we show that the upper part of an LKPS-lattice contains the most complete but the least correct link keys, while the lower part of the LKPS-lattice contains the most correct but the least complete link keys.

Starting from this observation, we introduce an original pruning strategy combining a "bottom-up" and a "top-down" pruning strategies, while the most relevant link key candidates are lying "in the middle" and achieve the best compromise between completeness and correctness. Accordingly, we propose the Sandwich algorithm that traverse and prune the LKPS-lattice w.r.t. two measures estimating correctness and completeness. The input of this algorithm is an LKPS-lattice, a correctness measure and a threshold, and as well a completeness measure and a threshold. The output of the Sandwich algorithm is a poset of relevant link key candidates. To the best of our knowledge, this is the first time that such an algorithm is proposed for selecting the best link key candidates w.r.t. adapted measures. In addition, this is also an elegant way of taking advantage of the fact that link key candidates correspond to the closed sets of a given Galois connection which is made explicit in the following.

The organization of the paper is as follows. First we make precise definitions and notations, and we briefly present the problem of link key discovery in a pattern structure framework. Then we present the correctness and the completeness of link keys, and the Sandwich algorithm for pruning an LKPS-lattice. Finally we discuss on the benefit of our strategy.

## 2    The Discovery of Link Keys with Pattern Structures

### 2.1    A Definition of Link Keys

We aim to discover identity links among two RDF datasets $D_1$ and $D_2$. An identity link is a statement of the form $\langle s_1, \text{owl:sameAs}, s_2 \rangle$ expressing that the subject $s_1 \in S(D_1)$ and the subject $s_2 \in S(D_2)$ represent the same real-world entity. For example, given $D_1$ and $D_2$ in Fig. 1, the data interlinking task should discover the identity link $\langle a_1, \text{owl:sameAs}, b_1 \rangle$ because the subjects $a_1$ and $b_1$ both represent the same vaccine "Pfizer-BioNTech". For short, we write $\langle a_1, b_1 \rangle$ and we call this pair a *link*. A link key is used to generate such links.

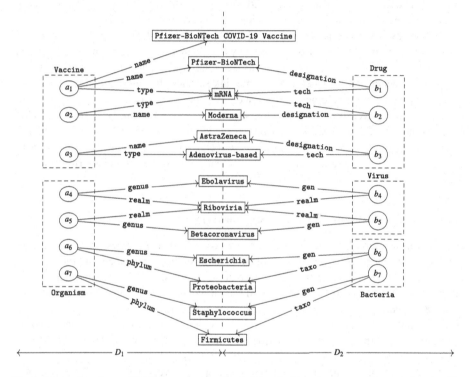

**Fig. 1.** Example of two RDF datasets. On the left-hand side, the dataset $D_1$ populated with instances of the classes: Vaccine and Organism. On the right-hand side, the dataset $D_2$ populated with instances of the classes: Drug, Virus and Bacteria.

Let us consider two RDF datasets $D_1$ and $D_2$, two non empty subsets of pairs of properties, namely $Eq$ and $In$, such that $Eq \subseteq P(D_1) \times P(D_2)$, $In \subseteq P(D_1) \times P(D_2)$, $Eq \subseteq In$, two class expressions –conjunction or disjunction– $C_1$ over $D_1$ and $C_2$ over $D_2$. Then $k = (Eq, In, \langle C_1, C_2 \rangle)$ is a "link key" over $D_1$ and $D_2$. An example of a link key is $k = (\{\langle \text{type}, \text{tech} \rangle\}, \{\langle \text{type}, \text{tech} \rangle, \langle \text{name}, \text{designation} \rangle\}, \langle \text{Vaccine}, \text{Drug} \rangle)$.

A link key may or may not generate links among two datasets as made precise here after. Let $k = (Eq, In, \langle C_1, C_2 \rangle)$ be a link key over $D_1$ and $D_2$. The link key $k$

generates a link $\langle s_1, s_2 \rangle \in S(C_1) \times S(C_2)$ iff $\langle s_1, s_2 \rangle$ verifies the link key $k$, i.e. (i) $p_1(s_1)$ and $p_2(s_2)$ should be non empty, (ii) $p_1(s_1) = p_2(s_2)$ for all $\langle p_1, p_2 \rangle \in Eq$, and (iii) $p_1(s_1) \cap p_2(s_2) \neq \emptyset$ for all $\langle p_1, p_2 \rangle \in In$.

The set of pairs of subjects $\langle s_1, s_2 \rangle \in S(C_1) \times S(C_2)$ verifying $k$ is called the link set of $k$ and denoted by $L(k)$. As the properties in RDF are not functional, we compare the values of subjects in two ways, (i) $Eq$ are pairs of properties for which two subjects share all their values, and (ii) $In$ are those pairs of properties for which two subjects share at least one value. Then $\langle a_1, b_1 \rangle$ verifies $k = (\{\langle \texttt{type}, \texttt{tech} \rangle\}, \{\langle \texttt{type}, \texttt{tech} \rangle, \langle \texttt{name}, \texttt{designation} \rangle\}, \langle \texttt{Vaccine}, \texttt{Drug} \rangle)$, because $\langle a_1, b_1 \rangle \in S(\texttt{Vaccine}) \times S(\texttt{Drug})$, and $\texttt{type}(a_1) = \texttt{tech}(b_1)$, and $\texttt{name}(a_1) \cap \texttt{designation}(b_1) \neq \emptyset$.

Algorithms for link key discovery [1,2,4] discover firstly the so-called "link key candidates" and then evaluate each candidate using quality measures. The relevant link key candidates are selected to generate identity links between datasets.

A link key candidate is defined in [4] as the intent of a formal concept computed within a particular formal context for link key candidate discovery, given pair of classes $\langle C_1, C_2 \rangle \in Cl(D_1) \times Cl(D_2)$. Actually, link keys are equivalent w.r.t. their link set, i.e. if two link keys $k_1$ and $k_2$ generate the same link set then they are equivalent. Link key candidates are maximal elements of their equivalence classes and thus correspond to closed sets. Moreover, the set of links must not be empty for a link key candidate. This explains the use of Formal Concept Analysis [6] in the discovery of link key candidate.

However, the pair of classes $\langle C_1, C_2 \rangle \in Cl(D_1) \times Cl(D_2)$ is not always known in advance and thus a generalization of the existing algorithms based on Pattern Structures was proposed in [1], as explained in the following.

### 2.2  A Pattern Structure for Link Key Discovery

In a pattern structure designed for the discovery of link key candidates over two datasets [1], the set of objects is the set of pairs of subjects from the two datasets and the descriptions of objects are potential link keys over these datasets. A link key candidate corresponds to an intent of a pattern concept in the lattice generated from this pattern structure. Moreover the link set of a link key candidate corresponds to the extent of the formal concept. In the following, we do not provide any definition but we recall some important results from [1]. Moreover, for simplicity, we consider only the $In$ set of pairs of properties in a link key, i.e. $k = (In, \langle C_1, C_2 \rangle)$ (as $Eq \subseteq In$).

The $Lk$-pattern structure for the datasets in Fig. 1 is given in Table 1. The associated concept lattice, called an LKPS-Lattice, is displayed in Fig. 2. An example of link key candidate is given by $k_2 = (\{\langle \texttt{name}, \texttt{designation} \rangle, \langle \texttt{type}, \texttt{tech} \rangle\}, \langle \texttt{Vaccine}, \texttt{Drug} \rangle)$, and the related link set is $L(k_2) = \{\langle a_1, b_1 \rangle, \langle a_2, b_2 \rangle, \langle a_3, b_3 \rangle\}$.

## 3  The Pruning of an LKPS-lattice

### 3.1  Correction and Completeness of a Link Key Candidate

Given an LKPS-lattice, we aim at identifying the most relevant link key based on a set of adapted interest measures. Link key relevance depends on two main criteria, namely

**Table 1.** The *Lk*-pattern structure over the datasets $D_1$ and $D_2$ represented in Fig. 1.

| $S(D_1) \times S(D_2)$ | *In* | $\langle C_1, C_2 \rangle$ |
|---|---|---|
| $\langle a_1, b_1 \rangle$ | $\{\langle \texttt{name}, \texttt{designation} \rangle, \langle \texttt{type}, \texttt{tech} \rangle\}$ | $\langle \texttt{Vaccine}, \texttt{Drug} \rangle$ |
| $\langle a_1, b_2 \rangle$ | $\{\langle \texttt{type}, \texttt{tech} \rangle\}$ | $\langle \texttt{Vaccine}, \texttt{Drug} \rangle$ |
| $\langle a_2, b_2 \rangle$ | $\{\langle \texttt{name}, \texttt{designation} \rangle, \langle \texttt{type}, \texttt{tech} \rangle\}$ | $\langle \texttt{Vaccine}, \texttt{Drug} \rangle$ |
| $\langle a_2, b_1 \rangle$ | $\{\langle \texttt{type}, \texttt{tech} \rangle\}$ | $\langle \texttt{Vaccine}, \texttt{Drug} \rangle$ |
| $\langle a_3, b_3 \rangle$ | $\{\langle \texttt{name}, \texttt{designation} \rangle, \langle \texttt{type}, \texttt{tech} \rangle\}$ | $\langle \texttt{Vaccine}, \texttt{Drug} \rangle$ |
| $\langle a_4, b_4 \rangle$ | $\{\langle \texttt{genus}, \texttt{gen} \rangle, \langle \texttt{realm}, \texttt{realm} \rangle\}$ | $\langle \texttt{Organism}, \texttt{Virus} \rangle$ |
| $\langle a_4, b_5 \rangle$ | $\{\langle \texttt{realm}, \texttt{realm} \rangle\}$ | $\langle \texttt{Organism}, \texttt{Virus} \rangle$ |
| $\langle a_5, b_5 \rangle$ | $\{\langle \texttt{genus}, \texttt{gen} \rangle, \langle \texttt{realm}, \texttt{realm} \rangle\}$ | $\langle \texttt{Organism}, \texttt{Virus} \rangle$ |
| $\langle a_5, b_4 \rangle$ | $\{\langle \texttt{realm}, \texttt{realm} \rangle\}$ | $\langle \texttt{Organism}, \texttt{Virus} \rangle$ |
| $\langle a_6, b_6 \rangle$ | $\{\langle \texttt{genus}, \texttt{gen} \rangle, \langle \texttt{phylum}, \texttt{taxo} \rangle\}$ | $\langle \texttt{Organism}, \texttt{Bacteria} \rangle$ |
| $\langle a_7, b_7 \rangle$ | $\{\langle \texttt{genus}, \texttt{gen} \rangle, \langle \texttt{phylum}, \texttt{taxo} \rangle\}$ | $\langle \texttt{Organism}, \texttt{Bacteria} \rangle$ |

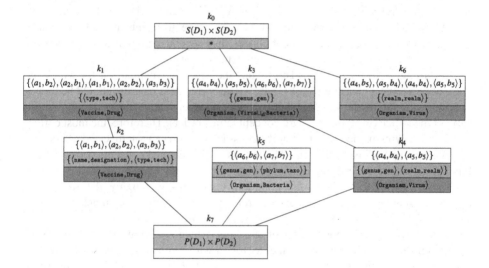

**Fig. 2.** The LKPS-lattice generated from the *Lk*-pattern structure in Table 1. The different colors distinguish the different pairs of classes.

"correctness" and "completeness". The correctness of a link key is its ability to generate correct links, while its completeness is its ability to generate all the correct links.

Firstly, we start from the hypothesis that the higher the number of pairs of properties for which two subjects $s_1$, $s_2$ share a value is, the greater is the probability that the link $\langle s_1, s_2 \rangle$ is correct. Then the *size* of the link key candidate $k = (In, \langle C_1, C_2 \rangle)$ is $|k| = |In|$. Moreover, the size of link keys is monotone w.r.t the intents of pattern concept intents in an LKPS-lattice.

In [2], the measure of coverage was proposed to evaluate the completeness of a link key candidate. The coverage of a link key $k = (In, \langle C_1, C_2 \rangle)$ is defined as follows:

$co(k) = |\pi_1(L(k)) \cup \pi_2(L(k))|/|S(C_1) \cup S(C_2)|$, where $\pi_1(L(k)) = \{s_1|\langle s_1, s_2\rangle \in L(k)\}$ and $\pi_2(L(k)) = \{s_2|\langle s_1, s_2\rangle \in L(k)\}$. The coverage is locally monotone, i.e. when the link key candidates are associated with the same pairs of classes, the coverage is monotone w.r.t extents of these candidates in the LKPS-lattice.

### 3.2 Sandwich: An Algorithm for Selecting the Most Relevant Link Key Candidates

Given an LKPS-lattice, we propose the Sandwich algorithm for identifying the relevant link key candidates (intents) and discarding the irrelevant candidates. The input of Sandwich is a LKPS-lattice, a correctness measure $\sigma_{cor}$ and a minimum threshold $\mu_{cor}$, and a completeness measure $\sigma_{comp}$ and a minimum threshold $\mu_{comp}$. The output of Sandwich is the poset of all relevant link key candidates. It should be noticed that a link key candidate may be relevant for a given pair of classes and not relevant for another pair, i.e. a given link key candidate may generate all the correct links over a pair of classes and no correct link over another pair. Accordingly, it is more appropriate to identify relevant link keys associated with each pair of classes. Thus, in a first step, Sandwich splits the lattice into sub-lattices where all intents are link key candidates associated with the same pairs of classes. In a second step, Sandwich prunes the sub-lattices based on correctness and completeness measures.

Regarding correctness, Sandwich retains the link key candidates $k$ for which the score of correctness measure $\sigma_{cor}(k) \geq \mu_{cor}$. The correctness measure should be monotone w.r.t. the intents in the LKPS-lattice, and the larger intents are at in the "bottom part" of the lattice (w.r.t. the standard concept lattice order). Therefore, the most correct link keys are lying in the lower part of the considered given LKPS-lattice and a "bottom-up pruning strategy" is carried out. The intents of the retained concepts correspond to link key candidates $k$ verifying $\sigma_{cor}(k) \geq \mu_{cor}$.

The strategy for retaining the complete kink keys is roughly the same, i.e., Sandwich retains link key candidates $k$ for which the score of completeness measure $\sigma_{comp}(k) \geq \mu_{comp}$. However, by contrast, the most complete link keys are having the better covering w.r.t. the extents of concepts, which are lying in the "upper part" of a given LKPS-sub-lattice. This time, a "top-down pruning strategy" is carried out, and the extents of the retained concepts correspond to link key candidates $k$ verifying $\sigma_{comp}(k) \geq \mu_{comp}$.

Finally, the Sandwich algorithm retains the concepts which are selected at the same time by both pruning strategies.

For illustrating the pruning strategy[1], let us consider the example of LKPS$(D_3, D_4)$ displayed in Fig. 3. The correctness measure which is monotone w.r.t. intents is the size of the link key and the threshold is set to $\mu_{cor} = 3$ (minimum size). The completeness measure which is monotone w.r.t extents is the coverage of a link key and the threshold is set to $\mu_{comp} = 0.9$ (minimum coverage). For the pair of classes $\langle$Person, Personne$\rangle$, the bottom-up pruning strategy returns all the pattern concepts whose intent size is greater than 3, i.e., $\{k_5, k_6, k_7\}$. The top-down pruning strategy returns all the pattern concepts

---

[1] The datasets and the implementation generating the lattice can be checked at https://gitlab.inria.fr/nabbas/sandwich_algorithm.

whose coverage is above 0.9, i.e., $\{k_4, k_5\}$. Finally, the best link key w.r.t. to both strategies is $k_5$, and it can be used to find identity links over the RDF datasets $D_3$ and $D_4$.

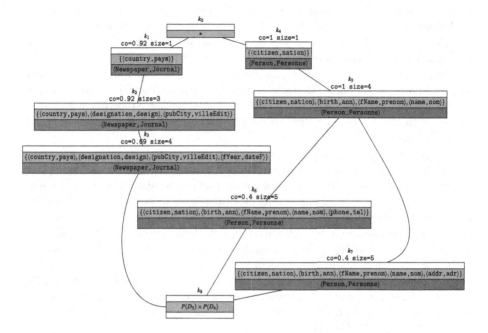

**Fig. 3.** The LKPS-lattice of link key candidate over the datasets $D_3$ and $D_4$.

# 4  Discussion and Conclusion

In this paper, we have studied the problem of the discovery of link keys in an FCA framework. We have proposed an algorithm based on pattern structures which returns a pattern concept lattice where the intent of a pattern concept corresponds to a link key candidate and the associated extent corresponds to the set of links related with the link key candidate. Indeed, FCA and pattern structures are well adapted to study the discovery of link keys above two datasets as link keys can be considered as "closed sets". Actually they are introduced as the maximum element in an equivalence class in [4]. Making a parallel with equivalence classes of itemsets this emphasizes the fact that a link key corresponds to a closed set.

Then, given the pattern concept lattice including all link key candidates, one crucial problem is to select the best link keys, in the same way as one could be interested in some "good concepts" extracted from the lattice to be checked by a domain expert. For that we introduce a set of quality measures that can be used for checking two main properties, namely correctness and completeness of the link keys. As a (pattern) concept lattice is based on duality and two anti-isomorphic orders, the measures of correctness and completeness are also behaving in a dual way. Intuitively, the correctness of a link key measures the capability to generate correct links, and in this way, the largest link key will be among the best link keys (w.r.t. a reference pair of classes). Dually, the completeness of a link key measures the capability to generating the largest set of links, and such a link key will also be among the best link keys (w.r.t. a reference pair of classes).

Furthermore, following the duality principle, designing an algorithm able to discover the best link keys, i.e. reaching the best compromise between correctness and completeness, amounts to exploring the pattern concept lattice in both and dual ways, namely top-down and bottom-up. This is precisely the work of the Sandwich algorithm which combines two pruning strategies, a top-down traversal and a bottom-up traversal, for reaching the most complete set of links and at the same time the most correct link key candidates. This is a straightforward illustration of the duality principle in a (pattern) concept lattice.

Now, the Sandwich is generic and is able to work with different quality measures as soon as they are monotone or locally monotone. For the next step, we should improve the present research work in a number of directions. First, we should enlarge the collection of measures and include more current measures used in the FCA and data mining communities, such as "stability" or "lift" for example. They could provide interesting directions of investigations for characterizing link keys. Second, we should improve the global traversal strategy and combine the characterization of link key candidates at the construction of the pattern concept lattice if possible. In this way, each pattern concept could be tagged with its characteristics w.r.t. a pair of classes and some quality measures. Finally, we have run a complete set of experiments for validating the current proposal on a real-world basis, and as well check the foundations and improve the algorithmic part of the Sandwich algorithm.

## References

1. Abbas, N., David, J., Napoli, A.: Discovery of Link Keys in RDF data based on pattern structures: preliminary steps. In: Proceedings of CLA, pp. 235–246. CEUR Workshop Proceedings 2668 (2020)
2. Atencia, M., David, J., Euzenat, J.: Data interlinking through robust linkkey extraction. In: Proceedings of ECAI. pp. 15–20. IOS Press (2014)
3. Atencia, M., David, J., Euzenat, J.: What can FCA do for database linkkey extraction? In: Proceedings of FCA4AI Workshop. pp. 85–92. CEUR Workshop Proceedings 1257 (2014)
4. Atencia, M., David, J., Euzenat, J., Napoli, A., Vizzini, J.: Link key candidate extraction with relational concept analysis. Discrete Appl. Math. **273**, 2–20 (2020)
5. Ganter, B., Kuznetsov, S.: Pattern Structures and Their Projections. In: Delugach, H.S., Stumme, G. (eds.) ICCS-ConceptStruct 2001. LNCS (LNAI), vol. 2120, pp. 129–142. Springer, Heidelberg (2001). https://doi.org/10.1007/3-540-44583-8_10
6. Ganter, B., Wille, R.: Formal Concept Analysis. Springer, Heidelberg (1999). https://doi.org/10.1007/978-3-642-59830-2
7. Heath, T., Bizer, C.: Linked Data: Evolving the Web into a Global Data Space. Morgan & Claypool Publishers, Synthesis Lectures on the Semantic Web (2011)
8. Nentwig, M., Hartung, M., Ngomo, A.N., Rahm, E.: A survey of current link discovery frameworks. Semant. Web **8**(3), 419–436 (2017)

9. Ngomo, A.N., Auer, S.: LIMES - a time-efficient approach for large-scale link discovery on the web of data. In: Proceedings of IJCAI, pp. 2312–2317. IJCAI/AAAI (2011)
10. Symeonidou, D., Armant, V., Pernelle, N.: BECKEY: understanding, comparing and discovering keys of different semantics in knowledge bases. Knowl Based Syst.**195**, 105708 (2020)
11. Volz, J., Bizer, C., Gaedke, M., Kobilarov, G.: Silk - a link discovery framework for the web of data. In: Proceedings of the WWW 2009 Workshop on Linked Data on the Web (LDOW). CEUR Workshop Proceedings 538 (2009)

# Decision Concept Lattice vs. Decision Trees and Random Forests

Egor Dudyrev[✉][iD] and Sergei O. Kuznetsov[✉][iD]

National Research University Higher School of Economics, Moscow, Russia
eodudyrev@edu.hse.ru, skuznetsov@hse.ru

**Abstract.** Decision trees and their ensembles are very popular models of supervised machine learning. In this paper we merge the ideas underlying decision trees, their ensembles and FCA by proposing a new supervised machine learning model which can be constructed in polynomial time and is applicable for both classification and regression problems. Specifically, we first propose a polynomial-time algorithm for constructing a part of the concept lattice that is based on a decision tree. Second, we describe a prediction scheme based on a concept lattice for solving both classification and regression tasks with prediction quality comparable to that of state-of-the-art models.

**Keywords:** Concept lattice · Decision trees · Random forest

## 1 Introduction

In this work we propose an approach to combining the ideas based on concept lattices and decision trees, which are extensively used in practical machine learning (ML), in order to create a new ML model which generates good classifiers and regressors in polynomial time.

Formal Concept Analysis (FCA) is a mathematically-founded theory well suited for developing models of knowledge discovery and data mining [8,11,18]. One of the serious obstacles to the broad use of FCA for knowledge discovery is that the number of formal concepts (i.e. patterns found in a data) can grow exponentially in the size of the data [15]. Sofia algorithm [7] offers a solution to this problem by constructing only a limited amount of most stable concepts.

Learning decision trees (DT) [5] is one of the most popular supervised machine learning approaches. Most famous methods based on ensembles of decision trees – aimed at increasing the accuracy of a single tree – are random forest (RF) [6] and gradient boosting over decision trees [9]. Both algorithms are considered among the best in terms of accuracy [17].

There are a number of papers which highlight the connection between the concept lattice and the decision tree. The work [4] states that a decision tree can be induced from a concept lattice. In [16] the author compares the ways the concept lattice and the decision tree can be used for supervised learning.

© Springer Nature Switzerland AG 2021
A. Braud et al. (Eds.): ICFCA 2021, LNAI 12733, pp. 252–260, 2021.
https://doi.org/10.1007/978-3-030-77867-5_16

Finally, in [12] the authors provide a deep mathematical explanation on the connection between the concept lattice and the decision tree.

In this paper we develop the previous work in a more practical way. We show that the decision tree (and its ensembles) can induce a subset of concepts of the concept lattice. We propose a polynomial-time algorithm to construct a supervised machine learning model based on a concept lattice with prediction quality comparable to that of the state-of-the-art models.

## 2   Basic Definitions

For standard definitions of FCA and decision trees we refer the reader to [10] and [5], respectively.

In what follows we describe algorithms for binary attributes, numerical data can be processed by means of interval pattern structures or can be scaled to binary contexts [13].

## 3   Construct a Concept Lattice via a Set of Decision Trees

**Definition 1 (Classification rule).** *Let $M$ be a set of attributes of a context $\mathbb{K}$ and $Y$ be a set of "target" values. A pair $(\rho, \hat{y}_\rho), \rho \subseteq M, \hat{y}_\rho \in Y$ is a classification rule where $\rho$ is called a premise and $\hat{y}_\rho$ is a target prediction.*

Applied to object $g \subseteq G$ it can be interpreted as "if the description of $g$ falls under the premise $\rho$, then object $g$ should have the target value $\hat{y}_\rho$" or "if $\rho \subseteq g' \Rightarrow \hat{y}_\rho$".

In the case of classification task $Y$ can be represented either as a set $\{0,1\}$: $Y = \{y \in \{0,1\}\}_{i=1}^{|G|}$ or a set of probabilities of a positive class: $Y = \{y \in [0,1]\}_{i=1}^{|G|}$. In the case of regression task target value $Y$ is a set of real valued numbers: $Y = \{y \in \mathbb{R}\}_{i=1}^{|G|}$.

We can define a decision tree $DT$ as a partially ordered set (poset) of classification rules:

$$DT \subseteq \{(\rho, \hat{y}_\rho) \mid \rho \subseteq M, \hat{y}_\rho \in Y\} \tag{1}$$

where by *the order of classification rules* we mean the inclusion order on their premises:

$$(\rho_1, \hat{y}_{\rho_1}) \leq (\rho_2, \hat{y}_{\rho_2}) \Leftrightarrow \rho_1 \subseteq \rho_2 \tag{2}$$

Here we assume that a decision tree is a binary tree, i.e. its node is either a leaf (has no children) or has exactly 2 children nodes.

The other property of a decision tree is that each premise of its classification rules describes its own unique subset of objects:

$$\forall(\rho_1, \hat{y}_{\rho_1}) \in DT, \nexists(\rho_2, \hat{y}_{\rho_2}) \in DT : \rho_1' = \rho_2' \tag{3}$$

These simple properties result in an idea that 1) we can construct a concept lattice by closing premises of a decision tree, 2) join semilattice of such concept lattice is isomorphic to a decision tree.

**Proposition 1.** *Let* $\mathbb{K} = (G, M, I)$ *be a formal context,* $L(\mathbb{K})$ *be a lattice of the context* $\mathbb{K}$*. A subset of formal concepts* $L_{DT}(\mathbb{K})$ *forming a lattice can be derived from the decision tree* $DT(\mathbb{K})$ *constructed from the same context as:*

$$L_{DT} = \{(\rho', \rho'') \mid \forall(\rho', \hat{y}_\rho) \in DT(\mathbb{K})\} \cup \{(M', M)\} \tag{4}$$

**Proposition 2.** *Join-semilattice of a concept lattice* $L_{DT}$ *is isomorphic to the decision tree* $DT$*.*

*Proof.* Given two classification rules $(\rho_1, \hat{y}_{\rho_1}), (\rho_1, \hat{y}_{\rho_1}) \in DT$ let us consider two cases:

1. $\rho_1 \subseteq \rho_2 \Rightarrow (\rho_1', \rho_1'') \leq (\rho_2', \rho_2'')$
2. $\rho_1 \nsubseteq \rho_2, \rho_2 \nsubseteq \rho_1 \Rightarrow \exists m \in M : m \in \rho_1, \neg m \in \rho_2 \Rightarrow (\rho_1', \rho_1'') \nleq (\rho_2', \rho_2''), (\rho_2', \rho_2'') \nleq (\rho_1', \rho_1'')$

Thus the formal concepts from the join-semilattice of $L_{DT}$ possess the same partial order as the classification rules from $DT$.

Since we can construct a concept lattice from a decision tree and there is a union operation for concept lattices then we can construct a concept lattice which will correspond to a "union" of a number of independent decision trees (i.e. a random forest).

**Proposition 3.** *Let* $\mathbb{K} = (G, M, I)$ *be a formal context,* $L(\mathbb{K})$ *be a lattice of the context* $\mathbb{K}$*. A subset of formal concepts* $L_{RF}(\mathbb{K})$ *of the concept lattice* $L(\mathbb{K})$ *forming a lattice can be obtained from a random forest, i.e. from a set of* $m$ *decision trees constructed on subsets of a formal context* $DT_i(K_i), i = 1, ..., m, K_i \subseteq \mathbb{K}$:

$$L_{RF}(\mathbb{K}) = \bigcup_{i=1}^{m} L_{DT_i}(K_i) \tag{5}$$

The size of the lattice $L_{RF}$ is close to the size of the underlying random forest $RF$: $|L_{RF}| \sim |RF| \sim O(mG \log(G))$, where $m$ is the number of trees in $RF$ [3]. According to [2] the time complexity of constructing a decision tree is $O(MG^2 \log(G))$. Several algorithms for constructing decision trees and random forests are implemented in various libraries and frameworks like Sci-kit learn[1], H2O[2], Rapids[3]. The latter is even adapted to be run on GPU.

Thus, our lattice construction algorithm has two steps:

1. Construct a random forest $RF$
2. Use random forest $RF$ to construct a concept lattice $L_{RF}$ (by Eq. 3)

Both strong and weak side of this algorithm is that it relies on a supervised machine learning model, so it can be applied only if target labels $Y$ are given. In addition, the result set of concepts may not be optimal w.r.t. any concept interestingness measure [14]. Though it is natural to suppose that such set of concepts should be reasonable for supervised machine learning tasks.

---

[1] https://scikit-learn.org/stable/modules/ensemble.html#random-forests.
[2] http://h2o-release.s3.amazonaws.com/h2o/master/1752/docs-website/datascience/rf.html.
[3] https://docs.rapids.ai/api/cuml/stable/api.html#random-forest.

## 4    Decision Lattice

Given a formal concept $(A, B)$ we can use its intent $B$ as a premise of a classification rule $(B, \hat{y}_B)$.

The target prediction $\hat{y}_B$ of such classification rule $(B, \hat{y}_B)$ can be estimated via an aggregation function over the set $\{y_g \mid \forall g \in A\}$. In what follows we use the average aggregation function:

$$\hat{y}_B = \frac{1}{|A|} \sum_{\forall g \in A} y_g \tag{6}$$

Let us define a decision lattice (DL) as a poset of classification rules.

**Definition 2.** *Let $M$ be a set of attributes of a formal context $\mathbb{K}$ and $Y$ be a set of target values. Then a poset of classification rules is called a decision lattice DL if a premise of each classification rule of DL describes its own unique subset of objects (similar to DT in Eq. 3).*

Decision lattice $DL$ can be constructed from a concept lattice $L$ as follows:

$$DL = \{(B, \hat{y}_B) \mid (A, B) \in L\} \tag{7}$$

where $\hat{y}_B$ can be computed in various ways (we use the Eq. 6).

To get a final prediction $\hat{y}_g$ for an object $g$ a decision tree $DT$ firstly selects all the classification rules $DT^g$ describing the object $g$. Then it uses the target prediction of the maximal classification rule from $DT^g$

$$DT^g = \{(\rho, \hat{y}_\rho) \in DT \mid \rho \subseteq g'\} \tag{8}$$

$$DT^g_{max} = \{(\rho, \hat{y}_\rho) \in DT^g \mid \nexists(\rho_1, \hat{y}_{\rho_1}) \in DT^g : \rho \subset \rho_1\} \tag{9}$$

$$\hat{y}_g = \hat{y}_\rho, \quad (\rho, \hat{y}_\rho) \in DT^g_{max} \tag{10}$$

We use the same algorithm to get a final prediction $\hat{y}_g$ for an object $g$ by a decision lattice $DL$. The only difference is that when the subset $DT^g_{max}$ always contains only one classification rule a subset $DL^g_{max}$ may contain many. In this case we average the predictions of maximal classification rules $DL^g_{max}$:

$$\hat{y}_g = \frac{1}{|DL^g_{max}|} \sum_{(\rho, \hat{y}_\rho) \in DL^g_{max}} \hat{y}_\rho \tag{11}$$

Let us consider the *fruit context* $\mathbb{K} = (G, M, I)$ and *fruit* label $Y$ presented in Table 1. We want to compare the way decision lattice makes an estimation of the label $y_{mango}$ of object *mango* when this object is included in the train or the test context.

Figure 1 represents decision lattices constructed upon *fruit context* with (on the left) and without (in the center) *mango* object. In both cases we show only the classification rules which cover (describe) *mango* object.

**Table 1.** *Fruit context* and *fruit* labels

| Objects G | Attributes M | | | | | | | | | Label Y |
| | Firm | Smooth | Color | | | | Form | | | Fruit |
| | | | Yellow | Green | Blue | White | Round | Oval | Cubic | |
| Apple | | X | X | | | | X | | | 1 |
| Grapefruit | | | X | | | | X | | | 1 |
| Kiwi | | | | X | | | | X | | 1 |
| Plum | | X | | | X | | | X | | 1 |
| Toy cube | X | X | X | | | | | | X | 0 |
| Egg | X | X | | | | X | | X | | 0 |
| Tennis ball | | | | | | X | X | | | 0 |
| Mango | | X | | X | | | | X | | 1 |

The left picture represents a decision lattice with 8 classification rules and 1 single maximal classification rule: ("color_is_green & form_is_oval & smooth", 1). Therefore we use this classification rule to predict the target label of *mango*.

The picture in the center shows a decision lattice with 6 classification rules and 2 maximal classification rules: ("color_is_green & form_is_oval", 1), ("form_is_oval & smooth", 1/2). We average the target predictions of these classification rules to get a final prediction of 3/4 as shown in the picture on the right.

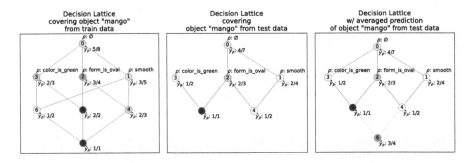

**Fig. 1.** Example of prediction of mango object

## 5    Experiments

We compare our decision lattice (DL) approach with the most popular machine learning models on real world datasets. One can reproduce the results by running Jupyter notebooks stored on GitHub [1]. Decision lattice models are implemented in open-source Python library for FCA which is called FCApy and located in the same GitHub repository.

We use ensembles of 5 and 10 decision trees to construct decision lattice models DL_RF_5 and DL_RF_10, respectively. The same ensemble of 5 decision trees is used by random forest models RF_5. Thus, we can compare prediction qualities of DL_RF_5 and RF_5 based on the same set of decision trees (and, consequently, classification rules).

The non-FCA model we use for the comparison are decision tree (DT), random forest (RF) and gradient boosting (GB) from sci-kit learn library, gradient boostings from LightGBM (LGBM), XGBoost (XGB), CatBoost (CB) libraries.

We also test Sofia algorithm [7] as a polynomial-time approach to construct a decision lattice DL_Sofia. We compute only 100 of most stable concepts by Sofia algorithm because of its time inefficiency.

Metadata of the datasets is given in Table 2.

**Table 2.** Description of the datasets

| Dataset name | Task type | # Instances | # Attributes |
|---|---|---|---|
| Adult[a] | Bin. class | 48842 | 14 |
| Amazon[b] | Bin. class | 32770 | 10 |
| Bank[c] | Bin. class | 45211 | 17 |
| Breast[d] | Bin. class | 569 | 32 |
| Heart[e] | Bin. class | 303 | 75 |
| Kick[f] | Bin. class | 72984 | 34 |
| Mammographic[g] | Bin. class | 961 | 6 |
| Seismic[h] | Bin. class | 2584 | 19 |
| Boston[i] | Regression | 506 | 14 |
| Calhouse[j] | Regression | 20640 | 8 |
| Diabetes[k] | Regression | 442 | 10 |

[a] https://archive.ics.uci.edu/ml/datasets/Adult
[b] https://www.kaggle.com/c/amazon-employee-access-challenge/data
[c] https://archive.ics.uci.edu/ml/datasets/bank+marketing
[d] https://archive.ics.uci.edu/ml/datasets/Breast+Cancer+Wisconsin+(Diagnostic)
[e] https://archive.ics.uci.edu/ml/datasets/heart+Disease
[f] https://www.kaggle.com/c/DontGetKicked/data?select=training.csv
[g] http://archive.ics.uci.edu/ml/datasets/mammographic+mass
[h] https://archive.ics.uci.edu/ml/datasets/seismic-bumps
[i] https://archive.ics.uci.edu/ml/machine-learning-databases/housing
[j] https://scikit-learn.org/stable/datasets/real_world.html#california-housing-dataset
[k] https://scikit-learn.org/stable/datasets/toy_dataset.html#diabetes-dataset

**Table 3.** Weighted average percentage error (best model delta)

| Model | Boston | | Calhouse | | Diabetes | | Mean delta | |
|---|---|---|---|---|---|---|---|---|
| | Train | Test | Train | Test | Train | Test | Train | Test |
| DL_RF_5 | 0.02 | 0.06 | 0.14 | 0.05 | 0.05 | 0.00 | 0.07 | 0.04 |
| DL_RF_10 | 0.01 | 0.07 | 0.13 | 0.04 | 0.01 | 0.01 | 0.05 | 0.04 |
| DL_Sofia | 0.29 | 0.20 | | | 0.40 | | 0.11 | 0.16 |
| DT | 0.00 | 0.05 | 0.00 | 0.09 | 0.00 | 0.12 | 0.00 | 0.09 |
| RF_5 | 0.05 | 0.01 | 0.14 | 0.04 | 0.15 | 0.02 | 0.12 | 0.03 |
| RF | 0.04 | 0.00 | 0.06 | 0.02 | 0.12 | 0.00 | 0.07 | 0.01 |
| GB | 0.05 | 0.00 | 0.17 | 0.02 | 0.16 | 0.00 | 0.13 | 0.01 |
| LGBM | 0.04 | 0.01 | 0.13 | 0.00 | 0.11 | 0.00 | 0.09 | 0.01 |
| CB | 0.02 | 0.00 | 0.13 | 0.00 | 0.06 | 0.00 | 0.07 | 0.00 |
| Best result | 0.00 | 0.14 | 0.00 | 0.21 | 0.00 | 0.31 | 0.00 | 0.22 |

For each dataset we use 5-fold cross-validation. We compute F1-score to measure the predictive quality of classification and weighted average percentage error (WAPE) to that of regression. In Tables 3, 4 we show the difference between the metric value of the model and the best obtained metric value among all methods.

As can be seen from Tables 3, 4 DL_RF model does not always show the best result among all the tested models, though its prediction quality is comparable to the state-of-the-art.

Table 4. F1 score (best model delta)

| Model | Adult | | Amazon | | Bank | | Breast | | Heart | | Kick | | Mamm. | | Seismic | | Mean delta | |
|---|---|---|---|---|---|---|---|---|---|---|---|---|---|---|---|---|---|---|
| | Train | Test | Train | Test | Train | Test | Train | Test | Train | Test | Train | Test | Train | Test | Train | Test | Train | Test |
| DL_RF_5 | −0.35 | −0.06 | −0.01 | −0.00 | −0.41 | −0.16 | −0.01 | −0.01 | −0.03 | −0.02 | −0.59 | −0.03 | −0.03 | −0.02 | −0.24 | −0.15 | −0.21 | −0.05 |
| DL_RF_10 | −0.33 | −0.05 | −0.01 | −0.00 | −0.37 | −0.14 | −0.00 | −0.00 | −0.01 | 0.00 | −0.58 | −0.03 | −0.01 | −0.02 | −0.13 | −0.15 | −0.18 | −0.05 |
| DL_Sofia | | | | | | | −1.00 | −0.95 | −0.33 | −0.27 | | | −0.87 | −0.72 | −1.00 | −0.15 | −0.80 | −0.52 |
| DT | 0.00 | −0.10 | −0.00 | −0.01 | 0.00 | −0.24 | 0.00 | −0.06 | 0.00 | −0.15 | 0.00 | −0.05 | −0.00 | −0.08 | 0.00 | 0.00 | −0.00 | −0.09 |
| RF_5 | −0.35 | −0.05 | −0.01 | −0.00 | −0.41 | −0.12 | −0.01 | −0.01 | −0.05 | −0.07 | −0.60 | −0.02 | −0.04 | −0.02 | −0.36 | −0.07 | −0.23 | −0.05 |
| RF | −0.00 | −0.04 | 0.00 | 0.00 | 0.00 | −0.11 | 0.00 | 0.00 | 0.00 | −0.00 | −0.00 | −0.01 | 0.00 | −0.03 | −0.00 | −0.12 | −0.00 | −0.04 |
| GB | −0.36 | −0.02 | −0.01 | −0.00 | −0.47 | 0.00 | 0.00 | −0.01 | −0.07 | −0.01 | −0.62 | −0.00 | −0.07 | 0.00 | −0.42 | −0.09 | −0.25 | −0.02 |
| LGBM | −0.31 | −0.00 | −0.01 | −0.00 | −0.32 | −0.04 | 0.00 | −0.02 | 0.00 | −0.03 | −0.60 | −0.00 | −0.03 | −0.02 | −0.00 | −0.11 | −0.16 | −0.03 |
| CB | −0.31 | 0.00 | −0.01 | −0.00 | −0.31 | −0.05 | 0.00 | −0.01 | −0.01 | −0.01 | −0.59 | 0.00 | −0.04 | −0.01 | −0.33 | −0.13 | −0.20 | −0.02 |
| Best result | 1.00 | 0.65 | 0.98 | 0.97 | 1.00 | 0.48 | 1.00 | 0.95 | 1.00 | 0.76 | 1.00 | 0.35 | 0.95 | 0.81 | 1.00 | 0.15 | 0.99 | 0.64 |

DL_Sofia model shows the worst results. There may be 2 reasons for this. First, it uses only a hundred of concepts. Second, we use Sofia algorithm to find one of the most stable concepts, but not the ones which minimize the loss.

Figure 2 shows the time needed to construct a lattice by the sets of 5 (*DL_RF_5*) and 10 (*DL_RF_10*) decision trees and by Sofia algorithm (*DL_Sofia*). The lattice can be constructed in a time linear in the number of objects in the given data.

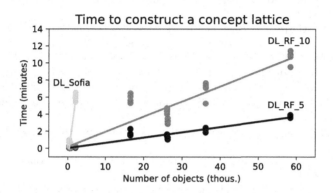

**Fig. 2.** Time needed to construct a lattice

## 6  Conclusions

In this paper we have introduced a new concept-based method to classification and regression. The proposed method constructs concept-based classifiers obtained with decision trees and random forests. This method is quite efficient and can be used for big datasets. We have shown that our approach is non-inferior to the predictive quality of the state-of-the-art competitors.

In the future work we plan to extend the algorithm for constructing decision trees in the case of data given by pattern structures.

**Acknowledgments.** The work of Sergei O. Kuznetsov on the paper was carried out at St. Petersburg Department of Steklov Mathematical Institute of Russian Academy of Science and supported by the Russian Science Foundation grant no. 17-11-01276

## References

1. Experiments source code. https://github.com/EgorDudyrev/FCApy/tree/main/notebooks/DecisionLattice_evaluation
2. Sci-kit learn description of decision trees. https://scikit-learn.org/stable/modules/tree.html
3. Sci-kit learn description of random forest. https://scikit-learn.org/stable/modules/ensemble.html#parameters

4. Belohlavek, R., De Baets, B., Outrata, J., Vychodil, V.: Inducing decision trees via concept lattices (2007)
5. Breiman, L., Friedman, J., Stone, C., Olshen, R.: Classification and Regression Trees. Taylor & Francis, New York (1984)
6. Breiman, L.: Random forests. Machine Learning (2001)
7. Buzmakov, A., Kuznetsov, S., Napoli, A.: Sofia: How to make FCA polynomial? In: FCA4AI@IJCAI (2015)
8. Buzmakov, A., Egho, E., Jay, N., Kuznetsov, S., Napoli, A., Rassi, C.: FCA and pattern structures for mining care trajectories. In: CEUR Workshop Proceedings, vol. 1058 (2013)
9. Drucker, H., Cortes, C.: Boosting decision trees, vol. 8, pp. 479–485 (1995)
10. Ganter, B., Wille, R.: Formal Concept Analysis: Mathematical Foundations. Springer, Berlin Heidelberg (1999). https://doi.org/10.1007/978-3-642-59830-2
11. Kaytoue, M., Kuznetsov, S., Napoli, A., Duplessis, S.: Mining gene expression data with pattern structures in formal concept analysis. Inf. Sci. **181**, 1989–2001 (2011)
12. Krause, T., Lumpe, L., Schmidt, S.: A link between pattern structures and random forests. In: CLA (2020)
13. Kuznetsov, S.: Pattern structures for analyzing complex data, vol. 5908, pp. 33–44 (2009)
14. Kuznetsov, S., Makhalova, T.: On interestingness measures of formal concepts. Inf. Sci. **442** (2016)
15. Kuznetsov, S., Obiedkov, S.: Comparing performance of algorithms for generating concept lattices. J. Exp. Theor. Artif. Intell. **14**, 189–216 (2002)
16. Kuznetsov, S.O.: Machine learning and formal concept analysis. In: Eklund, P. (ed.) ICFCA 2004. LNCS (LNAI), vol. 2961, pp. 287–312. Springer, Heidelberg (2004). https://doi.org/10.1007/978-3-540-24651-0_25
17. Prokhorenkova, L., Gusev, G., Vorobev, A., Dorogush, A.V., Gulin, A.: Catboost: unbiased boosting with categorical features (2019)
18. Wille, R.: Restructuring lattice theory: an approach based on hierarchies of concepts. In: Ferré, S., Rudolph, S. (eds.) ICFCA 2009. LNCS (LNAI), vol. 5548, pp. 314–339. Springer, Heidelberg (2009). https://doi.org/10.1007/978-3-642-01815-2_23

# Exploring Scale-Measures of Data Sets

Tom Hanika[1,2] and Johannes Hirth[1,2(✉)]

[1] Knowledge and Data Engineering Group, University of Kassel, Kassel, Germany
{tom.hanika,hirth}@cs.uni-kassel.de
[2] Interdisciplinary Research Center for Information System Design,
University of Kassel, Kassel, Germany

**Abstract.** Measurement is a fundamental building block of numerous scientific models and their creation in data driven science. Due to the high complexity and size of modern data sets, it is necessary to develop understandable and efficient scaling methods. A profound theory for scaling data is scale-measures from formal concept analysis. Recent developments indicate that the set of all scale-measures for a given data set constitutes a lattice. In this work we study the properties of said lattice and propose a novel and efficient scale-measure exploration algorithm, motivating multiple applications for (semi-)automatic scaling.

**Keywords:** Measurements · Data scaling · Formal concepts · Lattices

## 1 Introduction

An inevitable step of any data-based knowledge discovery process is *measurement* [14] and the associated (explicit or implicit) *scaling* of the data [17]. The latter is particularly constrained by the underlying mathematical formulation of the data representation, e.g., real-valued vector spaces or weighted graphs, the requirements of the data procedures, e.g., the presence of a distance function, and, more recently, the need for human understanding of the results. Considering the scaling of data as part of the analysis itself, in particular formalizing it and thus making it controllable, is a salient feature of formal concept analysis (FCA) [5]. This field of research has spawned a variety of specialized scaling methods, such as logical scaling [15], and in the form of *scale-measures* links the scaling process with the study of *continuous mappings* between *closure systems*.

Recent results by the authors [11] revealed that the set of all scale-measures for a given data set constitutes a lattice. Furthermore, it was shown that any scale-measure can be expressed in simple propositional terms using disjunction, conjunction and negation. Among other things, the previous results allow a computational transition between different scale-measures, which we may call *scale-measure navigation*, as well as their *interpretability* by humans. Despite these advances, the question of how to identify appropriate and meaningful scale-measures for a given data set with respect to a human data analyst remains

---

Authors are given in alphabetical order. No priority in authorship is implied.

© Springer Nature Switzerland AG 2021
A. Braud et al. (Eds.): ICFCA 2021, LNAI 12733, pp. 261–269, 2021.
https://doi.org/10.1007/978-3-030-77867-5_17

unanswered. In this paper, we propose an answer by adapting the well-known *attribute exploration algorithm* from FCA to present a method for exploring scale measures. Very similar to the original algorithm does *scale-measure exploration* inquire a (human) scaling expert for how to aggregate, separate, omit, or introduce data set features. Our efforts do finally result in a (semi-)automatic scaling framework. Please note, some proofs are outsourced to the journal version [10].

## 2  Scales and Measurement

Formalizing and understanding the process of *measurement* is, in particular in data science, an ongoing discussion, for which we refer the reader to *Representational Theory of Measurement* [13,18] as well as *Numerical Relational Structure* [14], and *algebraic (measurement) structures* [16, p. 253]. Formal concept analysis (FCA) is well equipped to handle and comprehend data scaling tasks. Within this work we use the standard notation, as introduced by B. Ganter and R. Wille [5]. A fundamental approach to comprehensible scaling, in particular for nominal and ordinal data as studied in this work, is the following.

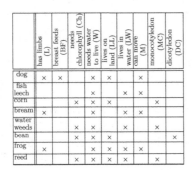 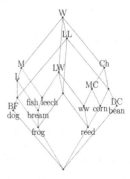

**Fig. 1.** This figure shows the *Living Beings and Water* context in the top. Its concept lattice is displayed at the bottom and contains nineteen concepts.

**Definition 1 (Scale-Measure (cf. Definition 91, [5])).** *Let* $\mathbb{K} = (G, M, I)$ *and* $\mathbb{S} = (G_\mathbb{S}, M_\mathbb{S}, I_\mathbb{S})$ *be formal contexts. The map* $\sigma : G \to G_\mathbb{S}$ *is called an* $\mathbb{S}$*-measure of* $\mathbb{K}$ *into the scale* $\mathbb{S}$ *iff the preimage* $\sigma^{-1}(A) := \{g \in G \mid \sigma(g) \in A\}$ *of every extent* $A \in Ext(\mathbb{S})$ *is an extent of* $\mathbb{K}$.

This definition resembles the idea of *continuity between closure spaces* $(G_1, c_1)$ and $(G_2, c_2)$. We say that the map $f : G_1 \to G_2$ is *continuous* if and only if for all $A \in \mathcal{P}(G_2)$ we have $c_1(f^{-1}(A)) \subseteq f^{-1}(c_2(A))$. In the light of the definition above we understand $\sigma$ as an interpretation of the objects from $\mathbb{K}$ in $\mathbb{S}$. Therefore we view the set $\sigma^{-1}(Ext(\mathbb{S})) := \bigcup_{A \in Ext(\mathbb{S})} \sigma^{-1}(A)$ as the set of extents that is *reflected* by the scale context $\mathbb{S}$.

We present in Fig. 2 the scale-context for some scale-measure and its concept lattice, derived from our *running example* context *Living Beings and Water* $\mathbb{K}_W$, cf. Fig. 1. This scaling is based on the original object set $G$, however, the attribute set is comprised of nine, partially new, elements, which may reflect species taxons. We observe in this example that the concept lattice of the scale-measure context reflects twelve out of the nineteen concepts from $\mathfrak{B}(\mathbb{K}_W)$.

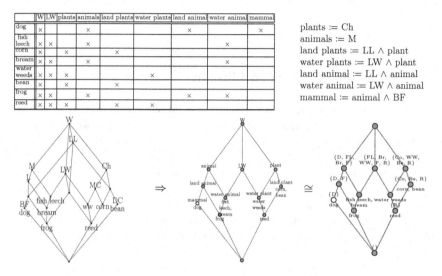

plants := Ch
animals := M
land plants := LL ∧ plant
water plants := LW ∧ plant
land animal := LL ∧ animal
water animal := LW ∧ animal
mammal := animal ∧ BF

**Fig. 2.** A scale context (top), its concept lattice (bottom middle) for which $\mathrm{id}_G$ is a scale-measure of the context in Fig. 1. The reflected extents by the scale $\sigma^{-1}(\mathrm{Ext}(\mathbb{S}))$ of the scale-measure are indicated in gray in the contexts concept lattice (bottom left). The concept lattice on the bottom right is the output scale of the scale-measure exploration algorithm and is discussed in Sect. 4. The employed object order is: Be > Co > D > WW > FL > Br > F > R

In our work [11] we derived a *scale-hierarchy* on the set of scale-measures, i.e., $\mathfrak{S}(\mathbb{K}) := \{(\sigma, \mathbb{S}) \mid \sigma \text{ is a } \mathbb{S}-\text{measure of } \mathbb{K}\}$, from a natural order of scales introduced by Ganter and Wille [5, Definition 92]). We say for two scale-measures $(\sigma, \mathbb{S}), (\psi, \mathbb{T})$ that $(\psi, \mathbb{T})$ is coarser then $(\sigma, \mathbb{S})$, iff $\psi^{-1}(\mathrm{Ext}(\mathbb{T})) \subseteq \sigma^{-1}(\mathrm{Ext}(\mathbb{S}))$, denoted $(\psi, \mathbb{T}) \leq (\sigma, \mathbb{S})$. This also yields a natural equivalence relation $\sim$, which in turn, allowed for the definition [11] of a *scale-hierarchy* $\underline{\mathfrak{S}}(\mathbb{K}) = (\mathfrak{S}(\mathbb{K})/\sim, \leq)$. For any given context $\mathbb{K}$, the scale-hierarchy is lattice ordered [11] and isomorphic to the set of all sub-closure systems of $\mathrm{Ext}(\mathbb{K})$ ordered by $\subseteq$. To show this, we defined a *canonical representation* of scale-measures, using the so-called *canonical scale* $\mathbb{K}_{\mathcal{A}} := (G, \mathcal{A}, \in)$ for $\mathcal{A} \subseteq \mathrm{Ext}(\mathbb{K})$ with $\mathrm{Ext}(\mathbb{K}_{\mathcal{A}}) = \mathcal{A}$. In fact, for context $\mathbb{K}$ and $(\mathbb{S}, \sigma) \in \mathfrak{S}(\mathbb{K})$ a scale-measure, then $(\sigma, \mathbb{S}) \sim (\mathrm{id}, \mathbb{K}_{\sigma^{-1}(\mathrm{Ext}(\mathbb{S}))})$.

We argued in [11] that the canonical representation eludes human explanation to some degree. To remedy this issue by means of logical scaling [15] we used scales with logical attributes $M_{\mathbb{S}} \subseteq \mathcal{L}(M, \{\wedge, \vee, \neg\})$ ([11, Problem 1]). Let $m \in M$, then we define $I_m = I \cap G \times \{m\}$, $gI_{\phi_1 \wedge \phi_2}(\phi_1 \wedge \phi_2)$ iff $gI_{\phi_1}\phi_1 \wedge gI_{\phi_2}\phi_2)$, $gI_{\phi_1 \vee \phi_2}(\phi_1 \vee \phi_2)$ iff $gI_{\phi_1}\phi_1 \vee gI_{\phi_2}\phi_2$, and $gI_{\neg\phi}(\neg\phi)$ iff not $gI_\phi\phi$.

**Proposition 1 (Conjunctive Normalform (cf. Proposition 23, [11])).**
*Let $\mathbb{K}$ be a context, $(\sigma, \mathbb{S}) \in \mathfrak{S}(\mathbb{K})$. Then the scale-measure $(\psi, \mathbb{T}) \in \mathfrak{S}(\mathbb{K})$ given by $\psi = \mathrm{id}_G$ and $\mathbb{T} = \mid_{A \in \sigma^{-1}(Ext(\mathbb{S}))}(G, \{\phi = \wedge A^I\}, I_\phi)$ is equivalent to $(\sigma, \mathbb{S})$ and is called conjunctive normalform of $(\sigma, \mathbb{S})$.*

# 3    Ideals in the Lattice of Closure Systems

The goal for the main part is to identify *outstanding* and particularly interesting data scalings. This quest leads to the natural question for a structural understanding of the scale-hierarchy. In order to do this we rely on the isomorphism [11, Proposition 11] between a context's scale-hierarchy $\mathfrak{S}(\mathbb{K})$ and the lattice of all sub-closure systems of $\mathrm{Ext}(\mathbb{K})$. The latter forms an order ideal, denoted by $\downarrow_{\mathfrak{F}_G} \mathrm{Ext}(\mathbb{K})$, in the lattice $\mathfrak{F}_G$ of all closure systems on $G$. This ideal is well-studied [1]. We may omit $G$ in $\downarrow_{\mathfrak{F}_G} \mathrm{Ext}(\mathbb{K})$ to improve the readability.

Equipped with this structure we have to recall a few notions and definitions for a complete lattice $(L, \leq)$. In the following, we denote by $\prec$ the *cover relation* of $\leq$. Furthermore, we say $L$ is 1) *lower semi-modular* if and only if $\forall x, y \in L : x \prec x \vee y \implies x \wedge y \prec y$, 2) *join-semidistributive* iff $\forall x, y, z \in L : x \vee y = x \vee z \implies x \vee y = x \vee (y \wedge z)$, 3) *meet-distributive (lower locally distributive*, cf [1]) iff $L$ is join-semidistributive and lower semi-modular, 4) *join-pseudocomplemented* iff for any $x \in L$ the set $\{y \in L \mid y \vee x = \top\}$ has a least, 6) *ranked* iff there is a function $\rho : L \mapsto \mathbb{N}$ with $x \prec y \implies \rho(x) + 1 = \rho(y)$, 7) *atomistic* iff every $x \in L$ can be written as the join of atoms in $L$. In addition to the just introduced lattice properties, there are properties for elements in $L$ that we consider. An element $x \in L$ is 1) *neutral* iff every triple $\{x, y, z\} \subseteq L$ generates a distributive sublattice of $L$, 2) *distributive* iff the equalities $x \vee (y \wedge z) = (x \vee y) \wedge (x \vee z)$ and $x \wedge (y \vee z) = (x \wedge y) \vee (x \wedge z)$ for every $y, z \in L$ hold, 3) *meet irreducible* iff $x \neq \top$ and $\bigwedge_{y \in Y} y$ for $Y \subseteq L$ implies $x \in Y$, 4) *join irreducible* iff $x \neq \bot$ and $\bigvee_{y \in Y} y$ for $Y \subseteq L$ implies $x \in Y$. For the rest of this work, we denote by $\mathcal{M}(L)$ the set of all meet-irreducible elements of $L$.

We can derive from literature [1, Proposition 19] the following statement.

**Corollary 1.** *For $\mathbb{K} = (G, M, I)$, $\downarrow Ext(\mathbb{K}) \subseteq \mathfrak{F}_G$ and $\mathcal{R}, \mathcal{R}' \in \downarrow Ext(\mathbb{K})$ that: $\mathcal{R}' \prec \mathcal{R} \iff \mathcal{R}' \cup \{A\} = \mathcal{R}$ with $A$ is meet-irreducible in $\mathcal{R}$.*

Of special interest in lattices are the (meet-) join-irreducibles, since every element of a lattice can be represented as a (meet) join of these elements.

**Proposition 2.** *For $\mathbb{K}$, $\downarrow Ext(\mathbb{K}) \subseteq \mathfrak{F}_G$ and $\mathcal{R} \in \downarrow Ext(\mathbb{K})$: $\mathcal{R}$ is join-irreducible in $\downarrow Ext(\mathbb{K}) \iff \exists A \in Ext(\mathbb{K}) \setminus \{G\}: \mathcal{R} = \{G, A\}$*

Next, we investigate the meet-irreducibles of $\downarrow \operatorname{Ext}(\mathbb{K})$ using a similar approach as done for $\mathfrak{F}_G$ [1] based on propositional logic. We recall, that an (object) implication for some context $\mathbb{K}$ is a pair $(A, B) \in \mathcal{P}(G) \times \mathcal{P}(G)$, shortly denoted by $A \to B$. We say $A \to B$ is valid in $\mathbb{K}$ iff $A' \subseteq B'$. The set $\mathcal{F}_{A,B} := \{D \subseteq G : A \not\subseteq D \vee B \subseteq D\}$ is the set of all *models* of $A \to B$. Additionally, $\mathcal{F}_{A,B}|_{\operatorname{Ext}(\mathbb{K})} := \mathcal{F}_{A,B} \cap \operatorname{Ext}(\mathbb{K})$ is the set of all extents $D \in \operatorname{Ext}(\mathbb{K})$ that are models of $A \to B$. The set $\mathcal{F}_{A,B}$ is a closure system [1] and therefor $\mathcal{F}_{A,B}|_{\operatorname{Ext}(\mathbb{K})}$, too. Furthermore, we can deduce that $\mathcal{F}_{A,B}|_{\operatorname{Ext}(\mathbb{K})} \in \downarrow \operatorname{Ext}(\mathbb{K})$.

**Lemma 1.** *For context $\mathbb{K}$, $\downarrow \operatorname{Ext}(\mathbb{K}) \subseteq \mathfrak{F}_G$, $\mathcal{R} \in \downarrow \operatorname{Ext}(\mathbb{K})$ with closure operator $\phi_{\mathcal{R}}$ we find $\mathcal{R} = \bigcap \{\mathcal{F}_{A,B}|_{Ext(\mathbb{K})} \mid A, B \subseteq G \wedge B \subseteq \phi_{\mathcal{R}}(A)\}$.*

For $\mathcal{R} \in \downarrow \operatorname{Ext}(\mathbb{K})$ the set $\{\mathcal{F}_{A,B}|_{\operatorname{Ext}(\mathbb{K})} \mid A, B \subseteq G \wedge B \subseteq \phi_{\mathcal{R}}(A)\}$ contains only closure systems in $\downarrow \operatorname{Ext}(\mathbb{K})$ and thus possibly meet-irred. elements of $\downarrow \operatorname{Ext}(\mathbb{K})$.

**Proposition 3.** *For context $\mathbb{K}$, $\downarrow Ext(\mathbb{K}) \subseteq \mathfrak{F}_G$ and $\mathcal{R} \in \downarrow Ext(\mathbb{K})$: 1. $\mathcal{R}$ is meet-irreducible in $\downarrow Ext(\mathbb{K})$ 2. $\exists A \in Ext(\mathbb{K}), i \in G$ with $A \prec_{Ext(\mathbb{K})} (A \cup \{i\})''$ such that $\mathcal{R} = \mathcal{F}_{A,\{i\}}|_{Ext(\mathbb{K})}$*

Propositions 2 and 3 provide a characterization of irreducible elements in $\downarrow\operatorname{Ext}(\mathbb{K})$ and thereby in the scale-hierarchy of $\mathbb{K}$. Those may be of particular interest, since any element of $\downarrow\operatorname{Ext}(\mathbb{K})$ is representable by irreducible elements. Equipped with this characterization we look into counting the irreducibles.

**Proposition 4.** *For context $\mathbb{K}$, the number of meet-irreducible elements in the lattice $\downarrow Ext(\mathbb{K}) \subseteq \mathfrak{F}_G$ is equal to $|\prec_{\downarrow Ext(\mathbb{K})}|$.*

Next, we turn ourselves to other lattice properties of $\downarrow \operatorname{Ext}(\mathbb{K})$.

**Lemma 2 (Join Complement).** *For $\mathbb{K}$, $\downarrow Ext(\mathbb{K}) \subseteq \mathfrak{F}_G$ and $\mathcal{R} \in \downarrow Ext(\mathbb{K})$, the set $\hat{\mathcal{R}} = \bigvee_{A \in \mathcal{M}(Ext(\mathbb{K})) \setminus \mathcal{M}(\mathcal{R})} \{A, G\}$ is the inclusion minimum closure-system for which $\mathcal{R} \vee \hat{\mathcal{R}} = Ext(\mathbb{K})$.*

All the above results in the following statement about $\downarrow \operatorname{Ext}(\mathbb{K})$:

**Proposition 5.** *For any context $\mathbb{K}$, the lattice $\downarrow Ext(\mathbb{K}) \subseteq \mathfrak{F}_G$ is: i) join-semidistributive ii) lower semi-modular iii) meet-distributive iv) join-pseudo-complemented v) ranked vi) atomistic*

# 4    Recommending Conceptual Scale-Measures

For the task of efficiently determining a scale-measure, based on human preferences, we propose the following approach. Motivated by the representation of meet-irreducible elements in the scale-hierarchy through object implications of the context ((Proposition 3), we employ the dual of the *attribute exploration* algorithm [6] by Ganter. We modified said algorithm toward exploring scale-measures

---

**Algorithm 1:** Scale-measure Exploration: A modified Exploration with Background Knowledge

---

**Input**  : Context $\mathbb{K} = (G, M, I)$
**Output**: $(id_G, \mathbb{S}) \in \mathfrak{S}(\mathbb{K})$ and optionally $\mathcal{L}_\mathbb{S}$
Init Scale $\mathbb{S} = (G, \emptyset, \in)$
Init $A = \emptyset, \mathcal{L}_\mathbb{S} = $ CanonicalBase($\mathbb{K}$) (or $\mathcal{L}_\mathbb{S} = \{\}$ for larger contexts)
while $A \neq G$ do
    while $A \neq A^{I_\mathbb{S} I_\mathbb{S}}$ do
        **if** *Further differentiate objects having* $A^{I_\mathbb{S} I_\mathbb{S} I_\mathbb{K}}$
        *by attributes in* $A^{I_\mathbb{K}} \setminus A^{I_\mathbb{S} I_\mathbb{S} I_\mathbb{K}}$ ? **then**
            $\mathcal{L}_\mathbb{S} = \mathcal{L}_\mathbb{S} \cup \{A \to A^{I_\mathbb{S} I_\mathbb{S}}\}$
            Exit While
        **else**
            Enter $B \subseteq A^{I_\mathbb{K}} \setminus (A)^{I_\mathbb{S} I_\mathbb{S} I_\mathbb{K}}$ that should be considered
            Add attribute $B^{I_\mathbb{K}}$ to $\mathbb{S}$
    $A = $ Next_Closure($A, G, \mathcal{L}_\mathbb{S}$)
**return** : $(id_G, \mathbb{S})$ and optionally $\mathcal{L}$

---

and present its pseudo-code in Algorithm 1. In this depiction we highlighted our modifications with respect to the original exploration algorithm (Algorithm 19, [7]) with darker print. This algorithm semi-automatically computes a scale context $\mathbb{S}$ and its canonical base. In each iteration of the inner loop of our algorithm the query that is stated to the *scaling expert* is if an object implication $A \to B$ is true in the closure system of user preferences. If the implication holds, it is added to the implicational base of $\mathbb{S}$ and the algorithm continues with the next query. Otherwise a counter example in the form of a closed set $C \in \text{Ext}(\mathbb{K})$ with $A \subseteq C$ but $B \not\subseteq C$ has to be constructed. This closed set is then added as attribute to the scale context $\mathbb{S}$ with the incidence given by $\in$. If $C \notin \text{Ext}(\mathbb{K})$ the scale $\mathbb{S}$ would contradict the scale-measure property (Proposition 20, [11]).

The object implicational theory $\mathcal{L}_\mathbb{S}$ is initialized to the object canonical base of $\mathbb{K}$, which is an instance of according to attribute exploration with background knowledge [6]. This initialization can be neglected for larger contexts, however it may reduce the number of queries. The algorithm terminates when the implication premise of the query is equal to $G$. The returned scale-measure is in canonical form, i.e., the canonical representation $(id_G, (G, \text{Ext}(\mathbb{S}), \in))$. The motivation behind exploration queries is to determine if an implication holds in the unknown representational context of the learning domain. In contrast, the exploration of scale-measures determines if a given $\text{Ext}(\mathbb{K})$ can be coarsened by implications $A \implies B$, resulting in a smaller and thus more human-comprehensible concept lattice $\underline{\mathfrak{B}}(\mathbb{S})$, adjusted to the preferences of the scaling expert.

Querying object implications may be less intuitive compared to attribute implications, hence, we suggest to rather not test for $A \implies A^{I_\mathbb{S} I_\mathbb{S}}$ for $A \subseteq G$ but for the difference of the intents $A^{I_\mathbb{K}}$ and $(A^{I_\mathbb{S} I_\mathbb{S}})'$ in $\mathbb{K}$. Finally, as a post-processing, one may apply the *conjunctive normalform* [11, Proposition 23] of scale-measures to improve human-comprehension. Yet, deriving other human-comprehensible representations of scale-measures is deemed to be future work.

*(Semi-)Automatic Large Data Set Scaling.* To demonstrate the applicability of the presented algorithm, we have implemented it in the `conexp-clj` ([9]) software suite. For this, we apply the scale-measure exploration Algorithm 1 on our running example $\mathbb{K}_W$, see Fig. 1. The evaluation steps of this algorithm are displayed in more detail in the long version of this work. One such intermediate step is for example row two where the implication $\{\} \implies \{D, FL, Br, F\}$ is true in the so far generated scale $\mathbb{S}$ and it is queried if it should hold. All objects of the implication do have at least the attributes *can move* and *needs water to live*. The answer of the scaling expert envisioned by us is the attribute *lives on land*. Thus, the object counter example is the attribute-derivation the union $\{M, W, LL\}^{I_W} = \{D, F\}$. In our example of the scale-measure exploration the algorithm terminates after the scaling expert provided in total nine counter examples and four accepts. The output is a scale context in canonical representation with twelve concepts as depicted in Fig. 2 (bottom right).

## 5   Related Work

Measurement is an important field of study in many (scientific) disciplines that involve the collection and analysis of data. A framework to describe and analyze the measurement for Boolean data sets has been introduced in [8] and [4], called *scale-measures*. It characterizes the measurement based on object clusters that are formed according to common feature (attribute) value combinations. An accompanied notion of dependency has been studied [19], which led to attribute selection based measurements of boolean data. The formalism includes a notion of consistency enabling the determination of different views and abstractions, called *scales*, to the data set. Despite the expressiveness of scale-measure framework, as demonstrated in this work, it is so far insufficiently studied in the literature. In particular algorithmical and practical calculation approaches are missing. Comparable and popular machine learning approaches, such as feature compression techniques, e.g., *Latent Semantic Analysis* [2,3], have the disadvantage that the newly compressed features are not interpretable by means of the original data and are not guaranteed to be consistent with said original data. The methods presented in this paper do not have these disadvantages, as they are based on meaningful and interpretable features with respect to the original features. In particular preserving consistency, as we did, is not a given, which was explicitly investigated in the realm scaling many-valued formal contexts [15] and implicitly studied for generalized attributes [12].

## 6   Conclusion

With this work we have shed light on the hierarchy of scale-measures. By applying multiple results from lattice theory, especially concerning ideals, to said hierarchy, we were able to give a more thorough structural description of $\downarrow_{\mathfrak{F}_G} \mathrm{Ext}(\mathbb{K})$.

Our main theoretical result is Proposition 5, which in turn leads to our practical applications. In particular, based on this deeper understanding we were able to present an algorithm for exploring the scale-hierarchy of a binary data set $\mathbb{K}$. Equipped with this algorithm a data scaling expert may explore the lattice of scale-measures for a given data set with respect to her preferences and the requirements of the data analysis task. The practical evaluation and optimization of this algorithm is a promising goal for future investigations. Even more important, however, is the implementation and further development of the automatic scaling framework, as outlined in Sect. 4. This opens the door to empirical scale recommendation studies and a novel approach for data preprocessing.

# References

1. Caspard, N., Monjardet, B.: The lattices of closure systems, closure operators, and implicational systems on a finite set: a survey. Discrete Appl. Math. **127**(2), 241–269 (2003). http://www.sciencedirect.com/science/article/B6TYW-46VBMT3-1/2/d6bbb12c7831e113c5e91bebc320bee1

2. Codocedo, V., Taramasco, C., Astudillo, A.: Cheating to achieve formal concept analysis over a large formal context. In: Napoli, A., Vychodil, V. (eds.) CLA, vol. 959, pp. 349–362. CEUR-WS.org (2011)

3. Dumais, S.T.: Latent semantic analysis. Annu. Rev. Inf. Sci. Technol. **38**(1), 188–230 (2004). https://doi.org/10.1002/aris.1440380105

4. Ganter, B., Wille, R.: Conceptual scaling. In: Robert, F. (ed.) Applications of Combinatorics and Graph Theory to the Biological and Social Sciences. IMA, vol. 17, pp. 139–167. Springer-Verlag, Heidelberg (1989). https://doi.org/10.1007/978-1-4684-6381-1_6

5. Ganter, B., Wille, R.: Formal Concept Analysis: Mathematical Foundations. Springer-Verlag, Berlin (1999). https://doi.org/10.1007/978-3-642-59830-2

6. Ganter, Bernhard: Attribute exploration with background knowledge. Theor. Comput. Sci. **217**(2), 215–233 (1999). https://doi.org/10.1016/S0304-3975(98)00271-0

7. Ganter, B., Obiedkov, S.: More expressive variants of exploration. In: Conceptual Exploration, pp. 237–292. Springer, Heidelberg (2016). https://doi.org/10.1007/978-3-662-49291-8_6

8. Ganter, B., Stahl, J., Wille, R.: Conceptual measurement and many-valued contexts. In: Gaul, W., Schader, M. (eds.) Classification as a Tool of Research, pp. 169–176. North-Holland (1986)

9. Hanika, T., Hirth, J.: Conexp-clj - A research tool for FCA. In: Cristea, D., Le Ber, F., Missaoui, R., Kwuida, L., Sertkaya, B. (eds.) Supplementary Proceedings of ICFCA 2019, volume 2378 of CEUR WS Proceedings, pp. 70–75. CEUR-WS.org (2019). http://ceur-ws.org/Vol-2378/shortAT8.pdf

10. Hanika, T., Hirth, J.: Exploring scale-measures of data sets (2021)

11. Hanika, T., Hirth, J.: On the lattice of conceptual measurements. arXiv preprint arXiv:2012.05287 (2020)

12. Kwuida, L., et al.: Generalized pattern extraction from concept lattices. Ann. Math. Artif. Intell. 72(1-2), 151–168 (2014)

13. Duncan Luce, R., Krantz, D.H., Suppes, P., Tversky, A.: Foundations of Measurement - Representation, Axiomatization, and Invariance, vol. 3. Academic Press (1990)

14. Pfanzagl, J.: Theory of Measurement. Physica, Heidelberg (1971)
15. Prediger, S., Stumme, G.: Theory-driven logical scaling: Conceptual information systems meet description logics. In Enrico Franconi and Michael Kifer, editors, Proc. KRDB'99, volume 21, pages 46–49. CEUR-WS.org, 1999. URL http://ceur-ws.org/Vol-21/prediger.ps
16. Fred, S.R.: Measurement Theory. Cambridge University Press, Cambridge (1984)
17. Stevens, S.S.: On the theory of scales of measurement. Science **103**(2684), 677–680 (1946). ISSN 0036-8075. https://doi.org/10.1126/science.103.2684.677, https://science.sciencemag.org/content/103/2684/677
18. Suppes, P., Krantz, D.H., Duncan Luce, R., Tversky, A.: Foundations of Measurement - Geometrical, Threshold, and Probabilistic Representations, vol. 2. Academic Press (1989)
19. Wille, R.: Dependencies of many valued attributes. In: Bock, H.-H. (ed.) Classification and related methods of data analysis, pp. 581–586. North-Holland (1988)

# Filters, Ideals and Congruences
# on Double Boolean Algebras

Tenkeu Jeufack Yannick Léa[1]([✉]), Etienne Romuald Alomo Temgoua[2],
and Léonard Kwuida[3]

[1] Department of Mathematics, Faculty of Sciences, University of Yaoundé 1,
P.O. Box 47, Yaoundé, Cameroon
[2] Department of Mathematics, Ecole Normale Supérieure, Taoundé, Cameroon
[3] Bern University of Applied Sciences (BFH), Bern, Switzerland
leonard.kwuida@bfh.ch

**Abstract.** Double Boolean algebras (dBas) are algebras $\underline{D} := (D; \sqcap, \sqcup, \neg, \lrcorner, \bot, \top)$ of type $(2, 2, 1, 1, 0, 0)$, introduced by R. Wille to capture the equational theory of the algebra of protoconcepts. Boolean algebras form a subclass of dBas. Our goal is an algebraic investigation of dBas, based on similar results on Boolean algebras. In these notes, we describe filters, ideals and congruences, and show that principal filters as well as principal ideals of dBas form (non necessary isomorphic) Boolean algebras.

**Keywords:** Double Boolean algebra · Protoconcept algebra · Concept algebra · Formal concept

## 1 Introduction

In order to extend Formal Concept Analysis (FCA) to Contextual Logic, a negation has to be formalized [8,10]. There are many options; one of these wants to preserve the correspondence between negation and set complementation, and leads to the notions of semiconcept, protoconcept and preconcept [10]. To capture their equational theory, double Boolean algebras have been introduced by Rudolf Wille and coworkers. Wille proved that each double Boolean algebra "quasi-embeds" into an algebra of protoconcepts. Thus the equational axioms of double Boolean algebras generate the equational theory of the algebras of protoconcepts [10](Corollary 1).

To the best of our knowledge, the investigation of dBas has been so far concentrated on representation problem such as equational theory [10], contextual representation [9], and most recently topological representation [2,6]. Of course the prime ideal theorem [7] plays a central role in such representation.

For a better understanding of the structure of dBas, our goal is to start with purely algebraic notions such as filters, ideals, congruences, homomorphisms, etc. They are the cornerstone in structure theory, representation, decomposition as well as classification of algebraic structures. In these notes we describe filters (resp. ideals) generated by arbitrary subsets of a dBa and show that the set of

© Springer Nature Switzerland AG 2021
A. Braud et al. (Eds.): ICFCA 2021, LNAI 12733, pp. 270–280, 2021.
https://doi.org/10.1007/978-3-030-77867-5_18

principal filters (resp. ideals) of a dBa form a bounded sublattice of the lattice of its filters (resp. ideals), and are (non necessary isomorphic) Boolean algebras. For any congruence $\theta$ in any bounded lattice $L$ (particularly in any Boolean algebra), the congruence class of the smallest (resp. largest) element $\bot$ (resp. $\top$) is an ideal (resp. filter). This result is not more true for double Boolean algebras. We therefore give necessary and sufficient conditions for the congruence class $[\bot]_\theta$ (resp. $[\top]_\theta$) to be an ideal (resp. a filter). The rest of this contribution is organized as follows: In Sect. 2 we recall some basic notions and present protoconcept algebras as a rich source of examples for dBas. We present our results on filters and ideals of dBas in Sect. 3 and those on congruences in Sect. 4. We finish with a conclusion and further research in Sect. 5.

## 2   Concepts, Protoconcepts and Double Boolean Algebras

In this section, we provide the reader with some basic notions and notations. For more details we refer to [4,10].

A **formal context** is a triple $\mathbb{K} := (G, M, I)$ where $G$ is a set of objects, $M$ a set of attributes and $I \subseteq G \times M$, a binary relation to describe if an object of $G$ has an attribute in $M$. We write $gIm$ for $(g, m) \in I$. To extract clusters, the following derivation operators are defined on subsets $A \subseteq G$ and $B \subseteq M$ by:

$$A' := \{m \in M \mid gIm \text{ for all } g \in A\} \quad \text{and} \quad B' := \{g \in G \mid gIm \text{ for all } m \in B\}.$$

The maps $A \mapsto A'$ and $B \mapsto B'$ form a Galois connection between the powerset of $G$ and that of $M$. The composition $''$ is a closure operator. A **formal concept** is a pair $(A, B)$ with $A' = B$ and $B' = A$; We call $A$ the **extent** and $B$ the **intent** of the formal concept $(A, B)$. They are closed subsets with respect to $''$ (i.e. $X'' = X$). The set $\mathfrak{B}(\mathbb{K})$ of all formal concepts of the formal context $\mathbb{K}$ can be ordered by $(A_1, B_1) \le (A_2, B_2) : \iff A_1 \subseteq A_2$   (or equivalently, $B_2 \subseteq B_1$).

The poset $\underline{\mathfrak{B}}(\mathbb{K}) := (\mathfrak{B}(\mathbb{K}), \le)$ is a complete lattice, called the **concept lattice** of the context $\mathbb{K}$. Conversely each complete lattice is isomorphic to a concept lattice. This basic theorem on concept lattice (Theorem 3[4]) is a template for contextual representation problems. The lattice operations $\wedge$ (meet) and $\vee$ (join) can be interpreted as a logical conjunction and a logical disjunction for concepts, and are given by:

$$\text{meet:} \quad (A_1, B_1) \wedge (A_2, B_2) = \left(A_1 \cap A_2, (A_1 \cap A_2)'\right)$$
$$\text{join:} \quad (A_1, B_1) \vee (A_2, B_2) = \left((B_1 \cap B_2)', B_1 \cap B_2\right).$$

To extent FCA to contextual logic, we need to define the negation of a concept. Unfortunately, the complement of a closed subset is not always closed. To preserve the correspondence between set complementation and negation, the notions of concept is extended to that of protoconcept.

Let $\mathbb{K} := (G, M, I)$ be a formal context and $A \subseteq G, B \subseteq M$. The pair $(A, B)$ is called a **semi-concept** if $A' = B$ or $B' = A$, and a **protoconcept** if $A'' = B'$. The set of all semi-concepts of $\mathbb{K}$ is denoted by $\mathfrak{h}(\mathbb{K})$, and that of all protoconcepts by $\mathfrak{P}(\mathbb{K})$. Note that each semi-concept is a protoconcept; i.e. $\mathfrak{h}(\mathbb{K}) \subseteq \mathfrak{P}(\mathbb{K})$. Meet and join of protoconcepts are then defined, similar as above for concepts. A negation (resp. opposition) is defined by taking the complement on objects (resp. attributes). More precisely, for protoconcepts $(A_1, B_1)$, $(A_2, B_2)$, $(A, B)$ of $\mathbb{K}$ we define the operations:

$$\text{meet:} \quad (A_1, B_1) \sqcap (A_2, B_2) := (A_1 \cap A_2, (A_1 \cap A_2)')$$
$$\text{join:} \quad (A_1, B_1) \sqcup (A_2, B_2) := ((B_1 \cap B_2)', B_1 \cap B_2)$$
$$\text{negation:} \quad \neg(A, B) := (G \setminus A, (G \setminus A)')$$
$$\text{opposition:} \quad \lrcorner(A, B) := ((M \setminus B)', M \setminus B)$$
$$\text{nothing:} \quad \bot := (\emptyset, M)$$
$$\text{all:} \quad \top := (G, \emptyset).$$

The algebra $\mathfrak{P}(\mathbb{K}) := (\mathfrak{P}(\mathbb{K}), \sqcap, \sqcup, \neg, \lrcorner, \bot, \top)$ is called the **algebra of proto-concepts** of $\overline{\mathbb{K}}$. Note that applying any operation above on protoconcepts gives a semi-concept as result. Therefore $\mathfrak{H}(\mathbb{K})$ is a subalgebra of $\mathfrak{P}(\mathbb{K})$. For the structural analysis of $\mathfrak{P}(\mathbb{K})$, we split $\mathfrak{H}(\mathbb{K})$ in $\sqcap$-**semiconcepts** and $\sqcup$-**semiconcepts**, $\mathfrak{P}(\mathbb{K})_\sqcap := \{(A, \overline{A'}) \mid A \subseteq G\}$ and $\mathfrak{P}(K)_\sqcup := \{(B', B) \mid B \subseteq M\}$, and set $x \vee y := \neg(\neg x \sqcap \neg y)$, $x \wedge y = \lrcorner(\lrcorner x \sqcup \lrcorner y)$, $\overline{\top} := \neg \bot$ and $\bot := \lrcorner \top$ for $x, y \in \mathfrak{P}(\mathbb{K})$. $\mathfrak{P}(\mathbb{K})_\sqcap := (\mathfrak{P}(\mathbb{K})_\sqcap, \sqcap, \vee, \neg, \bot, \neg\bot)$ (resp. $\mathfrak{P}(\mathbb{K})_\sqcup := (\mathfrak{P}(\mathbb{K})_\sqcup, \wedge, \sqcup, \lrcorner, \lrcorner\top, \top))$) is a Boolean algebra isomorphic (resp. anti-isomorphic) to the powerset algebra of $G$ (resp. $M$).

**Theorem 1.** *[10] The following equations hold in $\mathfrak{P}(\mathbb{K})$:*

*(1a)* $(x \sqcap x) \sqcap y = x \sqcap y$

*(2a)* $x \sqcap y = y \sqcap x$

*(3a)* $x \sqcap (y \sqcap z) = (x \sqcap y) \sqcap z$

*(4a)* $\neg(x \sqcap x) = \neg x$

*(5a)* $x \sqcap (x \sqcup y) = x \sqcap x$

*(6a)* $x \sqcap (y \vee z) = (x \sqcap y) \vee (x \sqcap z)$

*(7a)* $x \sqcap (x \vee y) = x \sqcap x$

*(8a)* $\neg\neg(x \sqcap y) = x \sqcap y$

*(9a)* $x \sqcap \neg x = \bot$

*(10a)* $\neg\bot = \top \sqcap \top$

*(11a)* $\neg\top = \bot$

*(1b)* $(x \sqcup x) \sqcup y = x \sqcup y$

*(2b)* $x \sqcup y = y \sqcup x$

*(3b)* $x \sqcup (y \sqcup z) = (x \sqcup y) \sqcup z$

*(4b)* $\lrcorner(x \sqcup x) = \lrcorner x$

*(5b)* $x \sqcup (x \sqcap y) = x \sqcup x$

*(6b)* $x \sqcup (y \wedge z) = (x \sqcup y) \wedge (x \sqcup z)$

*(7b)* $x \sqcup (x \wedge y) = x \sqcup x$.

*(8b)* $\lrcorner\lrcorner(x \sqcup y) = x \sqcup y$

*(9b)* $x \sqcup \lrcorner x = \top$

*(10b)* $\lrcorner\top = \bot \sqcup \bot$

*(11b)* $\lrcorner\bot = \top$

*(12)* $(x \sqcap x) \sqcup (x \sqcap x) = (x \sqcup x) \sqcap (x \sqcup x)$

A **double Boolean algebra** (dBa) is an algebra $\underline{D} := (D; \sqcap, \sqcup, \neg, \lrcorner, \bot, \top)$ of type $(2, 2, 1, 1, 0, 0)$ that satisfies the equations in Theorem 1. Wille showed that these equations generate the equational theory of protoconcept algebras [10]. A dBa $\underline{D}$ is called **pure** if it satisfies $x \sqcap x = x$ or $x \sqcup x = x$, for all $x \in D$. In fact, $\mathfrak{H}(\mathbb{K})$ is a pure dBa. A quasi-order $\sqsubseteq$ is defined on dBas by: $x \sqsubseteq y :$ $\iff$ $x \sqcap y = x \sqcap x$ and $x \sqcup y = y \sqcup y$. A dBa is **regular** if $\sqsubseteq$ is an order relation. We set $D_\sqcap := \{x \in D : x \sqcap x = x\}$ and $D_\sqcup := \{x \in D : x \sqcup x = x\}$.

The algebra $\underline{D}_\sqcap := (D_\sqcap, \sqcap, \vee, \neg, \bot, \neg\bot)$ (resp. $\underline{D}_\sqcup = (D_\sqcup, \sqcup, \wedge, \lrcorner, \top, \lrcorner\top)$) is a Boolean algebra. In addition $x \sqsubseteq y$ iff $x \sqcap x \sqsubseteq y \sqcap y$ and $x \sqcup x \sqsubseteq y \sqcup y$, for all $x, y \in D$ [10]. A dBa $\underline{D}$ is called **complete** if and only if its Boolean algebras $\underline{D}_\sqcap$ and $\underline{D}_\sqcup$ are complete. Additional known properties of dBas, that we will need in these notes are put together in the next proposition.

**Proposition 1.** *[5, 7, 10] Let $\underline{D}$ be a dBa and $x, y, a \in D$. Then:*

1. $\bot \sqsubseteq x$ and $x \sqsubseteq \top$.
2. $x \sqcap y \sqsubseteq x, y \sqsubseteq x \sqcup y$.
3. $x \sqsubseteq y \implies \begin{cases} x \sqcap a \sqsubseteq y \sqcap a \\ x \sqcup a \sqsubseteq y \sqcup a. \end{cases}$
4. $\neg(x \vee y) = \neg x \sqcap \neg y$.
5. $\neg(x \sqcap y) = \neg x \vee \neg y$.
6. $x \sqsubseteq \lrcorner y \iff y \sqsubseteq \lrcorner x$.
7. $\neg\neg x = x \sqcap x$ and $\lrcorner\lrcorner x = x \sqcup x$.
8. $\neg x, x \vee y \in D_\sqcap$ and $\lrcorner x, x \wedge y \in D_\sqcup$.
9. $x \sqsubseteq y \iff \begin{cases} \neg y \sqsubseteq \neg x \\ \lrcorner y \sqsubseteq \lrcorner x. \end{cases}$
10. $\lrcorner(x \wedge y) = \lrcorner x \sqcup \lrcorner y$.
11. $\lrcorner(x \sqcup y) = \lrcorner x \wedge \lrcorner y$.
12. $\neg x \sqsubseteq y \iff \neg y \sqsubseteq x$.
13. $\neg x \sqcap \neg x = \neg x$ and $\lrcorner x \sqcup \lrcorner x = \lrcorner x$.

## 3    Filters and Ideals of a Double Boolean Algebra

In this section, we study filters and ideals of dBas with respect to their order structures and generating subsets. The goal is to extent some results from Boolean algebras to dBas. Let $\underline{D}$ be a dBa. A nonempty subset $F$ of $D$ is called **filter** if for all $x, y \in D$, it holds: $x, y \in F \implies x \sqcap y \in F$ and $(x \in F, x \sqsubseteq y) \implies y \in F$. **Ideals** of dBas are defined dually. We denote by $\mathcal{F}(\underline{D})$ (resp. $\mathcal{I}(\underline{D})$) the set of filters (resp. ideals) of the dBa $\underline{D}$. Both sets are each closed under intersection [7]. Note that $\mathcal{F}(\underline{D}) \cap \mathcal{I}(\underline{D}) = \{D\}$. For $X \subseteq D$, the smallest filter (resp. ideal) containing $X$, denoted by Filter$\langle X \rangle$ (resp. Ideal$\langle X \rangle$), is the intersection of all filters (resp. ideals) containing $X$, and is called the **filter** (resp. **ideal**) **generated by** $X$. A **principal filter** (resp. **ideal**) is a filter (resp. ideal) generated by a singleton. In that case we omit the curly brackets and set F$(x)$ := Filter$\langle \{x\} \rangle$, and I$(x)$ := Ideal$\langle \{x\} \rangle$. A set $F_0$ is a **basis of a filter** $F$ if $F_0 \subseteq F$ and $F = \{y \in D : x \sqsubseteq y$ for some $x \in F_0\}$. **Basis of ideals** are defined similarly.

**Lemma 1.** *[10] Let $F \in \mathcal{F}(\underline{D})$. Then $F \cap D_\sqcap$ and $F \cap D_\sqcup$ are filters of the Boolean algebras $\underline{D}_\sqcap$ and $\underline{D}_\sqcup$, respectively. Each filter of the Boolean algebra $\underline{D}_\sqcap$ is a base of some filter of $\underline{D}$, and each ideal of $\underline{D}_\sqcup$ is a base of some ideal of $\underline{D}$.*

To describe the ideal (resp. filter) generated by $X \subseteq D$, we need the following lemma, which is a slight generalization of Proposition 1. 3. in [7].

**Lemma 2.** *Let $\underline{D}$ be a dBa and $a, b, c, d \in D$. Then if $a \sqsubseteq b$ and $c \sqsubseteq d$, then $a \sqcap c \sqsubseteq b \sqcap d$ and $a \sqcup c \sqsubseteq b \sqcup d$. i.e. $\sqcap$ and $\sqcup$ are compatible with $\sqsubseteq$.*

*Proof.* We assume $a \sqsubseteq b$ and $c \sqsubseteq d$. Then

$$(a \sqcap c) \sqcap (b \sqcap d) = (a \sqcap b) \sqcap (c \sqcap d) = (a \sqcap a) \sqcap (c \sqcap c) = (a \sqcap c) \sqcap (a \sqcap c).$$
$$(a \sqcap c) \sqcup (b \sqcap d) = ((a \sqcap a) \sqcap c) \sqcup (b \sqcap d) = [(a \sqcap b) \sqcap c] \sqcup (b \sqcap d)$$
$$= [(a \sqcap b) \sqcap (c \sqcap c)] \sqcup (b \sqcap d) = [(a \sqcap b) \sqcap (c \sqcap d)] \sqcup (b \sqcap d)$$
$$= [(b \sqcap d) \sqcap (a \sqcap c)] \sqcup (b \sqcap d) = (b \sqcap d) \sqcup (b \sqcap d).$$

Thus $a \sqcap c \sqsubseteq b \sqcap d$. The rest, $a \sqcup c \sqsubseteq b \sqcup d$, is proved similarly. □

For $F \in \mathcal{F}(\underline{D})$, $I \in \mathcal{I}(\underline{D})$ and $a \in D$, we have $a \in F$ iff $a \sqcap a \in F$ and $a \in I$ iff $a \sqcup a \in I$. Therefore $\text{Filter}\langle a \rangle = \text{Filter}\langle a \sqcap a \rangle$ and $\text{Ideal}\langle a \rangle = \text{Ideal}\langle a \sqcup a \rangle$. To prove the prime ideal theorem for dBas, a description of filter (resp. ideal) generated by an element $w$ and a filter $F$ (resp. ideal $I$) was given in [7] as follows:

$$\text{Filter}\langle F \cup \{w\} \rangle = \{x \in D : w \sqcap b \sqsubseteq x \text{ for some } b \in F\},$$
$$\text{Ideal}\langle J \cup \{w\} \rangle = \{x \in D : x \sqsubseteq w \sqcup b \text{ for some } b \in J\}.$$

In the following proposition, we extent this description to arbitrary subsets.

**Proposition 2.** *Let $\underline{D}$ be a dBa, $\emptyset \neq X \subseteq D$, $F_1, F_2 \in \mathcal{F}(\underline{D})$ and $I_1, I_2 \in \mathcal{I}(\underline{D})$.*

*(a)* $\text{I}(a) = \{x \in D \mid x \sqsubseteq a \sqcup a\}$ *and* $\text{F}(a) = \{x \in D \mid a \sqcap a \sqsubseteq x\}$.
*(b)* $\text{Ideal}\langle \emptyset \rangle = \text{I}(\bot) = \{x \in D \mid x \sqsubseteq \bot \sqcup \bot\}$ *and*
      $\text{Filter}\langle \emptyset \rangle = \text{F}(\top) = \{x \in D \mid \top \sqcap \top \sqsubseteq x\}$.
*(c)* $\text{Ideal}\langle X \rangle = \{x \in D \mid x \sqsubseteq b_1 \sqcup \ldots \sqcup b_n \text{ for some } b_1, \ldots, b_n \in X, n \geq 1\}$.
*(d)* $\text{Filter}\langle X \rangle = \{x \in D \mid x \sqsupseteq b_1 \sqcap \ldots \sqcap b_n \text{ for some } b_1, \ldots, b_n \in X, n \geq 1\}$.
*(e)* $\text{Ideal}\langle I_1 \cup I_2 \rangle = \{x \in D \mid x \sqsubseteq i_1 \sqcup i_2 \text{ for some } i_1 \in I_1 \text{ and } i_2 \in I_2\}$.
*(f)* $\text{Filter}\langle F_1 \cup F_2 \rangle = \{x \in D \mid f_1 \sqcap f_2 \sqsubseteq x \text{ for some } f_1 \in F_1 \text{ and } f_2 \in F_2\}$.

*Proof.* (e) follows from (c), with the facts that $\sqcup$ is commutative, associative, and ideals are closed under $\sqcup$. (d) and (f) are similarly to (c) and (e). Let $\underline{D}$ de a dBa, and $\emptyset \neq X \subseteq D$.

(a) Let $a \in D$. We will show that the set $J := \{x \in D \mid x \sqsubseteq a \sqcup a\}$ is an ideal containing $a$, and is contained in any ideal containing $a$. From $a \sqsubseteq a \sqcup a$ we get $a \in J$. If $x, y$ are in $J$, then we have $x \sqsubseteq a \sqcup a$ and $y \sqsubseteq a \sqcup a$. Thus $x \sqcup y \sqsubseteq (a \sqcup a) \sqcup (a \sqcup a) = a \sqcup a$ (by Lemma 2 and Theorem 1(1b)). If $y \in J$ and $x \sqsubseteq y$, then $x \sqsubseteq y \sqsubseteq a \sqcup a$, and $x \sqsubseteq a \sqcup a$; i.e. $x \in J$. If $G$ is an ideal and $a \in G$, then $a \sqcup a \in G$. Each $x \in J$ satisfies $x \sqsubseteq a \sqcup a$, and is also in $G$. Thus $J = \text{I}(a)$. The description of $\text{F}(a)$ is obtained dually.

(b) From $\bot \sqsubseteq x$ for all $x \in D$ (Proposition 1.1.) and that each ideal is non empty, it follows that $\bot$ is in each ideal. Thus $\text{Ideal}\langle \emptyset \rangle = \text{I}(\bot) = \{x \in D \mid x \sqsubseteq \bot \sqcup \bot\}$. The last equality follows from (a). The equality $\text{Filter}\langle \emptyset \rangle = \text{F}(\top) = \{x \in D \mid \top \sqcap \top \sqsubseteq x\}$ is proved dually.

(c) We set $J := \{x \in D \mid x \sqsubseteq b_1 \sqcup \ldots \sqcup b_n \text{ for some } b_1, \ldots, b_n \in X, n \geq 1\}$. We will show that $J$ is the smallest ideal that contains $X$. For $x \in X$, we have $x \sqsubseteq x \sqcup x$, and $x \in J$. If $y \in D$, $x \in J$ and $y \sqsubseteq x$, then $y \sqsubseteq x \sqsubseteq b_1 \sqcup \ldots \sqcup b_n$ for an $n \geq 1$ and $b_1, \ldots, b_n \in X$. By transitivity of $\sqsubseteq$, we get $y \in J$. Now, if $a, b \in J$, then $a \sqsubseteq a_1 \sqcup \ldots \sqcup a_n$ and $b \sqsubseteq b_1 \sqcup \ldots \sqcup b_m$ for some $n, m \geq 1$ and $a_i, b_i \in J$, $1 \leq i \leq n, 1 \leq j \leq m$. Therefore $a \sqcup b \sqsubseteq a_1 \sqcup \ldots \sqcup a_n \sqcup b_1 \sqcup \ldots \sqcup b_m$, by Lemma 2. Thus $a \sqcup b \in J$. It is easy to see that any ideal containing $X$ also contains $J$. □

Recall that $\mathcal{F}(\underline{D})$ and $\mathcal{I}(\underline{D})$ are closure systems, and thus complete lattices. The meet is then the intersection, and the join the filter and ideal generated by the union. For $F_1, F_2 \in \mathcal{F}(\underline{D})$ we have $F_1 \wedge F_2 = F_1 \cap F_2$ and $F_1 \vee F_2 = \mathrm{Filter}\langle F_1 \cup F_2 \rangle$. Similarly we have $I_1 \wedge I_2 = I_1 \cap I_2$ and $I_1 \vee I_2 = \mathrm{Ideal}\langle I_1 \cup I_2 \rangle$ for $I_1, I_2 \in \mathcal{I}(\underline{D})$. We set $\underline{\mathcal{F}}(\underline{D}) := (\mathcal{F}(\underline{D}); \vee, \wedge, \mathrm{F}(\top), D)$ and $\underline{\mathcal{I}}(\underline{D}) := (\mathcal{I}(\underline{D}); \vee, \wedge, \mathrm{I}(\bot), D)$.

Now, we study some properties of the set $\mathcal{F}_p := \{\mathrm{F}(a) \mid a \in D\}$ of principal filters of $\underline{D}$, and the set $\mathcal{I}_p := \{\mathrm{I}(a) \mid a \in D\}$ of principal ideals of $\underline{D}$. Observe that $\mathcal{F}_p = \{\mathrm{F}(a) \mid a \in D_\sqcap\}$ and $\mathcal{I}_p = \{\mathrm{I}(a) \mid a \in D_\sqcup\}$, since $\mathrm{F}(a) = \mathrm{F}(a \sqcap a)$ and $\mathrm{I}(a) = \mathrm{I}(a \sqcup a)$ for any $a \in D$. The following proposition gives distributivity-like properties of dBas.

**Proposition 3.** *Let $\underline{D}$ be a dBa and $a, b, c, d \in D$. We have:*

(i) $a \vee (b \sqcap c) = (a \vee b) \sqcap (a \vee c)$.　　(iv) $a \wedge (a \sqcup b) = a \sqcup a$.

(ii) $a \wedge (b \sqcup c) = (a \wedge b) \sqcup (a \wedge c)$.　　(v) $(a \sqcap a) \vee (b \sqcap b) = a \vee b$.

(iii) $a \vee (a \sqcap b) = a \sqcap a$.　　(vi) $(a \sqcup a) \wedge (b \sqcup b) = a \wedge b$.

(vii) *The binary operations $\vee$ and $\wedge$ are compatible with $\sqsubseteq$, that is, if $a \sqsubseteq b$ and $c \sqsubseteq d$, then $a \vee c \sqsubseteq b \vee d$ (1)　and $a \wedge c \sqsubseteq b \wedge d$ (2) .*

*Proof.* (ii), (iv) and (vi) are the dual of (i), (iii) and (v), respectively.

(i)

$$a \vee (b \sqcap c) \overset{\mathrm{def}}{=} \neg(\neg a \sqcap \neg(b \sqcap c)) = \neg(\neg a \sqcap (\neg b \vee \neg c)) \text{ (by (5), Proposition 1)}$$
$$= \neg((\neg a \sqcap \neg b) \vee (\neg a \sqcap \neg c)) \text{ (by axiom 6a))}$$
$$= \neg(\neg(a \vee b) \vee \neg(a \vee c)) = \neg[\neg[(a \vee b) \sqcap (a \vee c)]]$$
$$= \neg\neg[(a \vee b) \sqcap (a \vee c)] = (a \vee b) \sqcap (a \vee c) \text{ (by axiom (8.a))}$$

(iii)

$$a \vee (a \sqcap b) = \neg(\neg a \sqcap \neg(a \sqcap b)) = \neg(\neg a \sqcap (\neg a \vee \neg b)) \text{ (by (5), Proposition 1)}$$
$$= \neg(\neg a \sqcap \neg a) = \neg(\neg a) = a \sqcap a = a \vee a$$

(v) $(a \sqcap a) \vee (b \sqcap b) = \neg(\neg(a \sqcap a) \sqcap \neg(b \sqcap b)) = \neg(\neg a \sqcap \neg b) = a \vee b$.

(vii) We assume $a \sqsubseteq b$ and $c \sqsubseteq d$. Then by the definition of $\sqsubseteq$ we have

$$a \sqcap b = a \sqcap a \text{ (1.1)}, \quad a \sqcup b = b \sqcup b \text{ (1.2)}, \quad c \sqcap d = c \sqcap c \text{ (1.3)}, \quad c \sqcup d = d \sqcup d \text{ (1.4)}.$$

We need to show that $a \vee c \sqsubseteq b \vee d$ (1) and $a \wedge c \sqsubseteq b \wedge d$ (2).

$$(a \vee c) \sqcap (b \vee d) = [(a \vee c) \sqcap b] \vee [(a \vee c) \sqcap d] \text{ (by axiom (6a))}$$
$$= [(b \sqcap a) \vee (b \sqcap c)] \vee [(d \sqcap a) \vee (d \sqcap c)]$$
$$= [(a \sqcap a) \vee (b \sqcap c)] \vee [(a \sqcap d) \vee (c \sqcap c)] \text{ (by (1.1) and (1.3))}$$
$$= (a \sqcap a) \vee (c \sqcap c) \vee (b \sqcap c) \vee (a \sqcap d) \text{ ($\vee$ associative in $D_\sqcap$)}$$
$$= (a \vee c) \vee (b \sqcap c) \vee (a \sqcap d) \text{ (by (v) )}$$
$$= (a \vee c) \vee (b \sqcap c \sqcap c) \vee (a \sqcap a \sqcap d) \text{ (by axiom (1.a))}$$
$$= (a \vee c) \vee (b \sqcap c \sqcap d) \vee (a \sqcap (b \sqcap d))$$
$$= (a \vee c) \vee ([(b \sqcap d) \sqcap c] \vee [(b \sqcap d) \sqcap a])$$
$$= (a \vee c) \vee [(b \sqcap d) \sqcap (a \vee c)] \text{ (by axiom (6a))}$$
$$= (a \vee c) \sqcap (a \vee c) \text{ (by (iii))}$$

Therefore (1) holds. The proof of (2) is similarly to (1).　　□

**Lemma 3.** *Let $\underline{D}$ be a dBa and $a, b$ in $D$. Then*

*(i)* $a \sqsubseteq b \implies F(a) \supseteq F(b)$. *The converse holds if* $a, b \in D_\sqcap$.
*(ii)* $a \sqsubseteq b \implies I(a) \subseteq I(b)$. *The converse holds if* $a, b \in D_\sqcup$.
*(iii)* $F(a \sqcup b) \subseteq F(a) \cap F(b) \subseteq F(a), F(b) \subseteq F(a \sqcap b)$
*(iv)* $I(a \sqcap b) \subseteq I(a) \cap I(b) \subseteq I(a), I(b) \subseteq I(a \sqcup b)$.

*Proof.* (ii) and (iv) are dual of (i) and (iii). Let $a, b \in D$.

(i) We assume that $a \sqsubseteq b$. Then $a \sqcap a \sqsubseteq b \sqcap b \sqsubseteq b$. If $x \in F(b)$ then $b \sqcap b \sqsubseteq x$. Since $\sqsubseteq$ is transitive, we get $a \sqcap a \sqsubseteq x$, and $x \in F(a)$. Thus $F(b) \subseteq F(a)$. Conversely, if $F(b) \subseteq F(a)$ then $b \in F(a)$ and $a \sqcap a \sqsubseteq b$, which is equivalent to $a \sqsubseteq b$ if $a \in D_\sqcap$.
(iv) $a \sqcap b \sqsubseteq a, b \sqsubseteq a \sqcup b \implies F(a \sqcup b) \subseteq F(a) \cap F(b) \subseteq F(a), F(b) \subseteq F(a \sqcap b)$. $\quad\square$

Note that $F(\bot) = D$ and $I(\top) = D$. The next result shows that the principal filters form a bounded sublattice of the lattice of all filters.

**Proposition 4.** *Let $\underline{D}$ be a dBa, and $a_1, \ldots, a_n, a, b, c \in D$. Then:*

(1) $\bigvee\limits_{i=1}^{n} F(a_i) = F(\bigsqcap\limits_{i=1}^{n} a_i) = $
Filter$\langle \{a_1, \ldots, a_n\} \rangle$.

(2) $\bigcap\limits_{i=1}^{n} F(a_i) = F(\bigvee\limits_{i=1}^{n} a_i)$.

(3) $\bigvee\limits_{i=1}^{n} I(a_i) = I(\bigsqcup\limits_{i=1}^{n} a_i) = $
Ideal$\langle \{a_1, \ldots, a_n\} \rangle$.

(4) $\bigcap\limits_{i=1}^{n} I(a_i) = I(\bigsqcap\limits_{i=1}^{n} a_i)$.

(5) $b \in F(\neg a) \implies \neg b \in I(a)$.

(6) $b \in I(\lrcorner a) \implies \lrcorner b \in F(a)$.

(7) $F(a) \wedge F(\neg a) = F(\top)$.

(8) $I(a) \vee I(\lrcorner a) = I(\top)$.

(9) $F(a) \vee F(\neg a) = F(\bot)$.

(10) $I(a) \wedge I(\lrcorner a) = I(\bot)$.

(11) $F(a) \wedge (F(b) \vee F(c)) = (F(a) \wedge F(b)) \vee (F(a) \wedge F(c))$.
(12) $I(a) \vee (I(b) \wedge I(c)) = (I(a) \vee I(b)) \wedge (I(a) \vee I(c))$.

*Proof.* (1)–(4) can be proved by induction. We will give a proof for $n = 2$. The rest is obtained dually. Let $a, b, x \in D$.

(1) From $a \sqcap b \sqsubseteq a, b$ we get $F(a), F(b) \subseteq F(a \sqcap b)$ and $F(a) \vee F(b) \subseteq F(a \sqcap b)$. If $x \in F(a \sqcap b)$, then $a \sqcap b \sqsubseteq x$. But $a \in F(a)$ and $b \in F(b)$ imply $a \sqcap b \in F(a) \vee F(b)$, and yields $x \in F(a) \vee F(b)$. Thus $F(a \sqcap b) = F(a) \vee F(b)$.
(2) $x \in F(a \vee b) \iff (a \vee b) \sqcap (a \vee b) \sqsubseteq x \iff a \vee b \sqsubseteq x$

$\iff (a \sqcap a) \vee (b \sqcap b) \sqsubseteq x \iff a \sqcap a \sqsubseteq x$ and $b \sqcap b \sqsubseteq x$

$\iff x \in F(a)$ and $x \in F(b) \iff x \in F(a) \cap F(b)$.

(5) $b \in F(\neg a) \iff \neg a \sqcap \neg a \sqsubseteq b \iff \neg a \sqsubseteq b \iff \neg b \sqsubseteq a \implies \neg b \in I(a)$.
(7) $F(a) \wedge F(\neg a) = F(a \vee \neg a) = F(\neg(\neg a \sqcap \neg\neg a)) = F(\neg \bot) = F(\top \sqcap \top) = F(\top)$.
(9) $F(a) \vee F(\neg a) = F(a \sqcap \neg a) = F(\bot)$.
(11) $F(a) \vee (F(b) \wedge F(c)) = F(a) \vee (F(b \vee c)) = F(a \sqcap (b \vee c))$

$= F((a \sqcap b) \vee (a \sqcap c)) = F(a \sqcap b) \wedge F(a \sqcap c)$

$= (F(a) \vee F(b)) \wedge (F(a) \vee F(c))$. $\quad\square$

It is known that if $\underline{L}$ is a complete lattice, then its set $\mathcal{I}(\underline{L})$ (resp. $\mathcal{F}(\underline{L})$) of ideals (resp. filters) is a lattice and its set of principal ideals (resp. filters) is a sublattice of $\mathcal{I}(\underline{L})$ (resp. $\mathcal{F}(\underline{L})$) isomorphic to $\underline{L}$. Similar results hold for dBas. For $a \in D$ we set, $F(a)^c := F(\neg a)$, $I(a)^c := I(\lrcorner a)$, $\underline{\mathcal{F}}_p(\underline{D}) := (\mathcal{F}_p(\underline{D}); \vee, \wedge,^c, F(\top), F(\bot))$ and $\underline{\mathcal{I}}_p(\underline{D}) := (\mathcal{I}_p(\underline{D}); \vee, \wedge,^c, I(\bot), I(\top))$.

**Theorem 2.** *Let $\underline{D}$ be a double Boolean algebra. Then $\underline{\mathcal{F}}_p(\underline{D})$ and $\underline{\mathcal{I}}_p(\underline{D})$ are Boolean algebras, and bounded sublattices of $\underline{\mathcal{F}}(\underline{D})$ and $\underline{\mathcal{I}}(\underline{D})$, respectively. The map $\varphi : D_\sqcap \rightarrow \mathcal{F}_p(\underline{D}), a \mapsto F(a)$ is an anti-isomorphism, and $\psi : D_\sqcup \rightarrow \mathcal{I}_p(\underline{D}), a \mapsto I(a)$ is an isomorphism of Boolean algebras.*

*Proof.* From Proposition 4, we know that $\mathcal{F}_p(\underline{D})$ is a bounded sublattice of $\mathcal{F}(\underline{D})$, is distributive, and $F(\neg a)$ is a complement of $F(a)$ for each $a \in D$. Moreover, $\varphi(\neg a) = F(\neg a) = F(a)^c = \varphi(a)^c$, $\varphi(a \sqcap b) = F(a \sqcap b) = F(a) \vee F(b) = \varphi(a) \vee \varphi(b)$, and $\varphi(a \vee b) = F(a \vee b) = F(a) \wedge F(b) = \varphi(a) \wedge \varphi(b)$. By definition $\varphi$ is onto. Let $a, b \in D_\sqcap$ with $\varphi(a) = \varphi(b)$. Then $F(a) = F(b)$ and $a = b$. Thus $\varphi$ is an anti-isomorphism. The proof for $\psi$ follows dually. □

**Corollary 1.** *If $\underline{D}$ is a complete dBa, then $\mathcal{F}(\underline{D})$ and $\mathcal{I}(\underline{D})$ are Boolean algebras.*

Let $L$ be a lattice. An element $a \in L$ is compact iff whenever $\bigvee A$ exists and $a \leq \bigvee A$ for $A \subseteq L$, then $a \leq \bigvee B$ for some finite $B \subseteq A$. $L$ is compactly generated iff every element in $L$ is a supremum of compact elements. $L$ is algebraic if $L$ is complete and compactly generated [3].

**Proposition 5.** *The compact elements of $\mathcal{F}(\underline{D})$ (resp. $\mathcal{I}(\underline{D})$) are the finitely generated filters (resp. ideals). Moreover, $\mathcal{F}(\underline{D})$ and $\mathcal{I}(\underline{D})$ are algebraic lattices.*

*Proof.* Let $X := \{a_1, \ldots, a_n\} \subseteq D$ and $F := \text{Filter}\langle X \rangle \subseteq \bigvee_{i \in I} F_i$. Each $a_i$ is in a $F_{j_i}$, $1 \leq i \leq n$ and $j_i \in I$. Thus $F(a_i) \subseteq F_{j_i}$ and $F = \bigvee_{i=1}^{i=n} F(a_i) \subseteq \bigvee_{i=1}^{i=n} F_{j_i}$.

Let $G$ be a compact element of $\mathcal{F}(\underline{D})$. From $G \subseteq \bigvee_{a \in G} F(a)$ we get $G \subseteq \bigvee\{F(a_1), \ldots, F(a_n) : a_1, \ldots, a_n \in G\} = \text{Filter}\langle \{a_1, \ldots, a_n\} \rangle = G$.

Let $(F_i)_{i \in I}$ be a family of filters of $\underline{D}$. Then $\bigvee_{i \in I} F_i = \bigcup_{(i_1, \ldots, i_n) \in I^*} (F_{i_1} \vee \ldots \vee F_{i_n})$ where $I^*$ is the set of finite tuple $(i_1, \ldots, i_n) \in I^n, n \geq 1, n \in \mathbb{N}$. Dually one can see that $\mathcal{I}(\underline{D})$ is algebraic. □

In Boolean algebras there are several equivalent definitions of prime filters. These definitions can be carried over to dBas. To solve the equational theory problem for protoconcet algebras, Rudolf Wille introduced in [10] the set $\mathfrak{F}_P(\underline{D})$ of filters $F$ of $\underline{D}$ whose intersections $F \cap D_\sqcap$ are prime filters of the Boolean algebra $\underline{D}_\sqcap$, and the set $\mathfrak{I}_P(\underline{D})$ of ideals $I$ of $\underline{D}$ whose intersections $I \cap D_\sqcup$ are prime ideals of the Boolean algebra $\underline{D}_\sqcup$. To prove the prime ideal theorem for double Boolean algebras, Léonard Kwuida introduced in [7] **primary filters** as proper filters $F$ for which $x \in F$ or $\neg x \in F$, for each $x \in D$. Dually a **primary** ideal is a proper ideal $I$ for which $x \in I$ or $\lrcorner x \in I$, for each $x \in D$. We denote by $\mathfrak{F}_{pr}(\underline{D})$ the set of primary filters of $\underline{D}$, and by $\mathfrak{I}_{pr}(\underline{D})$ its set of primary ideals. In [5] Prosenjit Howlader and Mohua Banerjee showed that $\mathfrak{F}_P(\underline{D}) = \mathfrak{F}_{pr}(\underline{D})$ and $\mathfrak{I}_P(\underline{D}) = \mathfrak{I}_{pr}(\underline{D})$. The following theorem extends the above results.

**Theorem 3.** *Let $\underline{D}$ be a double Boolean algebra. Then $\mathcal{F}(\underline{D}_\sqcap)$ and $\mathcal{F}(\underline{D})$ are isomorphic lattices. Dually, $\mathcal{I}(\underline{D}_\sqcup)$ and $\mathcal{I}(\underline{D})$ are isomorphic lattices.*

*Proof.* The maps $\Phi : \mathcal{F}(\underline{D}_\sqcap) \to \mathcal{F}(\underline{D})$, $E \mapsto \Phi(E) := \{x \in D, \exists u \in E, u \sqsubseteq x\}$ and $\Psi : I(\underline{D}_\sqcup) \to I(\underline{D}), I \mapsto \Psi(I) = \{x \in D, \exists x_0 \in I, x \sqsubseteq x_0\}$ are isomorphisms.

## 4    Congruence on Double Boolean Algebras

**Definition 1.** *[7] A double Boolean algebra is trivial iff $\top \sqcap \top = \bot \sqcup \bot$.*

Let $\underline{D}$ be a dBa and $\sqsubseteq$ the quasi-order defined above. Let $\tau$ be the equivalence relation defined on $D$ by $x\tau y : \iff x \sqsubseteq y$ and $y \sqsubseteq x$. It is known that $\tau$ is a congruence on $\underline{D}$ and from [10] $\tau = \{(x,x), x \in D\} \cup \{(x_{\sqcap\sqcup}, x_{\sqcup\sqcap}), x \in D\} \cup \{(x_{\sqcup\sqcap}, x_{\sqcap\sqcup}), x \in D\}$. $(D/\tau; \tilde{\sqcup}, \tilde{\sqcap}, \tilde{\lnot}, \tilde{\lrcorner}, \tilde{\top}, \tilde{\bot})$ is the quotient algebra of $\underline{D}$ by $\tau$. Let $\tilde{\sqsubseteq}$ be the quasi-order defined on $D/\tau$. One can show that for any $x, y \in D$, $x\tilde{\sqsubseteq}y$ iff there exists $u \in [x]_\tau, v \in [y]_\tau$ such that $u \sqsubseteq v$.

It is known that for every boolean algebra $B$ and a congruence $\theta$ on $B$, the set $[\bot]_\theta$ is an ideal on $B$ and the set $[\top]_\theta$ is a filter on $B$ (see [3]). We consider the above congruence $\tau$ defined by $x\tau y$ iff $x \sqcap x = y \sqcap y$ and $x \sqcup x = y \sqcup y$. On the double Boolean algebra $\underline{D_3}$ of **Kwuida**, where $D_3 = \{\bot, a, \top\}$, $\bot \sqsubseteq a \sqsubseteq \top$, $a \sqcap a = a \sqcup a = a = \bot \sqcup \bot = \top \sqcap \top$. We have $[\bot]_\tau = \{\bot\}$ and $[\top]_\tau = \{\top\}$, it follows that $\bot \sqcup \bot = a \notin [\bot]_\tau$, so $[\bot]_\tau$ is not an ideal of $\underline{D_3}$. We have also that $\top \sqcap \top = a \notin [\top]_\tau$, so $[\top]_\tau$ is not a filter of $\underline{D_3}$.

**Observation 3.** A dBa of Kwuida shows that in a dBa $\underline{D}$, giving a congruence $\theta$, the class $[\bot]_\theta$ need not be an ideal and the class $[\top]_\theta$ need not be a filter. The above **Observation 3** allows us to ask the following questions: in which conditions $[\bot]_\theta$ is an ideal? $[\top]_\theta$ is a filter?

Now we are going to give some properties of congruences on a double Boolean algebras.

**Proposition 6.** *Let $\underline{D}$ be a dBa and $\theta \in Con(\underline{D})$. The following hold:*

(1) For any $x, y \in [\bot]_\theta$, $x \sqcap y \in [\bot]_\theta$.

(2) For any $x, y \in [\top]_\theta$, $x \sqcup y \in [\top]_\theta$.

(3) If $x \sqsubseteq y$ and $y \in [\bot]_\theta$, then $x \sqcap y \in [\bot]_\theta$.

(4) If $x \in [\top]_\theta$ and $x \sqsubseteq y$, then $x \sqcup y \in [\top]_\theta$.

(5) If $x \in [\bot]_\theta$, then for any $a \in D$, $(x \sqcap a, \bot) \in \theta$.

(6) If $x \in [\top]_\theta$, then for any $a \in D$, $(x \sqcup a, \top) \in \theta$.

(7) If $(\bot, \top) \in \theta$, then $D_\sqcap \cup D_\sqcup \subseteq [\bot]_\theta$.

(8) If $\underline{D}$ is a pure double Boolean algebra such that $(\bot, \top) \in \theta$, then $\theta = D^2$.

(9) If $\underline{D}$ is a trivial double Boolean algebra, then $\nabla$ is the only congruence on $D$ such that $[\bot]_\theta$ is an ideal and $[\top]_\theta$ is a filter.

(10) $[\top]_\theta$ is a filter if and only if $\top \sqcap \top \in [\top]_\theta$ and for any $x \in [\top]_\theta$, $[x, \top]_\theta \subseteq [\top]_\theta$.

(11) $[\bot]_\theta$ is an ideal if and only if $\bot \sqcup \bot \in [\bot]_\theta$ and for any $x \in [\bot]_\theta$, $[\bot, x] \subseteq [\bot]_\theta$.

*Proof.* The proofs of (2),(4),(6) and (11) are dual to those of (1),(3),(5) and (10), respectively. Here we give the proofs for (1),(3),(5) ,(7), (8) and (10). Let $x, y, a \in D$.

(1) Proof follows from $\sqcap$ preserves $\theta$ and $\bot \sqcap \bot = \bot$.

(3) Assume that $x \sqsubseteq y$ and $y \in [\bot]_\theta$, then $(y \sqcap x, \bot \sqcap x) = (x \sqcap x, \bot)$ due to $x \sqsubseteq y$ ,$\theta$ reflexive and $(y, \bot) \in \theta$.

(5) Assume that $x \in [\bot]_\theta$. Since $x \sqcap a \sqsubseteq x$ and $x \in [\bot]_\theta$, using (3) we get $(x \sqcap a, \bot) \in \theta$.

(7) Assume that $(\bot, \top) \in \theta$ and $x \in D_\sqcap \cup D_\sqcup$. If $x \in D_\sqcap$, then $x \sqsubseteq \top$ and $\top \in [\bot]_\theta$, so using (3), we get $x \sqcap \top = x \sqcap x = x \in [\bot]_\theta$. If $x \in D_\sqcup$, then $\bot \sqsubseteq x$ and $\bot \sqcup x = x \sqcup x = x \in [\top]_\theta = [\bot]_\theta$. Thus $D_\sqcap \cup D_\sqcup \subseteq [\bot]_\theta$.

(8) It is a consequence of (7).

(9) Assume that $\underline{D}$ is trivial, $[\bot]_\theta$ an ideal and $[\top]_\theta$ a filter. Since $[\bot]_\theta$ is an ideal and $[\top]_\theta$ a filter, we have $\bot \sqcup \bot = \top \sqcap \top \in [\top]_\theta \cap [\bot]_\theta$, so $(\bot, \top) \in \theta$ and by using (7) $D_\sqcap \cup D_\sqcup \subseteq [\bot]_\theta = [\top]_\theta$. It follows that $\theta = \nabla$.

(10) Assume that $[\top]_\theta$ is a filter, then $\top \sqcap \top \in [\top]_\theta$. For $x, y \in [\top]_\theta$ we have $(x \sqcap y, \top \sqcap \top) \in \theta$ due to the fact that $\theta$ is compatible with $\sqcap$, and by the transitivity of $\theta$ we ge $(x \sqcap y, \top) \in \theta$. Let $x \in [\top]_\theta$ and $y \in D$ such that $x \sqsubseteq y$, then $y \in [\top]_\theta$ due to $[\top]_\theta$ is a filter. Conversely, assume that $\top \sqcap \top \in [\top]_\theta$ and for any $x \in [\top]_\theta$, $[x, \top] \subseteq [\top]_\theta$. If $x, y \in [\top]_\theta$, then $(x \sqcap y, \top \sqcap \top) \in \theta$ and the transitivity of $\theta$ yields $(x \sqcap y, \top) \in \theta$, hence (10) holds.

## 5    Conclusion

We have described filters (ideals) generated by arbitrary subsets of a double Boolean algebras, and shown that its principal filters(ideals) form a bounded sublattice of the lattice of filters (resp. ideals) and are (non necessary isomorphic) Boolean algebras. We have also shown that $\mathcal{F}(\underline{D})$ (resp. $\mathcal{I}(\underline{D})$) and $\mathcal{F}(\underline{D_\sqcap})$ (resp. $\mathcal{I}(\underline{D_\sqcup})$) are isomorphic lattices and give some properties of congruences on dBas.

## References

1. Balbiani, P.: Deciding the word problem in pure double Boolean algebras. J. Appl. Logic **10**(3), 260–273 (2012)
2. Breckner, B.E., Sacarea, C.: A Topological representation of double Boolean lattices. Stud. Univ. Babes-Bolyai Math. **64**(1), 11–23 (2019)
3. Burris, S., Sankappanavar, H.P.: A Course in Universal Algebra. Graduate Texts in Mathematics. Springer, New York (1981)
4. Ganter, B., Wille, R.: Formal Concept Analysis: Mathematical Foundations. Springer, Heidelberg (1999). https://doi.org/10.1007/978-3-642-59830-2

5. Howlader, P., Banerjee, M.: Remarks on prime ideal and representation theorem for double Boolean algebras. In: The 15th International Conference on Concept Lattices and Their Applications (CLA 2020), pp. 83–94. Tallinn (2020)
6. Howlader, P., Banerjee, M.: Topological Representation of Double Boolean Algebras. ArXiv:2103.11387. (March 2021)
7. Kwuida, L.: Prime ideal theorem for double Boolean algebras. Discus. Math. Gen. Algebra Appl. **27**(2), 263–275 (2005)
8. Kwuida, L.: Dicomplemented lattices. A contextual generalization of Boolean algebras. Dissertation, TU Dresden, Shaker Verlag (2004)
9. Vormbrock, B., Wille, R.: Semiconcept and protoconcept algebras: the basic theorem. In Ganter, B., Stumme, G., Wille, R. (eds.) Formal Concept Analysis: Foundations and Applications. pp. 34–48. Springer, Berlin Heidelberg (2005)
10. Wille, R.: Boolean concept logic. In Ganter, B, Mineau, G.W. (eds.) Conceptual Structures: Logical Linguistic, and Computational Issue. pp. 317–331. Springer, Berlin Heidelberg (2000)

# Diagrammatic Representation
# of Conceptual Structures

Uta Priss[(✉)]

Ostfalia University, Wolfenbüttel, Germany
http://www.upriss.org.uk

**Abstract.** Conceptual exploration as provided by Formal Concept Analysis is potentially suited as a tool for developing learning materials for teaching mathematics. But even just a few mathematical notions can lead to complex conceptual structures which may be difficult to be learned and comprehended by students. This paper discusses how the complexity of diagrammatic representations of conceptual structures can potentially be reduced with Semiotic Conceptual Analysis. The notions of "simultaneous polysemy" and "observational advantage" are defined to describe the special kind of relationship between representations and their meanings which frequently occurs with diagrams.

## 1 Introduction

When learning mathematics, students need to acquire concepts using some kind of informal conceptual exploration where they mentally verify implications and identify counter examples to rule out misconceptions. The Formal Concept Analysis (FCA) method of conceptual exploration can formalise such a process by determining the relevant implicit knowledge that is contained in a domain. But concept lattices tend to contain too many concepts to be individually learned. Most likely some form of covering of the content of a concept lattice is required. Diagrams and visualisations often provide very concise representations of knowledge, because, as the proverb states, "a picture is worth 1000 words". But visualisations have both advantages as well as limits. Different students may be more or less adept in reading graphical representations. Thus not a single, but a variety of forms of representations may be required and students need to learn to switch between them.

A supportive theory for understanding the role of diagrams for conceptual learning is supplied by Semiotic Conceptual Analysis (SCA) – a mathematical formalisation of core semiotic notions that has FCA as a conceptual foundation (Priss 2017). A sign in SCA is a triple consisting of an interpretation, a representamen and a denotation (or meaning) where interpretations are partial functions from the set of representamens into the set of denotations. The capacity of a representamen (such as a diagram) to denote more than just one meaning is introduced in the notion of *simultaneous polysemy* in this paper. Simultaneous polysemy is contrasted with *ambiguous polysemy* which describes a representamen being mapped onto slightly different meanings in different usage

© Springer Nature Switzerland AG 2021
A. Braud et al. (Eds.): ICFCA 2021, LNAI 12733, pp. 281–289, 2021.
https://doi.org/10.1007/978-3-030-77867-5_19

contexts. Both forms of polysemy contribute to the efficiency of sets of representamens: a smaller set of representamens is capable of representing a larger set of denotations. For example, it is more efficient and easier to learn to read and write languages that use an alphabet as compared to Chinese. Letters of an alphabet are ambiguously polysemous because each letter expresses a small variety of different but similar phonemes in different usage contexts. Diagrams are usually simultaneously polysemous which is further explored in the notion of *observational advantage* adopted from the research about diagrams by Stapleton et al. (2017), although with a somewhat different formalisation.

The next section introduces a few non-standard FCA notions and the core SCA notions required for this paper. An introduction to FCA is not provided because FCA is the main topic of this conference. Section 3 discusses observational advantages of tabular, Euler and Hasse diagrams. Section 4 provides a small example of how to obtain an observationally efficient diagram. The paper finishes with a conclusion.

## 2   FCA and SCA Notions

This section repeats relevant notions from SCA, introduces some new SCA notions and a few non-standard FCA notions. A *supplemental concept* is a concept whose extension equals the union of the extensions of its proper subconcepts. In a Hasse diagram with minimal labelling, supplemental concepts are those that are not labelled by an object. Each supplemental concept corresponds to a clause because for such a concept $c$ with extension $ext(c)$ and intension $int(c)$ and the condition $\forall(o_i \in ext(c) : \exists(c_i < c : o_i \in ext(c_i)))$ it follows that $\bigwedge(a_i \in int(c)) \Rightarrow \bigvee(a_i \mid \exists c_i : c_i < c, a_i \in int(c_i), a_i \notin int(c))$ is a clause. Supplemental concepts are particularly interesting if the formal context is non-clarified and contains all objects that are known to be possible within a domain. In that case a clause presents not just information about the concept lattice but instead background knowledge about the domain.

In this paper the notion *conceptual class* is used as a placeholder for a formalisation of conceptual structures. A conceptual class is a structure consisting of sets, relations and functions. A *logical description of a conceptual class* $\mathcal{L}(\mathcal{C})$ is defined as a set of true statements according to the rules of some logical language $\mathcal{L}$. A logical description could be provided by a description logic, formal ontology or other formal language. Further details are left open so that SCA can be combined with a variety of conceptual formalisations.

The following definitions briefly summarise SCA. More details can be found in Priss (2017). A sign is a triple or element of a triadic relation:

**Definition 1.** For a set $R$ (called *representamens*), a set $D$ (called *denotations*) and a set $I$ of partial functions $i : R \twoheadrightarrow D$ (called *interpretations*), a *semiotic relation* $S$ is a relation $S \subseteq I \times R \times D$. A relation instance $(i, r, d) \in S$ with $i(r) = d$ is called a *sign*. For a semiotic relation, an equivalence relation $\approx_R$ on $R$, an equivalence relation $\approx_I$ on $I$, and a tolerance relation $\sim_D$ on $D$ are defined.

Additionally, the non-mathematical condition is assumed that signs are actually occurring in some communication event at a certain time and place. Representamens are entities that have a physical existence, for example as a sound wave, a neural brain pattern, a text printed in a book or on a computer screen or a state in a computer. Perceiving a sound wave or a pattern on a computer screen as a word is already an interpretation. Denotations represent meanings of signs usually in the form of concepts. Because of their physical existence, representamens tend to be at most equivalent instead of equal to each other. For example, two spoken words will never be totally equal but can be considered equivalent if they are sufficiently similar to each other. When referring to a sign, the notions "sign" and "representamen" are sometimes used interchangeably because a representamen is the perceptible part of a sign. For example, one might refer to "four" as a sign, word or representamen.

The tolerance relation in Definition 1 expresses similarity amongst meanings corresponding to synonymy. For example, "car" might be a synonym of "vehicle" and "vehicle" a synonym of "truck", but "truck" not a synonym of "car". In some domains (such as mathematics), equality of denotations is more important than similarity. The tolerance relation also serves the purpose of distinguishing polysemous signs (with similar meanings) from homographs which have totally unrelated meanings. For example $(i_1,$ "lead", [some metal]) and $(i_2,$ "lead", [to conduct]) are homographs whereas the latter is polysemous to $(i_3,$ "lead", [to chair]). An equivalence relation on interpretations can express a shared usage context of signs consisting of time, place and sign user. Equivalent interpretations belong to a single, shared usage context and differ only with respect to some further aspects that do not define a usage context. The first part of the next definition is repeated from Priss (2017) but the rest is new in this paper.

**Definition 2.** For a semiotic relation $S$, two signs $(i_1, r_1, d_1)$ and $(i_2, r_2, d_2)$ are *synonyms* $\Longleftrightarrow d_1 \sim_D d_2$. They are *polysemous* $\Longleftrightarrow r_1 \approx_R r_2$ and $d_1 \sim_D d_2$. Two polysemous signs are *simultaneously polysemous* if $i_1 \approx_I i_2$. Otherwise they are *ambiguously polysemous*.

Formally any sign is polysemous to itself, but in the remainder of this paper an individual sign is only called polysemous if another sign exists to which it is polysemous. Ambiguous polysemy refers to a representamen being used in different usage contexts with different meanings. The usage context disambiguates such polysemy. Ambiguous polysemy poses a problem if the ambiguity cannot be resolved by the usage context. A benefit of ambiguous polysemy is that a small number of representamens can refer to a much larger number of denotations because each representamen can occur in many usage contexts. Ambiguous polysemy involving different sign users expresses differences in understanding between the users. For example a student and a teacher might use the same terminology but not have exactly the same understanding of it.

**Definition 3.** For a semiotic relation $S$ with a sign $s := (i, r, d)$, the sets $S_{sp}(s) := \{s\} \cup \{(i_1, r_1, d_1) \mid \exists_{s_2 \in S_{sp}(s)} : (i_1, r_1, d_1) \text{ polysemous to } s_2 \text{ and } i_1 \approx_I i\}$ and $D_{sp}(s) := \{d_1 \mid \exists (i_1, r_1, d_1) \in S_{sp}(s)\}$ are defined. For $S_1 \subseteq S$ the sets $S_{sp}(S_1) := \bigcup_{s \in S_1} S_{sp}(s)$ and $D_{sp}(S_1) := \bigcup_{s \in S_1} D_{sp}(s)$ are defined.

The semiotic relation $S_{sp}(s)$ contains all signs that have interpretations that can be applied to $r$ within the same usage context. It should be noted that $|S_{sp}(s)| > |D_{sp}(s)|$ is possible because two interpretations of signs in $S_{sp}(s)$ can map $r$ onto the same denotation. For ambiguous polysemy we have in the past suggested to use neighbourhood lattices[1] (Priss and Old 2004). In the terminology of SCA, such neighbourhood lattices only consider a binary relation between representamens and denotations. A neighbourhood context is formed by starting with a denotation and finding all representamens that are in relation with it, then all other denotations which are in relation with one of the representamens and so on. Alternatively, one can start with a representamen. Depending on the sizes of the retrieved sets one can determine when to stop and whether to apply different types of restrictions (Priss and Old 2004). Definition 3 suggests a similar approach for simultaneous polysemy: $D_{sp}(s)$ and $D_{sp}(S_1)$ retrieve all denotations belonging to a representamen or set of representamens with equivalent interpretations.

In some cases, a sign $s$ has a representamen $r$ which has parts that are representamens of signs (e.g. $s_1$) themselves. In that case a mapping from $s$ to $s_1$ is called an *observation* if an additional non-mathematical condition is fulfilled that the relationship is based on some perceptual algorithm. For a diagram and the set $S_1$ of all signs that can be simultaneously observed from it, $D_{sp}(S_1)$ models an "observational advantage" in analogy to the notion of Stapleton et al. (2017). In domains such as mathematics, an *implication* can be considered to hold between signs if it is logically valid amongst the denotations of the signs. Observations, however, hold between signs based on representamens. Ideally observations amongst signs should imply implications and, thus, representamen-based relationships should correlate with or at least not disagree with denotation-based relationships.

**Definition 4.** For a semiotic relation $S$ with signs $s := (i, r, d)$, $s_1 := (i_1, r_1, d_1)$, $i \approx_I i_1$ and $d = d_1$, the sign $s$ has an *observational advantage* over $s_1$ if $|D_{sp}(s)| > |D_{sp}(s_1)|$. For two semiotic relations $S_1$ and $S_2$ whose interpretations all belong to the same equivalence class, $S_1$ has a *higher observational efficiency* over $S_2$ if $D_{sp}(S_1) = D_{sp}(S_2)$ and $|S_{sp}(S_1)| < |S_{sp}(S_2)|$. A semiotic relation $S$ has *maximal observational advantage* over a conceptual class if $D_{sp}(S)$ contains the set of true statements of the logical description of the conceptual class.

The last sentence of the definition extends the notion of observations to a relationship between signs and conceptual classes where a conceptual class is purely denotational and not a semiotic relation itself. Alternatively, conceptual

---

[1] Neighbourhood lattices were originally invented by Rudolf Wille in an unpublished manuscript.

classes could also be formalised as semiotic relations. Observational advantage and efficiency can be considered a measure of quality for representamens. A sign with a higher observational advantage might be better for certain purposes because it provides more information. A semiotic relation has a higher observational efficiency if it can express the same content with fewer signs. With respect to semiotic relations, the measure could be further refined, because otherwise just containing a single sign with an observational advantage is sufficient to cause a semiotic relation to have a higher observational efficiency.

Flower et al. (2008) distinguish concrete and abstract diagrams. A concrete diagram is actually drawn whereas an abstract diagram contains all the information that is required for producing a concrete diagram. For example, an abstract Hasse diagram describes nodes, edges, labels and their relationships to each other. A concrete diagram also contains x- and y- coordinates, fonts, colours and so. Such distinctions are, for example, relevant for the planarity of graphs: an abstract diagram is called planar if a concrete drawing without line crossings is possible. In SCA, an interpretation maps a concrete diagram onto an abstract diagram (as a denotation) which can then be considered a representamen and mapped onto a conceptual class. In general, it is always a matter of judgement to choose interpretations and denotations of a semiotic relation.

For example, one can argue that a (concrete or abstract) Hasse diagram of a concept lattice has maximal observational advantage over its concept lattice. This follows directly from how Hasse diagrams are defined, but still depends on how a conceptual class of concept lattices is defined. Formal contexts as diagrams might contain fewer representamens than Hasse diagrams of their concept lattices and could thus have a higher observational efficiency. But it can be argued that only a binary relation can be observed from a formal context. Constructing a lattice from a binary relation involves implications as well as observations. While it may be possible to read maximum rectangles from a formal context, most people would have great difficulty observing the complete conceptual ordering and, for example, the top and bottom concepts from a formal context. In any case, both formal context diagrams as well as Hasse diagrams of concept lattices have maximal observational advantage over the binary relation between objects and attributes.

## 3    Tabular, Euler and Hasse Diagrams

Euler diagrams are a form of graphical representation of subsets of a powerset that is similar to Venn diagrams but leaves off any zones that are known to be empty. Not all subsets of a powerset can be represented by an Euler diagram in a *well-formed* manner without including some supplemental zones (similar to supplemental concepts). Supplemental zones are often shaded in order to indicate that they are empty. Figure 1 shows an Euler diagram in the middle. It is slightly unusual because its curves are boxes instead of circles or ellipses. The correspondence between the Euler diagram and the Hasse diagram on the right should be evident from the letters. The reason for drawing the Euler diagram with boxes is because it allows a reduction to a diagram shown on the left which is called *tabular diagram* in this paper. It seems that there should be an established notion for "tabular diagrams" but there seem to be a variety of similar notions (mosaic plots/displays, contingency tables, Karnaugh maps) which all have slightly different additional meanings. In a sense, tabular diagrams are a 2-dimensional version of the "linear diagrams" invented by Leibniz (Chapman et al. 2014).

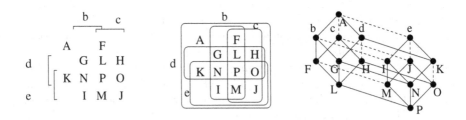

**Fig. 1.** Tabular, Euler and Hasse diagrams

Considered as single representamens, each diagram in Fig. 1 is simultaneously polysemous because it contains a large amount of information: objects, attributes, their binary relation and implications amongst attributes. The three types of diagrams can be considered to denote the same conceptual class and thus to have the same observational advantage. Instead of considering each diagram as one representamen, it can also be considered as a semiotic relation consisting of parts that are representamens. For the following theorem it is assumed that sets (or curves) with multiple labels are not allowed and that each diagram contains more than one set.

**Theorem:** Tabular, Euler and Hasse diagrams denote a shared conceptual class corresponding to partial orders of sets with labels and elements. If they exist, tabular diagrams have a higher or equal observational efficiency than Euler diagrams which have a higher observational efficiency than Hasse diagrams.

It should be noted that an Euler diagram might just be a partially ordered set, not a lattice. A translation from Euler to Hasse diagrams is discussed by Priss (2020) and shall be omitted here. The proof of the second half of the theorem is that tabular diagrams contain exactly one representamen for each object, at most two representamens for each attribute (the name of the attribute and a bracket) and nothing else. Euler diagrams also contain one representamen for each attribute and object, and one curve for each attribute. Because a bracket in a tabular diagram can be omitted if the attribute belongs to just one row or column and the outer curve may be omitted, tabular diagrams contain potentially fewer representamens than Euler diagrams. Hasse diagrams contain labels for the objects and attributes, one node for each attribute, but also some edges, thus more representamens than Euler diagrams.

The question arises as to which concept lattices and which Euler diagrams can be represented as tabular diagrams. Any tabular diagram can be converted into an Euler diagram by extending the brackets into boxes but the resulting Euler diagram may not be well-formed. For more than 4 elements, it may not be possible to construct a tabular diagram at all. A solution is to duplicate some of the row and column labels, if the diagram is not possible otherwise. Being able to embed a lattice into a direct product of two planar lattices, is not sufficient as a condition for a corresponding tabular diagram. Petersen's (2010) description of an "S-order" characterises lattices which correspond to 1-dimensional linear diagrams and could lead to a characterisation. Answering such questions may be as difficult as it is to determine which Euler diagrams are well-formed (Flower et al. 2008) which has only been solved by providing algorithms so far.

Observability as defined in the previous section is a formal condition. It does not imply that all users can actually observe the information. If a diagram gets too large and complex, users will have difficulties observing anything. Also, some people are more, some less skilled in reading information from graphical representations. Furthermore, we are not suggesting that tabular or Euler diagrams have a higher observational efficiency over Hasse diagrams with respect to all possible conceptual classes nor that they are in any other sense "superior" to lattices. In fact a comparison between Euler and Hasse diagrams shows varied results (Priss 2020).

One of the disadvantages of Euler and tabular diagrams is that some of the structural symmetry may be missing. The dashed lines in Fig. 1 indicate relationships between concepts that are neighbours in the Hasse diagram (and in the ordering relation), but are not neighbours in the tabular diagram. Thus observability has many aspects to it. Modelling with different semiotic relations and conceptual classes will produce different results. Psychological aspects relating to perception exist in addition to formal aspects.

## 4   Obtaining an Observationally Efficient Diagram

This section employs an example of a formal context and lattice from Ganter and Wille (1999, Sect. 2.2) consisting of seven prototypical types of triangles and their properties (Fig. 2). The example is discussed using a 3-step investigation: 1. conceptual exploration, 2. reduction of the concept lattice using background knowledge and 3. finding a diagrammatic representation that has a high observational efficiency.

The corresponding lattice is shown in Fig. 2 with empty nodes representing supplemental concepts and filled nodes for non-supplemental concepts. More than half of the concepts are supplemental – and adding more types of triangles as objects would not change that. If one removes the attribute "not equilateral" (and stores it as background knowledge), then the resulting lattice in the left half of Fig. 3 contains fewer supplemental concepts. Since there is only one object that has the attribute "equilateral", this information can still be observed from the lattice. Thus the complexity of the lattice can be reduced if some of the information is stored as background knowledge.

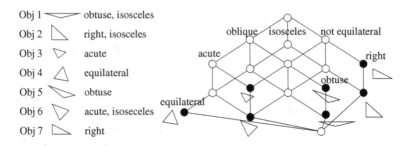

**Fig. 2.** Classification of triangles (according to Ganter and Wille (1999))

The third step consists of considering observational advantages and efficiency. By representing formal objects as diagrams of triangles, each object visually contains the information about which attributes it has. Thus, for example, the triangle for Obj 4 is simultaneously polysemous because the fact that it is equilateral, acute, oblique and isosceles can be observed from it if one knows what having such an attribute looks like. If one uses a string "equilateral triangle" for Obj 4 then only the attribute "equilateral" can be observed. The other attributes can be inferred but not observed. Representing formal objects as diagrams provides observational advantages over presenting them as strings. Of course, in most applications it will not be the case that objects can be represented as diagrams. Furthermore, in some disciplines diagrams are more suitable than in others.

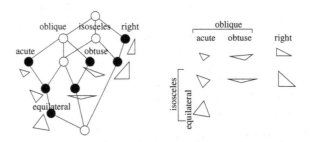

**Fig. 3.** Classification of triangles: Hasse and tabular diagram

The tabular diagram on the right of Fig. 3 provides a higher observational efficiency as explained in the previous section. If a student wants to memorise all existing types of triangles, then maybe this is the most suitable diagram. Apart from the bottom node of the lattice, the other supplemental concepts are still structurally present in the tabular diagram which becomes clear if the curves of the Euler diagram are added. If the attribute "not equilateral" was added then an Euler diagram would also have more empty zones, but those would not be visible in the tabular diagram. Again, we are not suggesting that in general hiding information is an advantage. It depends on the purpose of a diagram.

# 5    Conclusion

This paper discusses means of obtaining diagrammatic representations of conceptual structures that provide observational advantages and high observational efficiency. A background for this research is to find ways of developing teaching material that covers a topic area efficiently but also, if possible, by connecting to visual structures. Mathematical statements and proofs often present a combination of information that can be observed and information that must be known or inferred. We believe that in particular the notion of "seeing" information in representations is not yet fully understood, even though there is a long tradition of and large body of research in diagrammatic reasoning and information visualisation.

# References

Chapman, P., Stapleton, G., Rodgers, P., Micallef, L., Blake, A.: Visualizing sets: an empirical comparison of diagram types. In: Dwyer, T., Purchase, H., Delaney, A. (eds.) Diagrams 2014. LNCS (LNAI), vol. 8578, pp. 146–160. Springer, Heidelberg (2014). https://doi.org/10.1007/978-3-662-44043-8_18

Flower, J., Fish, A., Howse, J.: Euler diagram generation. J. Vis. Lang. Comput. **19**(6), 675–694 (2008)

Ganter, B., Wille, R.: Formal Concept Analysis. Mathematical Foundations. Springer, Berlin-Heidelberg-New York (1999)

Petersen, W.: Linear coding of non-linear hierarchies: revitalization of an ancient classification method. Adv. Data Anal., pp. 307–316. Springer, Data Handling and Business Intelligence (2010)

Priss, U., Old, L.J.: Modelling lexical databases with formal concept analysis. J. Univ. Comput. Sci. **10**, 8, 967–984 (2004)

Priss, U.: Semiotic-conceptual analysis: a proposal. Int. J. Gen. Syst. **46**(5), 569–585 (2017)

Priss, U.: Set visualizations with Euler and Hasse diagrams. In: Cochez, M., Croitoru, M., Marquis, P., Rudolph, S. (eds.) GKR 2020. LNCS (LNAI), vol. 12640, pp. 72–83. Springer, Cham (2020). https://doi.org/10.1007/978-3-030-72308-8_5

Stapleton, G., Jamnik, M., Shimojima, A.: What makes an effective representation of information: a formal account of observational advantages. J. Logic Lang. Inf. **26**(2), 143–177 (2017)

# Author Index

Printed in the United States
by Baker & Taylor Publisher Services